# RYPINS' INTENSIVE REV

**Series Editor**

## Edward D. Frohlich, MD, MACP, FACC

Alton Ochsner Distinguished Scientist
Vice President for Academic Affairs
Alton Ochsner Medical Foundation
Staff Member, Ochsner Clinic
Professor of Medicine and of Physiology
Louisiana State University of Medicine
Adjunct Professor of Pharmacology and
Clinical Professor of Medicine
Tulane University School of Medicine
New Orleans, Louisiana

RYPINS' INTENSIVE REVIEWS

# Anatomical Sciences

**Gross Anatomy**
**Embryology**
**Histology**
**Neuroanatomy**

Neal E. Pratt, PhD

Professor of Physical Therapy and Neurobiology/Anatomy
Allegheny University of the Health Sciences
Philadelphia, Pennsylvania

Dennis M. DePace, PhD

Associate Professor of Neurobiology and Anatomy
Allegheny University of the Health Sciences
Philadelphia, Pennsylvania

Lippincott - Raven
PUBLISHERS
Philadelphia • New York

Acquisitions Editor: Richard Winters
Developmental Editor: Mary Beth Murphy
Managing Editor: Susan E. Kelly
Manufacturing Manager: Kevin Watt
Supervising Editor: Mary Ann McLaughlin
Production Editor: Jenn Nagaj, Silverchair Science + Communications
Cover Designer: William T. Donnelly
Interior Designer: Susan Blaker
Design Coordinator: Melissa Olson
Indexer: Betty Hallinger
Compositor: Jennifer Whitlow, Silverchair Science + Communications
Printer: Courier/Kendallville



Printed in the United States of America

9 8 7 6 5 4 3 2 1

**Library of Congress Cataloging-in-Publication Data**

Pratt, Neal E.
Anatomical sciences : gross anatomy, embryology, histology, neuroanatomy / Neal E. Pratt, Dennis M. DePace.
p. cm. -- (Rypins' intensive reviews)
Includes bibliographical references and index.
ISBN 0-397-51552-9
1. Human anatomy--Examinations, questions, etc. 2. Histology--Examinations, questions, etc. 3. Embryology, Human--Examinations, questions, etc. 4. Neuroanatomy--Examinations, questions, etc. I. DePace, Dennis. II. Title. III. Series.
[DNLM: 1. Anatomy examination questions. QS 18.2 P916a 1999]
QM32.P68 1999
611'.0076--dc21
DNLM/DLC
for Library of Congress 98-43301
CIP

## Who Was "Rypins"?

Dr. Harold Rypins (1892–1939) was the founding editor of what is now known as the RYPINS' series of review books. Originally published under the title *Medical State Board Examinations,* the first edition was published by J. B. Lippincott Company in 1933. Dr. Rypins edited subsequent editions of the book in 1935, 1937, and 1939 before his death that year. The series that he began has since become the longest-running and most successful publication of its kind, having served as an invaluable tool in the training of generations of medical students. Dr. Rypins was a member of the faculty of Albany Medical College in Albany, New York, and also served as Secretary of the New York State Board of Medical Examiners. His legacy to medical education flourishes today in the highly successful *Rypins' Basic Sciences Review* and *Rypins' Clinical Sciences Review,* now in their 17th editions, and in the *Rypins' Intensive Reviews* series of subject review volumes. We at Lippincott–Raven Publishers take pride in this continuing success.

***—The Publisher***

# Series Preface

These are indeed very exciting times in medicine. Having made this statement, one's thoughts immediately reflect about the major changes that are occurring in our overall health care delivery system, utilization-review and shortened hospitalizations, issues concerning quality assurance, ambulatory surgical procedures and medical clearances, and the impact of managed care on the practice of internal medicine and primary care. Each of these issues has had a considerable impact on the approach to the patient and on the practice of medicine.

But even more mind-boggling than the foregoing changes are the dramatic changes imposed on the practice of medicine by fundamental conceptual scientific innovations engendered by advances in basic science that no doubt will affect medical practice of the immediate future. Indeed, much of what we thought of as having a potential impact on the practice of medicine of the future has already been perceived. One need only take a cursory look at our weekly medical journals to realize that we are practicing "tomorrow's medicine today." And consider that the goal a few years ago of actually describing the human genome is now near reality.

Reflect, then, for a moment on our current thinking about genetics, molecular biology, cellular immunology, and other areas that have impacted upon our current understanding of the underlying mechanisms of the pathophysiological concepts of disease. Moreover, paralleling these innovations have been remarkable advances in the so-called "high tech" and "gee-whiz" aspects of how we diagnose disease and treat patients. We can now think with much greater perspective about the dimensions of more specific biologic diagnoses concerned with molecular perturbations; gene therapy not only affecting genetic but oncological diseases; more specific pharmacotherapy involving highly specific receptor inhibition, alterations of intracellular signal transduction, manipulations of cellular protein synthesis; immunosuppresive therapy not only with respect to organ transplantations but also of autoimmune and other immune-related diseases; and therapeutic means for manipulating organ remodeling or the intravascular placement of stents. Each of these concepts has become inculcated into our everyday medical practice within the past decade. The reason why these changes have so rapidly promoted an upheaval in medical practice is continuing medical education, a constant awareness of the current medical literature, and a thirst for new knowledge.

To assist the student and practitioner in the review process, the publisher and I have initiated a new approach in the publication of *Rypins' Basic Sciences Review* and *Rypins' Clinical Sciences Review*. Thus, when I assumed responsibility to edit this long-standing board review series with the 13th edition of the textbook (first published in 1931), it was with a feeling of great excitement. I perceived that great changes would be coming to medicine, and I believed that this would be one ideal means of not only facing these changes head on but also for me personally to cope and keep up with these changes. Over the subsequent editions, this confidence was reassured and rewarded. The presentation for the updating of medical information was tremendously enhanced by the substitution of new authors, as the former authority "standbys" stepped down or retired from our faculty. Each of the authors who continue to be selected for maintaining the character of our textbook is an authority in his or her respective area and has had considerable pedagogic and formal examination experience. One dramatic recent example of the changes in author replacement just came about with the 17th edition. When I invited Dr. Peter Goldblatt to participate in the authorship of the pathology chapter of the textbook, his answer was "what goes around, comes around." You see, Dr. Goldblatt's father, Dr. Harry Goldblatt, a major contributor to the history of hypertensive disease, was the first author of the pathology chapter in 1931. What a satisfying experience for me personally. Other less human changes in our format came with the establishment of two soft cover volumes, the current basic and clinical sciences review volumes, replacing the single volume text of earlier years. Soon, a third supplementary volume concerned with questions and answers for the basic science volume appeared. Accompanying these more obvious changes was the constant updating of the knowledge base of each of the chapters, and this continues on into the present 17th edition.

And now we have introduced another major innovation in our presentation of the basic and clinical sciences reviews. This change is evidenced by the introduction of the *Rypins' Intensive Reviews* series, along with the 17th edition of *Rypins' Basic Sciences Review, Rypins' Clinical Sciences Review,* and the *Questions and Answers* third volume. These volumes are written to be used separately from the parent textbook. Each not only contains the material published in their respective chapters of the textbook, but is considerably "fleshed out" in the discussions, tables, figures, and questions and answers. Thus, the *Rypins' Intensive Reviews* series serves as an important supplement to the overall review process and also provides a study guide for those already in practice, in preparing for specific specialty board certification and recertification examinations.

Therefore, with continued confidence and excitement, I am pleased to present these innovations in review experience for your consideration. As in the past, I look forward to learning of your comments and suggestions. In doing so, we continue to look forward to our continued growth and acceptance of the *Rypins'* review experience.

Edward D. Frohlich, MD, MACP, FACC

# Dedication

This review is dedicated to the memory of J. Robert Troyer, PhD, a teacher with few peers, a dedicated mentor, an unflinchingly honest academic, a valued colleague, and a faithful friend. As Chairman of the Department of Anatomy at Temple University School of Medicine, he displayed his greatest gift by inspiring faculty members both to give and to achieve their best, an expertise that is seldom espoused but is the heart and soul of academe.

# Preface

Human anatomy is the study of the structure of the human body. It often is subdivided into gross anatomy, the study of structure as seen with the unaided eye; microscopic anatomy (histology), the study of structure as seen with the aid of a microscope; cytology, the study of the structure of cells, often included within the study of microscopic anatomy; embryology, the study of the origin, growth, and development of an organism from inception until birth; and neuroanatomy, the study of the structure of the nervous system.

Although these divisions of the anatomical sciences provide a convenient way to parcel the material for teaching purposes, remember that gross anatomy, microscopic anatomy, embryology, and neuroanatomy all look at the same organ systems in different ways. In general, an understanding of gross anatomy is valuable to the study of microscopic anatomy, embryology, and neuroanatomy.

This review is organized much like traditional anatomy courses, with emphasis on the important concepts found within each of the anatomical disciplines. Wherever possible, information is presented in tables, which serve as quick references. We hope this format will provide students with rapid access to information needed to review for U.S. Medical Licensing Examination Step 1 and for physician's assistant and physical therapy programs.

We thank our students past, present, and future.

Neal E. Pratt, PhD
Dennis M. DePace, PhD

# Introduction

## *Preparing for the USMLE*

### UNITED STATES MEDICAL LICENSING EXAMINATION (USMLE)

In August 1991 the Federation of State Medical Boards (FSMB) and the National Board of Medical Examiners (NBME) agreed to replace their respective examinations, the FLEX and NBME, with a new examination, the United States Medical Licensing Examination (USMLE). This examination will provide a common means for evaluating all applicants for medical licensure. It appears that this development in medical licensure will at last satisfy the needs for state medical boards licensure, the national medical board licensure, and licensure examinations for foreign medical graduates. This is because the 1991 agreement provides for a composite committee that equally represents both organizations (the FSMB and NBME) as well as a jointly appointed public member and a representative of the Educational Council for Foreign Medical Graduates (ECFMG).

As indicated in the USMLE announcement, "It is expected that students who enrolled in U.S. medical schools in the fall of 1990 or later and foreign medical graduates applying for ECFMG examinations beginning in 1993 will have access only to USMLE for purposes of licensure." The phaseout of the last regular examinations for licensure was completed in December 1994.

The new USMLE is administered in three steps. Step 1 focuses on fundamental basic biomedical science concepts, with particular emphasis on "principles and mechanisms underlying disease and modes of therapy." Step 2 is related to the clinical sciences, with examination on material necessary to practice medicine in a supervised setting. Step 3 is designed to focus on "aspects of biomedical and clinical science essential for the unsupervised practice of medicine."

Today Step 1 and Step 2 examinations are set up and scored as total comprehensive objective tests in the basic sciences and clinical sciences, respectively. The format of each part is no longer subject-oriented, that is, separated into sections specifically labeled Anatomy, Pathology, Medicine, Surgery, and so forth. Subject labels are therefore missing, and in each part questions from the different fields are intermixed or integrated so that the subject origin of any individual question is not immediately apparent, although it is known by the National Board office. Therefore, if necessary, individual subject grades can be extracted.

Step 1 is a two-day written test including questions in anatomy, biochemistry, microbiology, pathology, pharmacology, physiology, and the behavioral sciences. Each subject contributes to the examination a large number of questions designed to test not only knowledge of the subject itself but also "the subtler qualities of discrimination, judgment, and reasoning." Questions in such fields as molecular biology, cell biology, and genetics are included, as are questions to test the "candidate's recognition of the similarity or dissimilarity of diseases, drugs, and physiologic, behavioral, or pathologic processes." Problems are presented in narrative, tabular, or graphic form, followed by questions designed to assess the candidate's knowledge and comprehension of the situation described.

Step 2 is also a two-day written test that includes questions in internal medicine, obstetrics and gynecology, pediatrics, preventive medicine and public health, psychiatry, and surgery. The questions, like those in Step 1, cover a broad spectrum of knowledge in each of the clinical fields. In addition to individual questions, clinical problems are presented in the form of case histories, charts, roentgenograms, photographs of gross and microscopic pathologic specimens, laboratory data, and the like, and the candidate must answer questions concerning the interpretation of the data presented and their relation to the clinical problems. The questions are "designed to explore the extent of the candidate's knowledge of clinical situations, and to test his [or her] ability to bring information from many different clinical and basic science areas to bear upon these situations."

The examinations of both Step 1 and Step 2 are scored as a whole, certification being given on the basis of performance on the entire part, without reference to disciplinary breakdown. The grade for the examination is derived from the total number of questions answered correctly, rather than from an average of the grades in the component basic science or clinical science subjects. A candidate who fails will be required to repeat the entire examination. Nevertheless, as noted above, in spite of the interdisciplinary character of the examinations, all of the traditional disciplines are represented in the test, and separate grades for each subject can be extracted and reported separately to students, to state examining boards, or to those medical schools that request them for their own educational and academic purposes.

This type of interdisciplinary examination and the method of scoring the entire test as a unit have definite advantages, especially in view of the changing curricula in medical schools. The former type of rigid, almost standardized, curriculum, with its emphasis on specific subjects and a specified number of hours in each, has been replaced by a more liberal, open-ended curriculum, permitting emphasis in one or more fields and corresponding deemphasis in others. The result has been rather wide variations in the totality of education in different medical schools. Thus, the scoring of these tests as a whole permits accommodation to this variability in the curricula of different schools. Within the total score, weakness in one subject that has received relatively little emphasis in a given school may be balanced by strength in other subjects.

The rationale for this type of comprehensive examination as replacement for the traditional department-oriented examination in the basic sciences and the clinical sciences is given in the National Board Examiner:

> The student, as he [or she] confronts these examinations, must abandon the idea of "thinking like a physiologist" in answering a question labeled "physiology" or "thinking like a surgeon" in answering a question labeled "surgery." The one question may have been written by a biochemist or a pharmacologist; the other question may have been written by an internist or a pediatrician. The pattern of these examinations will direct the student to thinking more broadly of the basic sciences in Step 1 and to thinking of patients and their problems in Step 2.

Until a few years ago, the Part I examination could not be taken until the work of the second year in medical school had been completed, and the Part II test was given only to students who had completed the major part of the fourth year. Now students, if they feel they are ready, may be admitted to any regularly scheduled Step 1 or Step 2 examination during any year of their medical course without prerequisite completion of specified courses or chronologic periods of study. Thus, emphasis is placed on the acquisition of knowledge and competence rather than the completion of predetermined periods.

Candidates are eligible for Step 3 after they have passed Steps 1 and 2, have received the M.D. degree from an approved medical school in the United States or Canada, and subsequent to the receipt of the M.D. degree, have served at least six months in an approved hospital internship or residency. Under certain circumstances, consideration may be given to other types of graduate training provided they meet with the approval of the National Board. After passing the Step 3 examination, candidates will receive their diplomas as of the date of the satisfactory completion of an internship or residency program. If candidates have completed the approved hospital training prior to completion of Step 3, they will receive certification as of the date of the successful completion of Step 3.

The Step 3 examination, as noted above, is an objective test of general clinical competence. It occupies one full day and is divided into two sections, the first of which is a multiple-choice examination that relates to the interpretation of clinical data presented primarily in pictorial form, such as pictures of patients, gross and microscopic lesions, electrocardiograms, charts, and graphs. The second section, entitled Patient Management Problems, utilizes a programmed-testing technique designed to measure the candidate's clinical judgment in the management of patients. This technique simulates clinical situations in which the physician is faced with the problems of patient management presented in a sequential programmed pattern. A set of four to six problems is related to each of a series of patients. In the scoring of this section, candidates are given credit for correct choices; they are penalized for errors of commission (selection of procedures that are unnecessary or are contraindicated) and for errors of omission (failure to select indicated procedures).

All parts of the USMLE are given in many centers, usually in medical schools, in nearly every large city in the United States as well as in a few cities in Canada, Puerto Rico, and the Canal Zone. In some cities, such as New York, Chicago, and Baltimore, the examination may be given in more than one center.

The examinations of the National Board have become recognized as the most comprehensive test of knowledge of the medical sciences and their clinical application produced in this country.

## THE NATIONAL BOARD OF MEDICAL EXAMINERS

For years the National Board examinations have served as an index of the medical education of the period and have strongly influenced higher educational standards in each of the medical sciences. The Diploma of the National Board is accepted by 47 state licensing authorities, the District of Columbia, and the Commonwealth of Puerto Rico in lieu of the examination usually required for licensure and is recognized in the American Medical Directory by the letters DNB following the name of the physician holding National Board certification.

The National Board of Medical Examiners has been a leader in developing new and more reliable techniques of testing, not only for knowledge in all medical fields but also for clinical competence and fitness to practice. In recent years, too, a number of medical schools, several specialty certifying boards, professional medical societies organized to encourage their members to keep abreast of progress in medicine, and other professional qualifying agencies have called upon the National Board's professional staff for advice or for the actual preparation of tests to be employed in evaluating medical knowledge, effectiveness of teaching, and professional competence in certain medical fields. In all cases, advantage has been taken of the validity and effectiveness of the objective, multiple-choice type of examination, a technique the National Board has played an important role in bringing to its present state of perfection and discriminatory effectiveness.

Objective examinations permit a large number of questions to be asked, and approximately 150 to 180 questions can be answered in a 2½-hour period. Because the answer sheets are scored by machine, the grading can be accomplished rapidly, accurately, and impartially. It is completely unbiased and based on percentile ranking. Of long-range significance is the facility with which the total test and individual questions can be subjected to thorough and rapid statistical analyses, thus providing a sound basis for comparative studies of medical school teaching and for continuing improvement in the quality of the test itself.

## QUESTIONS

Over the years, many different forms of objective questions have been devised to test not only medical knowledge but also those subtler qualities of discrimination, judgment, and reasoning. Certain types of questions may test an individual's recognition of the similarity or dissimilarity of diseases, drugs, and physiologic or pathologic processes. Other questions test judgment as to cause and effect or the lack of causal relationships. Case histories or patient problems are used to simulate the experience of a physician confronted with a diagnostic problem; a series of questions then tests the individual's understanding of related aspects of the case, such as signs and symptoms, associated laboratory findings, treatment, complications, and prognosis. Case-history questions are set up purposely to place emphasis on correct diagnosis within a context comparable with the experience of actual practice.

It is apparent from recent certification and board examinations that the examiners are devoting more attention in their construction of questions to more practical means of testing basic and clinical knowledge. This greater realism in testing relates to an increasingly interdisciplinary approach toward fundamental material and to the direct relevance accorded practical clinical problems. These more recent approaches to questions have been incorporated into this review series.

Of course, the new approaches to testing add to the difficulty experienced by the student or physician preparing for board or certification examinations. With this in mind, the author of this review is acutely aware not only of the interrelationships of fundamental information within the basic science disciplines and their clinical implications but also of the necessity to present this material clearly and concisely despite its complexity. For this reason, the questions are devised to test knowledge of specific material within the text and identify areas for more intensive study, if necessary. Also, those preparing for examinations must be aware of the interdisciplinary nature of fundamental clinical material, the common multifactorial characteristics of disease mechanisms, and the necessity to shift back and forth from one discipline to another in order to appreciate the less than clear-cut nature separating the pedagogic disciplines.

The different types of questions that may be used on examinations include the completion-type question, in which the individual must select one best answer among a number of possible choices, most often five, although there may be three or four; the completion-type question in the negative form, in which all but one of the choices is correct and words such as *except* or *least* appear in the question; the true-false type of question, which tests an understanding of cause and effect in relationship to medicine; the multiple true-false type, in which the question may have one, several, or all correct choices; one matching-type question, which tests association and relatedness and uses four choices, two of which use the word, *both* or *neither;* another matching-type question that uses anywhere from three to twenty-six choices and may have more than one correct answer; and, as noted above, the patient-oriented question, which is written around a case and may have several questions included as a group or set.

Many of these question types may be used in course or practice exams; however, at this time the most commonly used types of questions on the USMLE exams are the completion-type question (one best answer), the completion-type negative form, and the multiple matching-type question, designating specifically how many choices are correct. Often included within the questions are graphic elements such as diagrams, charts, graphs, electrocardiograms, roentgenograms, or photomicrographs to elicit knowledge of structure, function, the course of a clinical situation, or a statistical tabulation. Questions then may be asked in relation to designated elements of the same. As noted above, case histories or patient-oriented questions are more frequently used on these examinations, requiring the individual to use more analytic abilities and less memorization-type data.

For further detailed information concerning developments in the evolution of the examination process for medical licensure (for graduates of both U.S. and foreign medical schools), those interested should contact the National Board of Medical Examiners at 3750 Market Street, Philadelphia, PA 19104; telephone 215-590-9500; or http://www.usmle.org.

## FIVE POINTS TO REMEMBER

In order for the candidate to maximize chances for passing these examinations, a few common sense strategies or guidelines should be kept in mind.

First, it is imperative to prepare thoroughly for the examination. Know well the types of questions to be presented and the pedagogic areas of particular weakness, and devote more preparatory study time to these areas of weakness. Do not use too much time restudying areas in which there is a feeling of great confidence and do not leave unexplored those areas in which there is less confidence. Finally, be well rested before the test and, if possible, avoid traveling to the city of testing that morning or late the evening before.

Second, know well the format of the examination and the instructions before becoming immersed in the challenge at hand. This information can be obtained from many published texts and brochures or directly from the testing service (National Board of Medical Examiners, 3750 Market Street, Philadelphia, PA 19104; telephone 215-590-9500). In addition, many available texts and self-assessment types of examination are valuable for practice.

Third, know well the overall time allotted for the examination and its components and the scope of the test to be faced. These may be learned by a rapid review of the examination itself. Then, proceed with the test at a careful, deliberate, and steady pace without spending an inordinate amount of time on any single question. For example, certain questions such as the "one best answer" probably should be allotted 1 to 1½ minutes each. The "matching" type of question should be allotted a similar amount of time.

Fourth, if a question is particularly disturbing, note appropriately the question (put a mark on the question sheet) and return to it later. Don't compromise yourself by so concentrating on a likely "loser" that several "winners" are eliminated because of inadequate time. One way to save this time on a particular "stickler" is to play your initial choice; your chances of a correct answer are always best with your first impression. If there is no initial choice, reread the question.

Fifth, allow adequate time to review answers, to return to the questions that were unanswered and "flagged" for later attention, and check every *n*th (e.g., 20th) question to make certain that the answers are appropriate and that you did not inadvertently skip a question in the booklet or answer on the sheet (this can happen easily under these stressful circumstances).

There is nothing magical about these five points. They are simple and just make common sense. If you have prepared well, have gotten a good night's sleep, have eaten a good breakfast, and follow the preceding five points, the chances are that you will not have to return for a second go-around.

Edward D. Frohlich, MD, MACP, FACC

# Series Acknowledgments

In no other writing experience is one more dependent on others than in a textbook, especially a textbook that provides a broad review for the student and fellow practitioner. In this spirit, I am truly indebted to all who have contributed to our past and current understanding of the fundamental and clinical aspects related to the practice of medicine. No one individual ever provides the singular "breakthrough" so frequently attributed as such by the news media. Knowledge develops and grows as a result of continuing and exciting contributions of research from all disciplines, academic institutions, and nations. Clearly, outstanding investigators have been credited for major contributions, but those with true and understanding humility are quick to attribute the preceding input of knowledge by others to the growing body of knowledge. In this spirit, we acknowledge the long list of contributors to medicine over the generations. We also acknowledge that in no century has man so exceeded the sheer volume of these advances than in the twentieth century. Indeed, it has been said by many that the sum of new knowledge over the past 50 years has most likely exceeded all that had been contributed in the prior years.

With this spirit of more universal acknowledgment, I wish to recognize personally the interest, support, and suggestions made by my colleagues in my institution and elsewhere. I specifically refer to those people from my institution who were of particular help and are listed at the outset of the internal medicine volume. But, in addition to these colleagues, I want to express my deep appreciation to my institution and clinic for providing the opportunity and ambience to maintain and continue these academic pursuits. As I have often said, the primary mission of a school of medicine is that of education and research; the care of patients, a long secondary mission to ensure the conduct of the primary goal, has now also become a primary commitment in these more pragmatic times. In contrast, the primary mission of the major multidisciplinary clinics has been the care of patients, with education and research assuming secondary roles as these commitments become affordable. It is this distinction that sets the multispecialty clinic apart from other modes of medical practice.

Over and above a personal commitment and drive to assure publication of a textbook such as this is the tremendous support and loyalty of a hard-working and dedicated office staff. To this end, I am tremendously grateful and indebted to Mrs. Lillian Buffa and Mrs. Caramia Fairchild. Their long hours of unselfish work on my behalf and to satisfy their own interest in participating in this major educational effort is appreciated no end. I am personally deeply honored and thankful for their important roles in the publication of the Rypins' series.

Words of appreciation must be extended to the staff of Lippincott–Raven Publishers. It is more than 25 years since I have become associated with this publishing house, one of the first to be established in our nation. Over these years, I have worked closely with Mr. Richard Winters, not only with the Rypins' editions but also with other textbooks. His has been a labor of commitment, interest, and full support—not only because of his responsibility to his institution, but also because of the excitement of publishing new knowledge. In recent years, we discussed at length the merits of adding the intensive review supplements to the parent textbook and together we worked out the details that have become the substance of our present "joint venture." Moreover, together we are willing to make the necessary changes to assure the intellectual success of this series. To this end, we are delighted to include a new member of our team effort, Ms. Susan Kelly. She joined our cause to ensure that the format of questions, the reference process of answers to those questions within the text itself, and the editorial process involved be natural and clear to our readers. I am grateful for each of these facets of the overall publication process.

Not the least is my everlasting love and appreciation to my family. I am particularly indebted to my parents who inculcated in me at a very early age the love of education, the respect for study and hard work, and the honor for those who share these values. In this regard, it would have been impossible for me to accomplish any of my academic pursuits without the love, inspiration, and continued support of my wife, Sherry. Not only has she maintained the personal encouragement to initiate and continue with these labors of love, but she has sustained and supported our family and home life so that these activities could be encouraged. Hopefully, these pursuits have not detracted from the development and love of our children, Margie, Bruce, and Lara. I assume that this has not occurred; we are so very proud that each is personally committed to education and research. How satisfying it is to realize that these ideals remain a familial characteristic.

Edward D. Frohlich, MD, MACP, FACC
New Orleans, Louisiana

# Contents

# PART I

# Gross Anatomy

# Chapter 1

# Back

## OSTEOLOGY OF THE VERTEBRAL COLUMN

### Vertebral Column as a Whole

**Support of vertebral column**. The static support of the normally aligned vertebral column is provided primarily by the ligaments as well as by the intervertebral disk. As soon as motion occurs, the muscles become important in controlling the overall posture, but the relationship between adjacent vertebrae is still maintained largely by the ligaments. Bony support is a factor at only certain levels. The **thoracic region** is greatly reinforced by the thoracic cage and, as a result, vertebral dislocations seldom occur there. At other levels, the bony support is derived from the articular facets that form the zygapophyseal joints. The amount of this support, however, is dependent on the orientation of the joint space. In the **cervical region**, these joint spaces are nearly horizontal; therefore, only minimal bony resistance prevents one vertebra from sliding anteriorly or posteriorly relative to an adjacent vertebra. As a result, cervical dislocation can occur with only soft tissue damage (no fracture). On the other hand, the planes of the **lumbar zygapophyseal joints** are vertically oriented, with the inferior articular facets overlapping the superior articular facets of the next lower vertebra. Any tendency toward dislocation is resisted by interlocking of the articular surfaces, and **fracture** usually accompanies dislocation.

The stability of the lower lumbar region (especially the **lumbosacral junction**) is particularly dependent on bony support. The body of L5 sits on the anteriorly inclined superior aspect of the sacrum, and this vertebra has a natural tendency to slide anteriorly. This tendency is resisted by the zygapophyseal joints between L5 and the sacrum. Occasional bony discontinuity of the lamina between the superior and inferior articular processes results in a condition called **spondylolysis**. When this discontinuity is present, the bony support normally provided by the zygapophyseal joint is lost, and anterior sliding of the L5 body is predisposed. Anterior subluxation of the vertebral body is referred to as **spondylolisthesis**.

**Normal and abnormal curves**. The anteroposterior curves of the vertebral column are compensatory adjustments to the bipedal posture. Normally each junctional area (lumbosacral, thoracolumbar, cervicothoracic, occipitocervical) is directly above the center of gravity of the body as a whole, thus requiring little or no muscular activity to hold the spine upright during quiet standing. These curves are such that the lumbar and cervical regions present posterior concavities (referred to as **lordotic curves**), whereas the thoracic and sacral regions present posterior convexities. Although lateral curvatures normally do not exist, such a curvature is called a **scoliosis**. An exaggerated thoracic curve is referred to as a **kyphosis**.

**Motion**. Motion of the vertebral column as a whole is the sum of the variable amounts of motion that occur between adjacent vertebrae. Although the **intervertebral disk** is easily distorted in any direction and thus permits motion in any direction, its thickness or height is a major determinant of the extent of motion. In addition, motion is limited by the tension of the vertebral column ligaments, bony features of certain vertebrae, and extrinsic factors such as the rib cage and soft tissue bulk. The direction of motion is determined largely by the orientation of the plane of the zygapophyseal joint.

## Individual Vertebrae and Regional Variation

**Typical vertebra.** Most of the component parts of the vertebral column are similar in construction. The two major portions are the anterior body and the posterior vertebral or neural arch. The **body** is in the form of a flattened cylinder that serves as the major weight-bearing portion of the vertebra. The periphery of the cylinder is formed of dense bone, and the core is entirely trabecular bone; the superior and inferior aspects are covered by hyaline cartilage plates. The **vertebral arch** is formed by the pedicles and laminae, has multiple projections, and together with the posterior aspect of the body forms the vertebral foramen. The paired **pedicles** project posteriorly from the posterolateral aspect of each body, and the **laminae** project posteromedially (to join in the midline) from the posterior aspects of the pedicles. The **transverse processes** extend laterally from the arch; the single **spinous process** is directed posteriorly in the midline. The pairs of superior and inferior articular processes arise from the vertebral arch at the junctions of the pedicles and laminae. Each articular process contains an articular facet, which is an articular surface of the zygapophyseal joint.

**Regional variation.** The cervical region is unique because it can be separated into upper (craniovertebral) and lower portions, which are both anatomically and biomechanically different. The upper part consists of vertebra C1 and the upper aspect of C2, and the lower part consists of the inferior aspect of C2 and the rest of the cervical vertebrae. Vertebra C1 (atlas) has no body, but rather anterior and posterior arches that interconnect two lateral masses. Vertebra C2 (axis) is transitional in that it interconnects the upper and lower aspects of the cervical spine. The **dens**, or **odontoid process**, extends superiorly from the body of C2 and occupies the anterior aspect of the vertebral foramen between the lateral masses of C1. Both the superior and inferior articular facets of C1, as well as the superior facets of C2, are positioned anterior to those of the rest of the cervical vertebrae. In the lower cervical region, the vertebrae are distinguished by bifid spinous processes and foramina in the transverse processes (C1–C6), which transmit the vertebral arteries. The spinous process of vertebra C7 is considerably longer than the other cervical spines and thus is a reasonably dependable bony landmark. The lateral aspects of the bodies of vertebrae C3–C7 have prominent upward flares called **uncinate processes**; the adjacent inferolateral aspects of the bodies above are beveled and match the orientation of the uncinate process. Thoracic vertebrae have articular facets or costal fovea (with which the ribs form synovial joints) on the posterolateral aspects of the bodies and the anterior aspects of the tips of the transverse processes, and very long spinous processes that are directed inferiorly so that they overlap the next lower vertebra. Lumbar vertebrae have very large, heavy bodies and short, strong spinous processes that are directed posteriorly. Their laminae are about half as high (superoinferiorly) as their bodies; hence, an interlaminar space exists between lumbar vertebrae.

# ARTICULATIONS BETWEEN ADJACENT VERTEBRAE AND MOTION PERMITTED

## Intervertebral Disks

The bodies are united (and separated) by an intervertebral disk. This is a cartilaginous joint, and thus permits limited motion. The disk is composed of a gelatinous core (**nucleus pulposus**), which is surrounded by a strong distensible envelope (**annulus fibrosus**) of fibrocartilage. The nucleus pulposus is separated from each vertebral body by a hyaline cartilage plate. Because of this combination of materials and the way they are arranged, the disk exhibits the properties of a closed fluid-elastic system. That is, any pressure within the disk is delivered equally and undiminished to all parts of the container, which, in this case, are the annulus fibrosus and the cartilaginous plates. The disk is also preloaded; it pushes the vertebral bodies apart and thus aids the alignment of adjacent vertebrae. As a result, the disk plays a major role in both permitting and limiting motion as well as in alignment.

## Zygapophyseal Joints

The vertebral arches are connected by the **intervertebral,** or **zygapophyseal, articulations**. These are the synovial joints between the superior and inferior articular facets of adjacent vertebrae. The articular surfaces are enclosed by a strong joint capsule that permits limited amounts of gliding between the surfaces. The planes of the joints are oriented differently in each region, and are the major determinant of the direction of motion that occurs between adjacent vertebrae.

In the lower cervical region (C2–C7), the planes of these joints are positioned obliquely between the coronal and transverse planes. As a result of this orientation, flexion, extension, rotation, and side-bending occur at all levels. In addition, side-bending and rotation are coupled and occur to the same side. The planes of the thoracic facets are nearly coronal; however, because of the rib cage, very little vertebral motion occurs. The lumbar facets are curved and oriented in both the sagittal and coronal planes. The upper facets are more sagittal in orientation, the lower more coronal. This orientation greatly limits rotation and side-bending but allows free flexion and extension.

## Specializations in the Cervical Region

Some articulations are unique to both the upper and lower portions of the cervical spine. The zygapophyseal joints in the upper cervical region are oriented differently than those in any other part of the vertebral column. The **atlanto-occipital joints** are formed by the inferiorly convex occipital condyles and the superiorly concave articular facets of the atlas. These joints permit flexion and extension and some side-bending of the occipital bone on vertebra C1.

Three articulations exist between the atlas and axis. The **lateral atlanto-axial joints** are formed between the obliquely (sloping inferiorly from medial to lateral) oriented inferior facets of C1 and superior facets of C2. The **middle atlanto-axial joint** is formed between the dens and the anterior arch of the atlas anteriorly and the transverse ligament of the atlas posteriorly. Rotation of the atlas around the dens occurs at all three of these joints simultaneously. Some flexion and extension also occur.

The **joints of Luschka** are found between the bodies of the lower cervical vertebrae (C2–C7), specifically between the uncinate processes of the lower vertebrae and the beveled inferolateral aspects of the bodies of the vertebrae above. These "joints" are thought to be clefts rather than true joints. The maintenance of the intervals (clefts) between these aspects of adjacent vertebral bodies depends on the presence of healthy intervertebral disks. When disk height is lost, approximation of the bony surfaces and increased contact with motion occur. This contact commonly causes hypertrophic changes, which can impinge on adjacent neural tissues.

## Ligaments of the Vertebral Column

### *Continuous Ligaments*

Two ligaments extend from the upper part of the cervical spine to the sacrum. The **anterior longitudinal ligament** reinforces the anterior and anterolateral aspects of the vertebral bodies and intervertebral disks. Superior to the anterior arch of the atlas, this ligament is continuous with the **anterior atlanto-occipital membrane**. The **posterior lon-**

gitudinal ligament attaches to the posterior aspects of the bodies and disks and is therefore within the vertebral canal. This ligament supports the posterior aspect of the disk in the midline, but adds little support posterolaterally. Superior to the body of the axis, this ligament is continuous with the **tectorial membrane**. Because the axis of flexion and extension is thought to pass through the posterior aspect of the intervertebral disk, the anterior longitudinal ligament limits extension and the posterior limits flexion.

### *Segmental Ligaments*

The segmental ligaments interconnect adjacent vertebrae of the entire vertebral column, except those of the upper cervical region. The **ligamentum flavum** interconnects the laminae and differs from the typical ligaments in that it contains a large number of elastic fibers. Because of its location, it is stretched with flexion. During extension, it shortens because of its elasticity and thereby does not bulge into the vertebral canal.

The other segmental ligaments are named by their locations (i.e., the **intertransverse**, **interspinous**, and **supraspinous ligaments**). All of these ligaments limit flexion; the intertransverse ligaments also limit side-bending. The supraspinous ligament extends posterior to the spinous processes and thus may span several vertebrae. This ligament is particularly well developed in the cervical region, where it is called the **ligamentum nuchae**.

# CONTENTS OF THE VERTEBRAL CANAL AND SPINAL NERVE

## Spinal Cord and Spinal Nerve

The spinal cord is a long cylindrical structure with a very small central opening called the **central canal**. The central canal is part of the **ventricular system** and is surrounded by the gray matter (cell bodies and terminal arborizations), which in turn is surrounded by the white matter (long ascending and descending cell processes). The diameter of the cord decreases from top to bottom, with the exceptions of the low cervical and the lumbosacral regions, where enlargements reflect the upper and lower extremities, respectively. The spinal cord terminates inferiorly at the inferior aspect of the first lumbar vertebra. This termination is in the form of an inverted cone and thus is called the **conus medullaris**.

The cord is segmented, each segment corresponding to that specific portion of the body wall (including extremities) that it innervates. Even though this segmentation is not grossly visible, the general extent of each segment can be determined by the origin of the rootlets that form the dorsal and ventral roots. Vertical lines of nerve rootlets attach to the anterolateral and posterolateral aspects of the cord. The rootlets from a single segment converge and form **anterior** and **posterior roots**. The two roots join in the intervertebral foramen to form the **spinal nerve** (Fig. 1-1). The spinal nerve, found in the intervertebral foramen, divides into **ventral** and **dorsal rami**.

## Meninges

Both the brain and spinal cord are surrounded by three membranes that have both trophic and protective functions (see Fig. 1-1). The outermost of these layers is the strong **dura mater**, sometimes referred to as the **pachymeninx**. This layer forms a long cylindrical sac that tapers to an end inferiorly at vertebral layer S2. An inferior extension of the dura, the **coccygeal ligament**, extends inferiorly through the sacral hiatus to attach to the coccyx. The dura is separated from the bones and ligaments that form the verte-

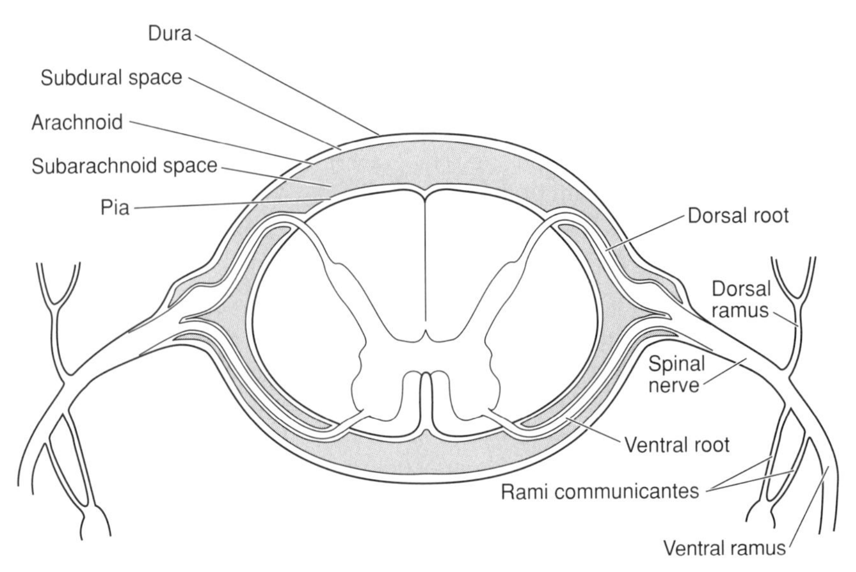


**Figure 1-1.** Cross-sectional diagram of the spinal cord, nerve roots, and meninges.

bral canal by an **epidural space,** which contains epidural fat and the **internal venous plexus** (Batson's plexus). This plexus is a valveless system of veins that extends the entire length of the vertebral canal and interconnects the body cavities and the cranial vault.

The two deeper layers, the **arachnoid** and **pia mater**, are much thinner than the dura and are called the **leptomeninges**. The innermost meninx is the pia mater, which contains the vessels that supply the central nervous system. This very thin membrane conforms closely to the contours of the spinal cord and is firmly attached to both the cord and the spinal nerve roots and rootlets.

There are two specializations of the pia. The **denticulate ligament** consists of a series of lateral extensions that attach to the dura mater to stabilize the spinal cord. The **filum terminale internum** extends inferiorly from the conus medullaris and joins the coccygeal ligament at the inferior extent of the dural sac. The intermediate meninx is the arachnoid, a thin filmy membrane that is loosely attached to the pia by numerous trabeculae. The arachnoid and pia mater are separated by the **subarachnoid space**, which is filled with cerebrospinal fluid. Because this fluid holds the arachnoid tightly against the dura, the subarachnoid space extends inferiorly as far as the dural sac and ends at vertebral level S2.

All three meningeal layers are continuous with the connective tissue coverings of the spinal nerve. This continuity occurs in the intervertebral foramen, where the dura and arachnoid join the pia mater as close investments of the spinal roots or nerve.

## Relationship of the Spinal Cord to the Vertebral Column

Because of the differential growth of the spinal cord and vertebral column, the spinal cord extends inferiorly only to the level of the lower border of the first lumbar vertebra. This means that only the very uppermost cervical cord segments are opposite the vertebra of the same name. Upper thoracic cord segments are one vertebral level higher than their correspondingly named vertebra, whereas the lumbar, sacral, and coccygeal cord segments lie opposite the last two thoracic and first lumbar vertebrae. Because each spinal nerve exits through its original intervertebral foramen, only the highest cervical spinal nerves are horizontally oriented, while each next lower nerve is more obliquely oriented as it travels farther to its intervertebral foramen. As a result of this incongruity between spinal cord and vertebral column, the symptoms resulting from a **spinal cord lesion** do not usually correspond to the vertebral level of the lesion. For example, a lesion at vertebral level T12 could logically be accompanied by symptoms that correspond to cord segments L3 and below.

The **cauda equina** is composed of lower lumbar, sacral, and coccygeal roots. These structures are the only neural elements in the subarachnoid space between the end of the spinal cord and the end of the dural sac. This area is the region of choice for spinal tap because risk to neural structures is minimal when a needle is inserted into the subarachnoid space. In addition, the interlaminar space between lumbar vertebrae allows easy access.

# INTERVERTEBRAL FORAMEN

## Normal Structure of the Intervertebral Foramen

The basic boundaries of this foramen are the same throughout the vertebral column. The superior and inferior aspects are the pedicles of the respective vertebrae. The anterior boundary consists of the intervertebral disk and portions of the adjacent vertebral bodies. Posteriorly, the superior and inferior articular facets form the zygapophyseal joint, and the lateral aspect of the ligamentum blends with the joint capsule to form the medial aspect of the boundary.

## Pathology Involving the Intervertebral Foramen

In the **cervical region**, the foramen is small, the intervertebral disk is positioned in the center of the anterior wall, and the relatively large spinal nerve almost fills the opening. **Protrusion** or **rupture of the disk** into the foramen will impinge on the nerve in the foramen; that is, rupture of the disk between cervical vertebrae 5 and 6 will involve spinal nerve C6. In addition, the size of the cervical intervertebral foramen can be reduced by **inflammation of the zygapophyseal joint** (arthritis) and by **bony projections** from the vertebral bodies. These posterolateral bony spurs are in the region of the posterior aspects of **Luschka's joints** on the posterolateral aspects of the vertebral bodies.

In the **lumbar region**, the disk forms the lower half of the anterior wall of the foramen, and the upper vertebral body forms the upper half. The opening is very large, and the relatively small spinal nerve exits in the upper part of the foramen opposite the vertebral body. Rupture of the disk at this level typically does not affect the spinal nerve in the same foramen, but rather the spinal nerve that is descending in the anterolateral aspect of the vertebral canal (across the posterolateral aspect of the disk) to exit from the next lower intervertebral foramen. Thus, a **rupture of the disk** between lumbar vertebrae 4 and 5 usually affects spinal nerve L5.

# MUSCLES OF THE BACK

The back is separated into superficial and deep regions by the **thoracolumbar fascia**. This fascia is well developed in the thoracic and lumbar regions, but gradually disappears in the cervical region. It has superficial and deep layers, and thus forms a sheath around the deep muscles of the back; it extends laterally from the spinous processes to the edge of the deep muscles and then medially to the transverse processes. The muscles superficial to this fascia are the superficial muscles of the back (Table 1-1). Because these muscles interconnect the bones of the shoulder girdle and axial skeleton, and thus control scapular position and motion, they are also called the **extrinsic**

**TABLE 1-1. Pectoral and Superficial Muscles of the Back**

| Muscle | Origin | Insertion | Action | Innervation |
|---|---|---|---|---|
| Trapezius | Cervical and thoracic spinous processes | Spine of scapula, acromion, lateral clavicle | Retraction and upward rotation of scapula | Accessory nerve, C3 and C4 |
| Latissimus dorsi | Thoracolumbar fascia (iliac crest, sacrum, and lumbar and thoracic vertebrae) | Intertubercular groove of humerus | Extension and medial rotation of humerus, shoulder depression | Thoracodorsal nerve |
| Serratus anterior | Anterolateral aspects of upper 8 or 9 ribs | Medial border of scapula, anterior aspect | Protraction, abduction, and upward rotation of scapula | Long thoracic nerve |
| Levator scapulae | Transverse processes—upper 3 or 4 cervical vertebrae | Superior angle of scapula | Elevation of scapula | Dorsal scapular nerve, C2 and C3 |
| Rhomboid major and minor | Spinous processes—lower cervical and upper thoracic vertebrae | Medial border of scapula | Downward rotation, retraction, and inferior rotation of scapula | Dorsal scapular nerve |
| Pectoralis major | Manubrium and body of sternum, lateral anterior clavicle | Crest of greater tubercle of humerus | Flexion, medial rotation, and horizontal adduction of humerus | Medial and lateral pectoral nerves |
| Pectoralis minor | Anterior aspects of ribs 2–5 | Coracoid process | Depression of scapula | Medial pectoral nerve |

**muscles of the shoulder**. They are mostly innervated by branches of the brachial plexus and have little involvement in movement of the vertebral column.

The deep or **intrinsic muscles of the back** consist of multiple groups of muscles that extend from the occipital bone to the sacrum, occupy the depression formed by the spinous and transverse processes, and are innervated segmentally by branches of the dorsal rami (Table 1-2). Bilateral contraction of these muscles produces extension of the vertebral column; unilateral contraction causes side-bending and rotation. In the upper cervical region, the **suboccipital muscles** (Table 1-3) are specifically designed to accommodate the unique kinematics of that region.

## BLOOD SUPPLY OF THE BACK

The blood supply of the back is provided predominantly by the segmental branches of the aorta. In the thoracic region these are the **posterior intercostal arteries** and in the lumbar region, the **lumbar arteries**. The cervical region is supplied by branches of the subclavian artery (**transverse cervical**, **thyrocervical trunk**).

## TABLE 1-2. Deep Muscles of the Back

| Muscle Group and Components | Origin → Orientation → Insertion | Regions Found |
|---|---|---|
| Spinotransversalis group<br>Splenius capitis<br>Splenius cervicis | Lower cervical and upper thoracic spinous processes → ascend and pass laterally → base of skull and upper cervical transverse processes | Cervical and upper thoracic |
| Erector spinae (sacrospinalis) group | | |
| Iliocostalis | Ilium and ribs → ascend → ribs | All levels |
| Longissimus | Transverse processes → ascend → transverse processes | Thoracic and cervical |
| Spinalis | Spinous processes → ascend → spinous processes | Thoracic and cervical |
| Transversospinal group | | |
| Semispinalis | All components: transverse processes → ascend obliquely for few segments → spinous processes | Cervical and thoracic |
| Multifidi | | All levels |
| Rotatores | | All levels |
| Segmental group<br>Interspinales<br>Intertransversarii | Interconnect adjacent spinous and transverse processes | All levels |

## TABLE 1-3. Suboccipital Muscles

| Muscle | Origin | Insertion | Action | Innervation |
|---|---|---|---|---|
| Rectus capitis posterior major | Spinous process of vertebra C2 | Occipital bone–lateral inferior nuchal line | Extension and ipsilateral rotation of occipital bone | Suboccipital nerve |
| Rectus capitis posterior minor | Posterior tubercle of vertebra C1 | Occipital bone–medial inferior nuchal line | Extension of occipital bone | Suboccipital nerve |
| Obliquus capitis superior | Transverse process of vertebra C1 | Occipital bone–lateral inferior nuchal line | Extension and ipsilateral side-bending of occipital bone | Suboccipital nerve |
| Obliquus capitis inferior | Spinous process of vertebra C2 | Transverse process of vertebra C1 | Ipsilateral rotation of upper cervical spine | Suboccipital nerve |

**TABLE 1-4. Lymphatics of the Back**

| Afferents From | Lymph Nodes | Efferents To |
|---|---|---|
| **Superficial back** | | |
| Cervical region | Cervical | Thoracic duct on left; right lymphatic trunk on right |
| Trunk (above umbilicus) | Axillary | Deep cervical nodes |
| Trunk (below umbilicus) | Superficial inguinal | External iliac |
| **Deep back** | | |
| Cervical region | Deep cervical | Thoracic duct or right lymphatic duct |
| Thoracic region | Mediastinal | Thoracic duct or right lymphatic duct |
| Lumbar and sacral regions | Aortic and sacral | Thoracic duct |

## LYMPHATICS OF THE BACK

The lymphatic drainage of the back is summarized in Table 1-4.

# Chapter 2

# Upper Limb

This chapter is organized with an initial discussion of the bones and articulations of the entire limb, followed by a consideration of each region (e.g., posterior cervical triangle, arm), and finally a discussion of the neurovascular structures.

## OSTEOLOGY

### Clavicle

The clavicle and scapula form the **shoulder** or **pectoral girdle**. The clavicle, through its articulations, is the only bony connection between the upper limb and the axial skeleton. As such, it keeps the limb away from the body, thereby enabling a large range of motion. Proximally directed force through the upper limb frequently causes **clavicular fracture**. These fractures are usually in the middle third of the bone because of its doubly curved shape. They are also easily diagnosed by palpation because of the clavicle's completely subcutaneous location. The clavicle is doubly curved with the medial half convex anteriorly and the lateral half convex posteriorly. The medial half is round in cross section and has an enlarged medial end, a small portion of which articulates with the sternum. Laterally the bone is flat and it has a small oval facet that articulates with the acromion.

### Scapula

The scapula is thin and flat and roughly **triangular**, creating three angles. Its apex is directed inferiorly and called the **inferior angle**; the **superior angle** is positioned superomedially; and the **lateral angle** is located superolaterally and formed by the glenoid fossa. Extending superomedially and superolaterally from the inferior angle is the medial or vertebral border and the lateral or axillary border. The superior border is the superior edge of the scapula. The anterior surface of the scapula, the subscapular fossa, is slightly concave, corresponding somewhat to the surface of the thoracic wall. The posterior surface is mildly convex and separated into supraspinatus and infraspinatus fossae by the obliquely oriented elevated ridge, the spine of the scapula. The spine extends from the medial border of the scapula (the base of the spine) to the lateral angle, where it extends laterally as the expanded acromion process. The **coracoid process** projects anteriorly from the superior border approximately 2 to 3 cm medial to the glenoid fossa.

### Humerus

The humerus is the **only bone in the arm**. Its expanded proximal end consists of the head and greater and lesser tubercles. The **humeral head** is entirely articular surface and directed superomedially; it joins the rest of the bone at the **anatomical neck**. The

**lesser tubercle** is directed anteriorly and is separated from the greater tubercle by the **intertubercular groove**, which houses the tendon of the long head of the biceps brachii muscle. The **greater tubercle** forms the entire superolateral aspect of the humerus and consists of three facets.

The humeral shaft is cylindrical. Proximally, the tapering region between the shaft and neck is the **surgical neck** of the humerus, so named because it is a common site of fracture. The **deltoid tuberosity** is located laterally at the midshaft level, and the **spiral** or **radial groove** curves posterolaterally around the bone. Distally, the shaft flattens and is inclined anteriorly.

As the distal end of the humerus widens, it consists of a medial and lateral column separated by depressions anteriorly and posteriorly, the **coronoid** and **olecranon fossae**, respectively. The distal end has two articular surfaces, the trochlea medially and the **capitulum** laterally. The medial epicondyle is large and separated from the trochlea by a deep notch; the lateral epicondyle is less prominent. Each condyle has a supracondylar ridge that extends superiorly, where it joins the shaft.

## Radius and Ulna

The radius and ulna are the **bones of the forearm**. The **radius** is the lateral bone of the forearm and bowed laterally to facilitate pronation. It has a small cylindrically shaped head proximally, a tapering neck, and a prominent **biceps tuberosity** just distal to the neck. The expanded distal part of the bone is the major forearm contribution to the wrist joint. The distal aspect consists of the lateral palpable **radial styloid**, which is the most distal bony prominence of the forearm, and the **dorsal radial tubercle** (of Lister), which is the most prominent bony landmark on the posterior aspect of the distal radius. There are two articular surfaces on the distal radius: the medially directed **ulnar (sigmoid) notch** and the distally directed aspect that consists of **articular facets** for the **scaphoid** (laterally) and lunate (medially).

The **ulna** is the reverse of the radius in that it has a large proximal extremity and is quite small distally. Proximally, the anteriorly directed **trochlear notch** is articular surface. It is bounded anteriorly by the coronoid process and posteriorly by the prominent **olecranon process**. Another articular surface, the radial notch, is found on the lateral aspect of the coronoid process. The distal aspect of the ulna is quite small, consisting of a small cylindrical **head** (articular surface) and a **styloid process** that extends distal to the head on its dorsomedial aspect.

## Bones of the Wrist

The bones of the wrist or carpus are the eight **carpal bones** that are arranged in two rows. From lateral to medial, the proximal row consists of the scaphoid, lunate, triquetrum, and pisiform bones. The distal row is composed of the trapezium, trapezoid, capitate, and hamate bones. These bones are arranged so the medial and lateral bones extend anterior to the more central bones, and thus form an anterior proximal-to-distal depression. This depression is bridged anteriorly by connective tissue to form the carpal tunnel.

## Bones of the Hand

The bones of the hand include the 5 **metacarpals** and 14 **phalanges**. The metacarpals consist of a base proximally and a head distally. The shafts are bowed dorsally to accommodate the soft tissue of the palm. The first metacarpal (of the thumb) is shorter and stouter than the others. Each of the digits is composed of three phalanges, except the thumb, which has only two. The phalanges are named by their positions: proximal, middle, and distal. Like the metacarpals, each phalanx has a proximal base and distal head, except the distal phalanges, which have only bases.

# ARTICULATIONS OF THE UPPER LIMB

## Shoulder Region

**Glenohumeral joint**. The shoulder joint is a **synovial joint** that is classified as a **ball-and-socket joint** between the head of the humerus and the glenoid fossa of the scapula (Fig. 2-1). Maximal mobility is available at the expense of stability. The glenoid cavity is deepened somewhat by the glenoid labrum, a fibrocartilaginous wedge that attaches to the periphery of the glenoid fossa. The joint capsule extends from the rim of the glenoid fossa and the labrum to the anatomical neck of the humerus; in the anatomical position, it is quite loose inferiorly. The tendons of the rotator cuff muscles (i.e., subscapularis, supraspinatus, infraspinatus, teres minor) blend with the lateral aspect of the joint capsule and are the major support of the joint. The capsule is also reinforced anteriorly and inferiorly by the glenohumeral and coracohumeral ligaments. The tendon of the long head of the biceps brachii ascends through the intertubercular groove and then passes across the superior aspect of the humeral head (between the fibrous and synovial portions of the capsule) to attach to the supraglenoid tubercle.

**Motion at the shoulder joint is free** and occurs around an infinite number of axes. The named motions occur mostly in the cardinal planes and are called **flexion** and **extension** (sagittal plane), **abduction** and **adduction** (coronal plane), and **medial** (internal) and **lateral** (external) **rotation** (axial rotation). The major muscles producing these motions at the glenohumeral joint are indicated in Table 2-1. Although shoulder joint motion is a major factor in allowing the hand to assume innumerable locations and positions, proper function of the entire shoulder complex is also necessary. Scapular motion, as well as motion of the clavicle, accompanies virtually every positional change of the upper limb. A loss of scapular, sternoclavicular, acromioclavicular, or glenohumeral range of motion can reduce the range of the entire upper limb.

**Sternoclavicular joint.** The medial end of the clavicle articulates with the superolateral aspect of the manubrium at the sternoclavicular joint. The bones are separated by an **intra-articular disk**, which supports the joint by resisting elevation and medial dislocation of the medial end of the clavicle. It also partitions the joint into two separate

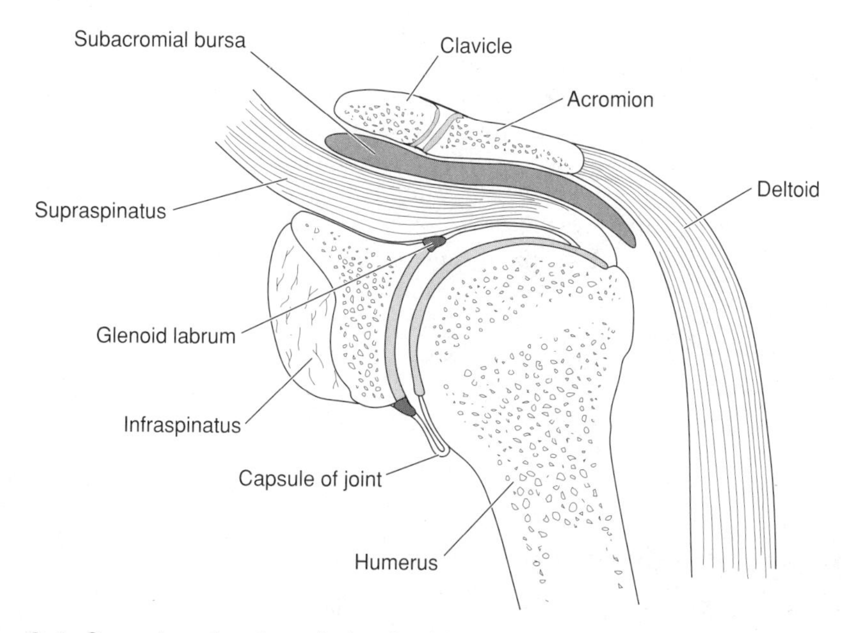


**Figure 2-1.** Coronal section through the shoulder joint and suprahumeral space.

**TABLE 2-1. Primary Muscles Producing Motion at the Glenohumeral Joint**

| Motion | Muscles Producing Motion |
|---|---|
| Flexion | Anterior deltoid, pectoralis major (clavicular part), coracobrachialis |
| Extension | Latissimus dorsi, teres major, posterior deltoid |
| Abduction | Middle deltoid, supraspinatus |
| Adduction | Latissimus dorsi, pectoralis major (sternal part) |
| Medial rotation | Pectoralis major, latissimus dorsi, teres major, anterior deltoid, subscapularis |
| Lateral rotation | Posterior deltoid, teres minor, infraspinatus |

synovial cavities. The joint also is reinforced by the articular capsule and sternoclavicular, interclavicular, and costoclavicular ligaments. The lateral end of the clavicle can be elevated and depressed, moved anteriorly and posteriorly, and undergo axial rotation.

**Acromioclavicular joint**. The lateral end of the clavicle articulates with the acromion at the acromioclavicular joint. The small oval articular surfaces are enclosed by a capsule, which is reinforced by the superior acromioclavicular ligament. The major supports for this joint, however, are the components of the coracoclavicular ligament (conoid and trapezoid ligaments), which extend between the inferior aspect of the clavicle and the coracoid process. These two ligaments also maintain a rather constant relationship between the scapula and clavicle.

**Thoracoscapular joint**. The thoracoscapular joint is **not a true articulation** but rather a pair of fascial planes that permit scapular movement between the scapula and thoracic wall. These planes are between the subscapularis and serratus anterior muscles, and between the serratus anterior and thoracic wall. Although the scapula has a bony link to the axial skeleton through the clavicle, it can move in a variety of ways. It can translate superiorly (**elevation**) or inferiorly (**depression**), and it can move laterally (**abduction**) or medially (**adduction**). It can also rotate around an anteroposterior axis so that the lateral angle moves superiorly (**upward or superior rotation**) or inferiorly (**downward or inferior rotation**). In addition, the scapula can rotate around a vertical axis, so its lateral aspect moves anteriorly (protraction) or posteriorly (retraction). The muscles that produce these motions of the scapula are listed in Table 2-2.

**Suprahumeral (subacromial) space**. This is **not a space** but rather an area packed with structures of various types. The suprahumeral space (see Fig. 2-1) is between the head of the humerus inferiorly and the arch formed by the acromion, coracoid, the intervening coracoacromial ligament, and the acromioclavicular joint. The major structures within this space are the superior portion of the capsule of the shoulder joint, the tendon and part of the supraspinatus muscle, the subdeltoid (subacromial) bursa, and the tendon of the long head of the biceps brachii muscle. Because humeral motion accompanies virtually every motion of the upper limb, the structures in this space are compressed continuously between the head of the humerus and the coracoacromial arch. As a result, **inflammation** of any of these structures can result in **pain** and **loss of range of motion** of the shoulder area.

## Elbow and Proximal Radioulnar Joints

**Elbow joint**. The elbow joint is formed between the trochlea of the humerus and the trochlear notch of the ulna medially, and between the capitulum of the humerus and

**TABLE 2-2. Muscles Controlling the Scapula**

| Motion | Muscles Moving Scapula |
|---|---|
| Elevation | Levator scapulae, upper trapezius |
| Depression | Lower trapezius, pectoralis minor, latissimus dorsi, pectoralis major |
| Abduction and protraction | Serratus anterior, pectoralis major and minor |
| Adduction and retraction | Rhomboids, trapezius |
| Upward rotation | Serratus anterior, upper and lower trapezius |
| Downward rotation | Rhomboids, pectoralis major and minor, latissimus dorsi |

the head of the radius laterally. The motion permitted at this joint—flexion and extension—is determined by the part of the joint between the humerus and ulna. A common joint capsule encloses the two portions of the elbow joint in addition to the proximal radioulnar joint. The capsule is thickened medially to form the ulnar (medial) collateral ligament and laterally to form the radial (lateral) collateral ligament. The major flexors at the elbow are the brachialis, biceps brachii, and brachioradialis. The major extensor is the triceps brachii.

**Proximal radioulnar joint**. The ulna and radius are united through synovial joints both proximally and distally, with the interosseous membrane interconnecting the two shafts between these synovial joints. The proximal radioulnar joint is formed by the radial head and the radial notch of the ulna. It shares an articular capsule with the elbow joint and is stabilized primarily by the **annular ligament**, which attaches to the edges of the radial notch of the ulna and surrounds the head of the radius.

**Pronation** (palm down) and **supination** (palm up) are the two motions that occur between the radius and ulna. When the hand is supinated, the two bones are parallel. When the hand is pronated, the radius is wrapped around the ulna. Pronation is produced primarily by the pronator teres and pronator quadratus; supination is produced by the biceps brachii and the supinator.

## Wrist and Distal Radioulnar Joints

**Distal radioulnar joint**. This articulation is formed between the head of the ulna and the ulnar notch of the radius, and between the head of the ulna and the intra-articular disk (triangular fibrocartilage) of the wrist. Even though the disk strongly reinforces the joint, it attaches to the ulnar styloid and thus permits the radius to rotate around the ulnar head. Because the joint capsule attaches to the intra-articular disk, the synovial cavity of the distal radioulnar joint is separate from that of the radiocarpal joint.

**Wrist joint**. The wrist consists of two separate articulations: the proximal radiocarpal (wrist) joint and the distal midcarpal (transverse carpal) joint. The radiocarpal joint is formed between the distal aspect of the radius and the intra-articular disk proximally, and the proximal row of carpal bones distally. Of the proximal row, the majority of the surface is formed by the scaphoid and lunate. The radiocarpal joint has its own synovial cavity, and it is separated from the distal radioulnar joint

**TABLE 2-3. Primary Muscles of the Wrist**

| Motion | Muscles Producing Motion |
|---|---|
| Flexion | Flexor carpi radialis, flexor carpi ulnaris |
| Extension | Extensor carpi radialis longus and brevis, extensor carpi ulnaris |
| Abduction (radial deviation) | Extensor carpi radialis longus and brevis, flexor carpi radialis, abductor pollicis longus, extensor pollicis brevis |
| Adduction (ulnar deviation) | Flexor carpi ulnaris, extensor carpi ulnaris |

by the intra-articular disk and from the midcarpal joint by the interosseous ligaments. The **interosseous ligaments** interconnect the adjacent bones in each of the carpal rows.

The **midcarpal articulation**, which also has a separate joint cavity, is found between the two rows of carpal bones. The capsules of both joints are reinforced by collateral, dorsal, and palmar radiocarpal ligaments. The motions that occur in the area of the wrist are contributed to by movement at both the radiocarpal and midcarpal joints. These motions and the muscles that produce them are listed in Table 2-3.

**TABLE 2-4. Muscles Producing Motion in the Hand**

| Motion | CMC Joint of Thumb | Four Medial MCP Joints | MCP Joint of Thumb | Four Medial PIP Joints | IP Joint of Thumb | DIP Joints |
|---|---|---|---|---|---|---|
| Flexion | Flexor pollicis longus and brevis | Lumbricals, interossei, flexor digitorum profundus and superficialis | Flexor pollicis longus and brevis | Flexor digitorum profundus and superficialis | Flexor pollicis longus | Flexor digitorum profundus |
| Extension | Extensor pollicis longus and brevis | Extensor digitorum, digit minimi, and indicis | Extensor pollicis longus and brevis | Lumbricals, interossei | Extensor pollicis longus | Lumbricals, interossei |
| Abduction | Abductor pollicis longus and brevis | Dorsal interossei | Abductor pollicis brevis | — | — | — |
| Adduction | Adductor pollicis | Palmar interossei | Adductor pollicis | — | — | — |
| Opposition | Opponens pollicis, flexor pollicis brevis | — | — | — | — | — |

*CMC, carpometacarpal; DIP, distal interphalangeal; IP, interphalangeal; MCP, metacarpophalangeal; PIP, proximal interphalangeal.*

## Joints of the Hand

**Carpometacarpal (CMC) joints.** The CMC articulations are formed between the distal row of carpal bones and the bases of the metacarpals. The articular surfaces forming the four medial joints are irregular in shape, and the bones are bound together by strong ligaments. As a result, little or no motion is available at these articulations even though they are synovial joints. The articular surfaces forming the CMC joint of the thumb, the trapezium and base of the first metacarpal, are both saddle-shaped, and the capsule and reinforcing ligaments are rather loose. Relatively free motion, therefore, is available at this joint and consists of flexion, extension, abduction, adduction, and, hence, circumduction. Circumduction is the essential ingredient of opposition.

**Metacarpophalangeal (MCP) joints.** The MCP joints are synovial in type and permit flexion, extension, abduction, adduction, and circumduction. Abduction and adduction are free only in extension as the collateral ligaments become taut in flexion and thereby reduce any side-to-side movement.

**Interphalangeal joints.** Both the **proximal interphalangeal (PIP)** and **distal interphalangeal (DIP)** joints are synovial joints formed between the bases and heads of adjacent phalanges. Due both to the tightness of the collateral ligaments and the bone architecture of the articular surfaces, these joints permit only flexion and extension.

The muscles producing motion in the hand are listed in Table 2-4.

# JUNCTIONAL REGIONS BETWEEN THE TRUNK AND UPPER LIMB

## Posterior Cervical Triangle

The posterior cervical triangle, or **posterior triangle of the neck,** is included in a discussion of the upper limb because this triangle houses the major neurovascular structures that supply most of the upper limb. In addition, common pathologies that affect the upper limb occur in this region.

**Boundaries.** The borders of the posterior triangle are easily defined on the surface and consist of the posterior border of the sternocleidomastoid muscle, the superior border of the trapezius muscle, and the middle third of the clavicle. The floor is entirely muscular and formed by the scalene and levator scapulae muscles.

**Contents.** The structures in the posterior triangle are embedded in a mass of loose connective tissue that usually contains a significant amount of fat. The **neurovascular structures** in the triangle include the roots or trunks, or both, of the brachial plexus, the first part of the subclavian artery and a variable number of branches, the phrenic nerve as it obliquely crosses the anterior scalene muscle, the accessory nerve (CN XI), and the cutaneous branches of the cervical plexus (i.e., supraclavicular, transverse cervical, greater auricular, and lesser occipital nerves). The accessory nerve passes superficially across the triangle along a course corresponding to a line between the earlobe and point of the shoulder. The cutaneous branches of the cervical plexus emerge from the triangle posterior to the middle third of the posterior border of the sternocleidomastoid muscle. A number of **lymph nodes** are also present in the triangle. The accessory nodes are associated with the accessory nerve, and the inferior deep cervical or scalene nodes are positioned on the anterior aspect of the anterior scalene muscle. The omohyoid muscle, one of the infrahyoid muscles, crosses the inferomedial aspect of the triangle.

## Pectoral Region

The pectoral region is the superolateral aspect of the trunk occupied by the **pectoral muscles**. It also contains the **mammary gland** in the subcutaneous tissue. The pec-

**TABLE 2-5. Muscles of the Arm**

| Muscle | Origin | Insertion | Action | Innervation |
|---|---|---|---|---|
| **Anterior compartment** | | | | |
| Biceps brachii | Supraglenoid tubercle, coracoid process | Radial tuberosity | Forearm flexion and supination | Musculocutaneous |
| Brachialis | Distal half of anterior humerus | Coronoid process of ulna | Forearm flexion | Musculocutaneous |
| Coracobrachialis | Coracoid process | Medial midhumerus | Flexion of arm | Musculocutaneous |
| **Posterior compartment** | | | | |
| Triceps brachii | Infraglenoid tubercle, posterior humerus | Olecranon process | Forearm extension | Radial |

toralis major and minor muscles are extrinsic muscles of the shoulder and are described in Table 1-1.

## Axillary Region

**Boundaries**. The axilla, or axillary region, is the junctional region through which the major neurovascular structures enter and leave the upper limb. Although irregular in shape, the axilla can be compared to a pyramid. Its base or floor is the concave axillary skin and fascia; its anterior wall is formed by the pectoral muscles, and its posterior wall by the latissimus dorsi, teres major, and subscapularis muscles. The medial wall is the serratus anterior muscle. The lateral wall is the thin strip of the humerus between the insertions of the pectoralis major and latissimus dorsi muscles. These walls converge superomedially toward an apex that is formed by the first rib, clavicle, and superior border of the scapula.

**Contents**. All of the contents are embedded in a mass of fatty, loose connective tissue. The **brachial plexus** and axillary vessels are encased in the fascial axillary sheath and extend from the apex to base where they enter the arm. The **intercostobrachial nerve**, the lateral cutaneous branch of the second intercostal nerve, passes through the axilla as it passes toward the skin of the medial arm. Several groups of axillary lymph nodes are also present. These nodes drain the entire upper limb, the pectoral region including the mammary gland, and the scapular region.

# ARM

The arm is divided into **anterior** and **posterior compartments** by the **medial** and **lateral intermuscular septa**, which extend from the investing brachial fascia to the humerus. The muscles in the anterior compartment are the **biceps brachii**, **brachialis**, and **coracobrachialis**. These muscles are innervated by the musculocutaneous nerve and produce flexion at the elbow and supination of the forearm. Only the **triceps brachii** is found in the posterior compartment. This elbow extensor is innervated by the radial nerve. These muscles are described in Table 2-5. At the junction of the medial intermuscular septum and the brachial fascia, the medial neurovascular bundle descends through the arm. This bundle contains the brachial artery and median, ulnar, and medial antebrachial cutaneous nerves.

## CUBITAL FOSSA

The cubital fossa is the depression anterior to the elbow and proximal part of the forearm. This fossa is bounded laterally by the brachioradialis muscle, medially by the pronator teres muscle, and proximally by a line interconnecting the medial and lateral epicondyles of the humerus. The floor of the fossa is the brachialis muscle. The tendon of the biceps brachii muscle disappears into this fossa as it passes toward the radial tuberosity; the bicipital aponeurosis (lacertus fibrosus) extends medially from the biceps tendon to the investing fascia of the forearm. This aponeurosis separates the superficially positioned median cubital vein from the deeper structures that pass through the fossa. The median cubital vein, commonly used for veni puncture, passes superficial to the aponeurosis. The deeper structures are the **brachial artery**, which passes medial to the biceps tendon, and the **median nerve**, which is medial to the brachial artery.

## FOREARM

The forearm is separated into **anterior** and **posterior compartments** by the **medial** and **lateral intermuscular septa**, which extend from the investing **antebrachial fascia** to the ulna and radius, respectively, and by the **interosseous membrane**, which interconnects the radius and ulna. The muscles in both the anterior and posterior compartments are

**TABLE 2-6. Muscles of the Anterior Compartment of the Forearm**

| Muscles | Origin | Insertion | Action | Innervation |
|---|---|---|---|---|
| **Superficial group** | | | | |
| Pronator teres | Medial epicondyle and ulna | Lateral midshaft of radius | Pronation | Median |
| Flexor carpi radialis | Medial epicondyle | Anterior base of second metacarpal | Flexion and abduction of hand | Median |
| Palmaris longus | Medial epicondyle | Palmar aponeurosis | Tense palmar fascia | Median |
| Flexor digitorum superficialis | Medial epicondyle, radius, and ulna | Anterior bases of middle phalanges | Flexion of middle phalanges | Median |
| Flexor carpi ulnaris | Medial epicondyle and ulna | Pisiform | Flexion and adduction of hand | Ulnar |
| **Deep group** | | | | |
| Flexor digitorum profundus | Anterior proximal two-thirds of ulna | Anterior bases of distal phalanges | Flexion of distal phalanges | 2, 3: median<br>4, 5: ulnar |
| Flexor pollicis longus | Ventral middle half of radius | Anterior base of thumb distal phalanx | Flexion of thumb distal phalanx | Median |
| Pronator quadratus | Distal anterior ulna | Distal anterior radius | Pronation | Median |

**TABLE 2-7. Posterior Muscles of the Forearm**

| Muscle | Origin | Insertion | Action | Innervation |
|---|---|---|---|---|
| **Superficial group** | | | | |
| Brachioradialis | Lateral supracondylar ridge of humerus | Distal lateral radius | Forearm flexion | Radial |
| Extensor carpi radialis longus | Lateral epicondyle of humerus | Dorsal base of second metacarpal | Extend and abduct hand | Radial |
| Extensor carpi radialis brevis | Lateral epicondyle of humerus | Dorsal base of third metacarpal | Extend and abduct hand | Radial |
| Extensor digitorum | Lateral epicondyle of humerus | Extensor aponeurosis of four medial digits | Extension of proximal phalanges | Radial |
| Extensor digiti minimi | Lateral epicondyle of humerus | Extensor aponeurosis of little finger | Extension of proximal phalanx | Radial |
| Extensor carpi ulnaris | Lateral epicondyle and posterior ulna | Dorsomedial base of fifth metacarpal | Extend and adduct hand | Radial |
| **Deep group** | | | | |
| Supinator | Lateral epicondyle and adjacent ulna | Anterolateral midshaft of radius | Supination | Radial |
| Abductor pollicis longus | Posterior proximal radius and ulna | Anterior base of first metacarpal | Abduct and extend thumb | Radial |
| Extensor pollicis longus | Dorsal midshaft of ulna | Lateral base of thumb distal phalanx | Extend thumb distal phalanx | Radial |
| Extensor pollicis brevis | Dorsal distal radius | Lateral base of thumb proximal phalanx | Extend thumb proximal phalanx | Radial |
| Extensor indicis | Dorsal distal ulna | Extensor aponeurosis of index finger | Extend index finger proximal phalanx | Radial |

divided into **superficial** and **deep groups**. The superficial groups cross the elbow and thus may contribute to motion at that joint. The superficial flexors are the pronator teres, flexor carpi radialis, palmaris longus, flexor carpi ulnaris, and flexor digitorum superficialis (which may be considered as an intermediate muscle). The superficial extensors are the brachioradialis, extensor carpi radialis longus, extensor carpi radialis brevis, extensor digitorum, extensor digiti minimi, and extensor carpi ulnaris. The deep groups arise from the radius, ulna, and interosseous membrane. The deep flexors are the flexor digitorum profundus, flexor pollicis longus, and pronator quadratus. The deep extensors are the supinator, abductor pollicis longus, extensor pollicis brevis, extensor pollicis longus, and extensor indicis. These muscles are described in Tables 2-6 and 2-7.

## WRIST

The wrist is the junctional region between the forearm and hand and consists of a very short segment of the upper limb. On the anterior surface, it begins at about the distal carpal skin crease and extends approximately 2 to 3 cm distally. This area contains all of the **carpal bones** and the **tendons** and **neurovascular structures** that enter the hand.

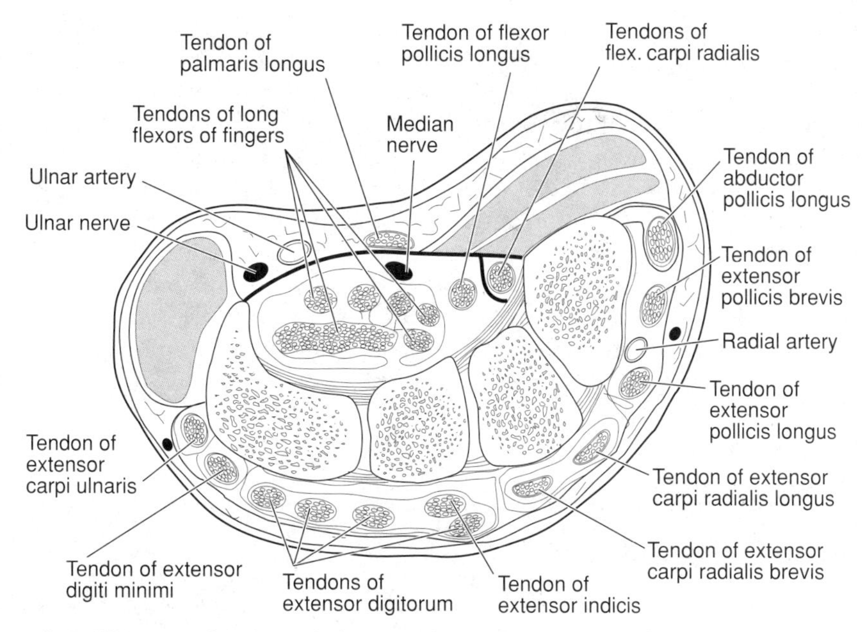


**Figure 2-2.** Cross section through the carpal tunnel.

## Volar Wrist

**Relationships. Four bony prominences** are palpable on the volar wrist. On the medial side, the **pisiform** is just distal to the distal wrist crease, and the **hook of the hamate** is just distal and lateral to the pisiform. On the lateral side, the **tubercles of the scaphoid** and **trapezium** are palpable in line with the tendon of the flexor carpi radialis muscle. From lateral to medial, the palpable tendons are those of the flexor carpi radialis, palmaris longus, flexor digitorum superficialis, and flexor carpi ulnaris. The radial artery is lateral to the tendon of the flexor carpi radialis, and the ulnar artery is lateral to the tendon of the flexor carpi ulnaris. The median nerve is between the tendons of the flexor carpi radialis and palmaris longus. The ulnar nerve is deep to the tendon of the flexor carpi ulnaris.

**Carpal tunnel**. The carpal tunnel interconnects the anterior compartment of the forearm and the palm of the hand (Fig. 2-2). It is a **fibro-osseous canal**, formed posteriorly and on both sides by the carpal bones and ventrally by the deep part of the flexor retinaculum (transverse carpal ligament). The structures that pass through this tunnel (canal) include the tendons of the flexor digitorum superficialis, flexor digitorum profundus, and flexor pollicis longus muscles, the synovial sheaths of those tendons, and the median nerve. Any of these structures, particularly the median nerve, can be compressed within the canal (**carpal tunnel syndrome**). The ulnar nerve and artery, tendon of the palmaris longus muscle, and palmar branch of the median nerve also cross the ventral aspect of the wrist but pass superficial to the deep part of the flexor retinaculum.

## Dorsal Wrist

**Relationships**. The most prominent bony landmarks of the dorsal aspect of the wrist are the dorsal radial tubercle (of Lister), which is about in line with the index finger, and the ulnar styloid on the most medial aspect of the wrist. The tendons crossing the dorsal aspect of the wrist are separated into six compartments, formed between the radius and ulna and the extensor retinaculum. These **compartments** are numbered from lateral to medial. The *first* compartment contains the tendons of the abductor pollicis longus and extensor pollicis brevis; the *second*, the extensor carpi radialis longus and brevis. Compartment *three* contains the extensor pollicis longus. The tendons of compartments one and three define a depression, the **anatomical snuff box**, on the dorsolateral aspect of the base of the thumb. The scaphoid can be palpated in the snuff box. Compartment *four* contains the tendons of two muscles, the extensor digitorum and extensor indicis. Compartments *five* and *six* contain one tendon each, those of the extensor digiti minimi and extensor carpi ulnaris, respectively.

Only cutaneous nerves cross the dorsal wrist. The superficial branch of the **radial nerve** crosses the dorsolateral aspect as it passes toward the lateral digits, and the dorsal branch of the **ulnar nerve** crosses the dorsomedial aspect. Only the radial artery crosses the wrist dorsally; it passes through the anatomical snuff box toward the first web space.

# HAND

## Retinacular System

The antebrachial fascia of the forearm continues into the hand where it attaches to the first and fifth metacarpals (Fig. 2-3). Distal to the flexor retinaculum of the wrist, the investing fascia is very thin laterally (thenar fascia) and medially (hypothenar fascia) and greatly thickened centrally as the palmar aponeurosis. This aponeurosis is very narrow proximally but widens distally, where it separates into four digital slips interconnected by the superficial transverse metacarpal ligament. The thenar septum extends from the junction of the palmar aponeurosis and thenar fascia to the first metacarpal; the hypothenar septum extends from the junction of the palmar aponeurosis and hypothenar septum to the fifth metacarpal. The anterior and posterior aspects of the five metacarpals are interconnected by the anterior adductor-interosseous and dorsal interosseous fasciae, respectively.

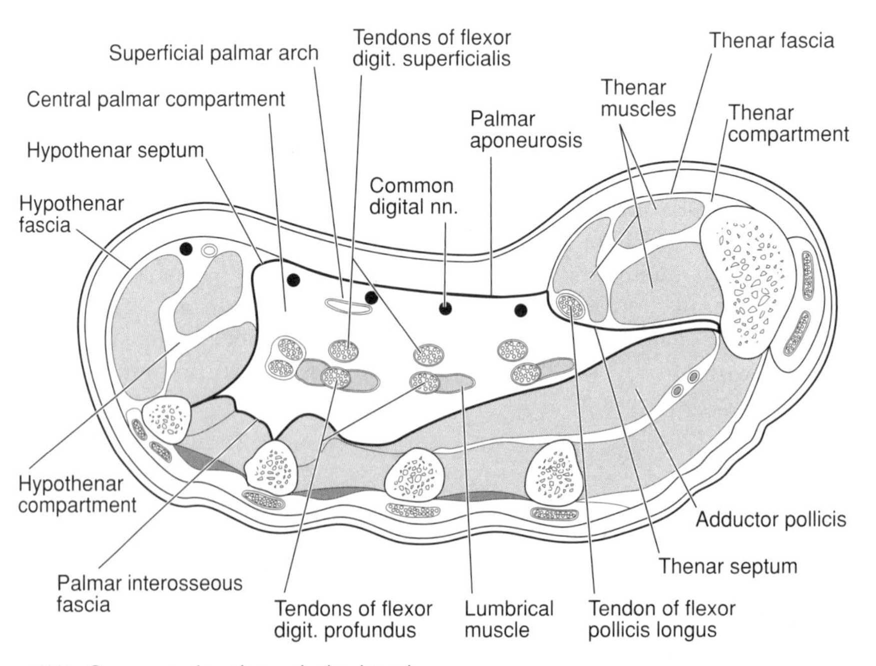


**Figure 2-3.** Cross section through the hand.

## Compartmentation and Intrinsic Muscles

The hand is separated into thenar, hypothenar, central, and adductor-interosseous compartments. The **thenar compartment** is associated with the first metacarpal and delineated by the thenar fascia and septum. It contains the abductor pollicis brevis, flexor pollicis brevis, and opponens pollicis muscles. The **hypothenar compartment** is associated with the fifth metacarpal and formed by the hypothenar fascia and septum. This compartment contains the abductor digiti minimi, flexor digiti, and opponens digiti minimi. The areas between the metacarpals are collectively called the **adductor-interosseous compartment**. This compartment is formed by the anterior adductor-interosseous and dorsal interosseous fasciae, and it contains all of the interossei muscles as well as the adductor pollicis. The remaining central area of the palm, delineated by the anterior adductor-interosseous fascia, the thenar and hypothenar septa, and the palmar aponeurosis, is the **central compartment**. This compartment contains the long digital flexor tendons to the four medial digits and the lumbrical muscles. It also contains the superficial palmar arterial arch, which is between the palmar aponeurosis and the long flexor tendons. The cutaneous branches of the median and ulnar nerves are distributed with the branches of the superficial arterial arch. The attachments, actions, and innervations of the intrinsic muscles are described in Table 2-8.

## Midpalmar and Thenar Spaces

These spaces or clefts are **potential spaces** where **blood** or **inflammatory material** can accumulate and thus produce a characteristically shaped swelling. This fascial plane is between the long digital flexor tendons and the interossei, where the central and adductor-interossei compartments meet. A septum extends from the palmar aponeurosis to the third metacarpal, thus separating the single fascial plane into the medial midpalmar space and the lateral thenar space.

## Digital Flexor Tendons and Tendon Sheaths

The long digital flexor tendons, the flexor digitorum superficialis and profundus muscles, pass through the carpal tunnel and central compartment and then into each of the four medial fingers. The profundus tendons are deep to those of the superficialis in the palm. At the level of the proximal phalanx, the superficialis tendons split and wrap around the profundus tendons. They then attach to the bases of the middle phalanges, and the profundus tendons continue to the bases of the distal phalanges.

These tendons are held in place and their movement facilitated by tendon sheaths. The radial and ulnar bursae are synovial tendon sheaths that surround the long flexor tendons of the digits. They both start proximal to the wrist, pass through the carpal tunnel with the tendons, and extend either partially or completely through the palm and into the fingers. The **radial bursa** is associated only with the tendon of the flexor pollicis longus. The **ulnar bursa** surrounds all four tendons of both the flexor digitorum superficialis and profundus muscles. That part of this bursa associated with the little finger usually extends through the palm and into the digit while the others terminate at midpalmar levels. Thus, the synovial portions of the digital tendon sheaths of the index, middle, and ring fingers are not continuous with the ulnar bursa. The fibrous part of the digital tendon sheath extends from the MCP joint to the DIP joint. It surrounds the synovial sheath and tendons and anchors the tendons to the phalanges and volar plates of the joint capsules. At certain critical points, the fibrous tendon sheath is thickened (to form annular and cruciform pulleys) to ensure that the position of the tendons is maintained.

**TABLE 2-8. Intrinsic Muscles of the Hand**

| Muscle | Origin | Insertion | Action | Innervation |
|---|---|---|---|---|
| **Thenar compartment** | | | | |
| Abductor pollicis brevis | TCL and trapezium | Ventral base thumb proximal phalanx | Abduct thumb | Recurrent branch of median |
| Flexor pollicis brevis | TCL and trapezium | Ventromedial thumb proximal phalanx | Flex MCP of thumb | Recurrent branch of median |
| Opponens pollicis | TCL and trapezium | Ventral shaft of first metacarpal | Opposition of thumb | Recurrent branch of median |
| **Hypothenar compartment** | | | | |
| Abductor digiti minimi | Pisiform | Medial base–fifth proximal phalanx | Abduct little finger | Deep branch of ulnar |
| Flexor digiti minimi | TCL and hook of hamate | Medial base–fifth proximal phalanx | Flex fifth proximal phalanx | Deep branch of ulnar |
| Opponens digiti minimi | Hook of hamate | Medial shaft of fifth metacarpal | Opposition of little finger | Deep branch of ulnar |
| **Central compartment** | | | | |
| Lumbricals | FDP tendons | Central and lateral bands of extensor aponeurosis | Flexion at MCP, extension at PIP and DIP joints | 2, 3: median<br>4, 5: ulnar |
| **Adductor-interosseous compartment** | | | | |
| Palmar interossei | Second metacarpal (medial), fourth and fifth metacarpal (lateral) | Central and lateral bands of extensor aponeurosis | Flexion at MCP, extension at PIP and DIP joints | Deep branch of ulnar |
| Dorsal interossei | Sides of adjacent metacarpals | Central and lateral bands of extensor aponeurosis | Flexion at MCP, extension at PIP and DIP joints | Deep branch of ulnar |
| Adductor pollicis | Lateral distal carpals, adjacent metacarpal bases, third metacarpal shaft | Ventromedial base–thumb proximal phalanx | Adduct thumb | Deep branch of ulnar |

*DIP, distal interphalangeal; FDP, flexor digitorum profundus; MCP, metacarpophalangeal; PIP, proximal interphalangeal; TCL, transverse carpal ligament.*

## Digital Extensor Tendons and Extensor Mechanism

The digital extensor tendons, from the extensor digitorum, extensor indicis, and extensor digiti minimi, do not have a single simple insertion, but rather give rise to an elaborate fibrous system on the dorsum of each of the four medial digits. As each of these tendons crosses the MCP joint, it blends with the extensor hood, a sheet of fibers that surrounds the joint on both sides and serves to hold the extensor tendon in place. Through its attachment to this hood, the extensor tendon extends the MCP joint. Distal to the joint, the extensor tendon splits into three bands, one central band and two lateral bands. The **central band** crosses the PIP joint and inserts into the dorsal base of the middle phalanx; the **lateral bands** diverge around the PIP and converge to cross the DIP joint and insert into the dorsal base of the distal phalanx. The lateral bands are stabilized at the level of the PIP joint by the triangular membrane and the retinacular ligament. The tendons of both the lumbrical and interossei (ventral and dorsal) muscles insert into both the central and lateral bands.

# NERVES OF THE UPPER LIMB

## Brachial Plexus

Most of the muscles of the upper limb are supplied by branches of the brachial plexus (Fig. 2-4). The plexus is formed by the ventral rami of spinal nerves C5, C6, C7, C8, and T1. The ventral rami (commonly called the **roots** of the plexus) of C5 and C6 unite to form the superior trunk, C7 continues as the middle trunk, and the inferior trunk is formed by the union of C8 and T1.

Each of the three trunks splits into **anterior** and **posterior divisions**. This separation into divisions determines the basic innervation pattern for the extremity; that is, the nerves formed from the anterior divisions innervate the muscles in the anterior compartments, and those from posterior divisions innervate posterior compartment muscles. The lateral cord is formed from the anterior divisions of the superior and middle trunks, and the anterior division of the inferior trunk continues as the medial cord.

The **posterior divisions** of all three trunks unite to form the posterior cord. The cords of the plexus receive their names from their relationships with the second part of the axillary artery. The terminal peripheral nerves are formed in the axilla from the cords. The median nerve (C6 to T1) is formed by contributions from both the medial and lateral cords. The remainder of the medial cord forms the ulnar nerve (C8 to T1), and the termination of the lateral cord is the musculocutaneous nerve (C5, C6). The posterior cord gives rise to the radial (C5 to C8) and axillary nerves (C5, C6).

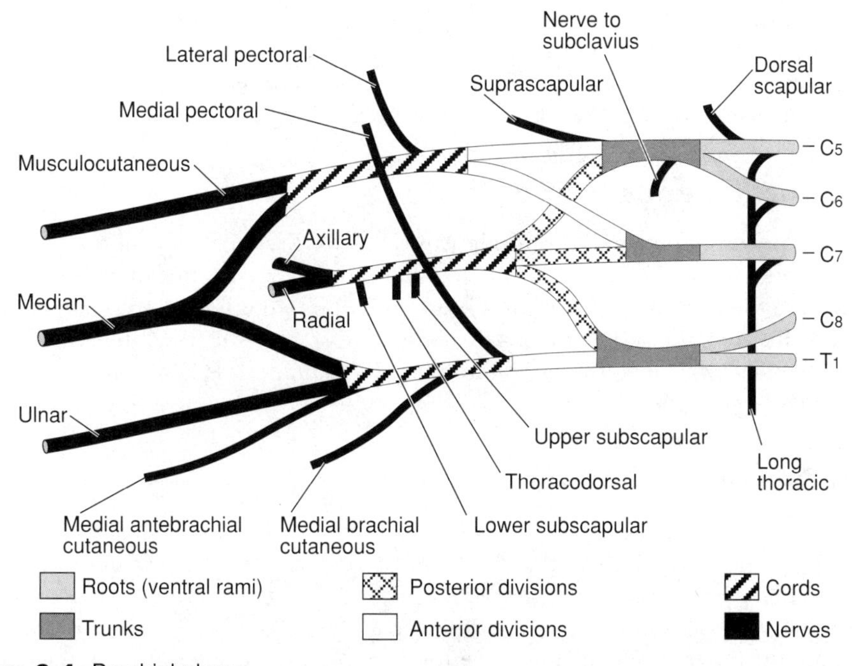


**Figure 2-4.** Brachial plexus.

Small branches from the plexus, called **collateral branches**, supply most of the extrinsic and intrinsic muscles of the shoulder. (Extrinsic muscles extend from the axial skeleton to the scapula, clavicle, or humerus; intrinsic muscles connect the scapula or clavicle with the humerus.) The branches from the ventral rami are the dorsal scapular nerve (C5), which supplies both rhomboids and part of the levator scapulae, and the long thoracic nerve (C5 to C7) to the serratus anterior.

Two branches arise from the superior trunk: the nerve to the subclavius (C5) and the suprascapular nerve (C5, C6), which innervates the supraspinatus and infraspinatus muscles. The medial (C8, T1) and lateral (C5 to C7) pectoral nerves branch from the medial and lateral cords, respectively. The medial pectoral nerve innervates the pectoralis major and minor; the lateral innervates only the major. The three subscapular nerves branch from the posterior cord. The upper and lower subscapular nerves (C5, C6) innervate the subscapularis and teres major muscles. The thoracodorsal nerve (middle subscapular) (C6 to C8) innervates the latissimus dorsi muscle.

## Median Nerve

The median nerve passes through the arm in the **medial neurovascular bundle**, which is located where the medial intermuscular septum joins the brachial fascia. In the distal part of the arm, it inclines laterally and passes through the cubital fossa just medial to the brachial artery. The median nerve exits from the cubital fossa by passing through the pronator teres muscle and descends in the forearm deep to the flexor digitorum superficialis.

The main trunk of the nerve supplies all of the superficial muscles except the flexor carpi ulnaris. The **anterior interosseous branch**, which arises in the proximal aspect of the forearm, supplies most of the deep muscles (i.e., flexor pollicis longus, pronator quadratus, lateral half of the flexor digitorum profundus). At the wrist, the median nerve is positioned deeply in the interval between the tendons of the flexor carpi radialis and the palmaris longus. It enters the hand by going through the carpal tunnel (see Fig. 2-2). Just distal to the deep part of the flexor retinaculum (transverse carpal ligament) the nerve branches into the **thenar** (recurrent, motor) and **digital** (common and proper) **branches**. The thenar branch innervates the **muscles** in the thenar compartment. The digital branches supply the two lateral lumbricals. They also supply the **skin** on the ventral and dorsal distal aspects of the lateral three and a half digits as well as the corresponding part of the palm. The lateral midpalmar skin is supplied by the palmar branch of the median nerve. This branch arises proximal to the wrist and does not pass through the carpal tunnel.

## Musculocutaneous Nerve

The musculocutaneous nerve supplies the biceps brachii, coracobrachialis, and brachialis muscles. From its origin in the axilla, this nerve inclines laterally, passing first through the coracobrachialis muscle and then between the biceps brachii and the brachialis. It enters the subcutaneous tissue in the distal lateral arm, after which it continues into the forearm as the lateral antebrachial cutaneous. It supplies the skin of the lateral aspect of the forearm.

## Ulnar Nerve

The ulnar nerve passes through the proximal half of the arm in the medial neurovascular bundle. It inclines posteriorly in the distal half of the arm, enters the posterior compartment and then passes posterior to the medial epicondyle of the **humerus** to enter the forearm. The nerve passes through the forearm deep to the flexor carpi ulnaris, supplying this muscle and the ulnar half of the flexor digitorum profundus. At the wrist, it is deep to the tendon of the same muscle. It enters the hand by passing superficial to the deep part of

**TABLE 2-9. Segmental Innervation of the Upper Limb**

| Spinal Cord Segment | Cutaneous Distribution | Muscular Distribution (Functions of Muscles) |
|---|---|---|
| C4 | Point of the shoulder | — |
| C5 | Lateral aspect of the arm | C5 and C6: intrinsic muscles of shoulder (abduction and internal rotation of arm); anterior compartment of arm (flexion and supination of forearm) |
| C6 | Lateral forearm and hand, including thumb and ring finger | — |
| C7 | Middle finger | C6 and C7 (8): pronators of forearm<br>C7 (6, 8): posterior compartment of arm (extension of arm); posterior compartment of forearm, superficial muscles (extension of hand and proximal phalanges) |
| C8 | Ring and little fingers, medial hand and wrist | C8 (7, T1): anterior compartment of forearm (flexion of hand and digits)<br>C8 and T1: intrinsic muscles of hand (adduction and abduction of digits, flexion at MP joints and extension at interphalangeal joints, opposition of thumb) |
| T1 | Medial aspect of forearm | — |
| T2 | Medial aspect of arm | — |

the flexor retinaculum and lateral to the pisiform. Just distal to the pisiform, it divides into deep and superficial branches. The superficial branch splits into common and proper digital nerves, which innervate the two medial lumbricals, the palmar skin of the medial one and a half digits, and the corresponding part of the palm. The deep branch passes through the hypothenar compartment and then sweeps laterally across the palm deep to the long flexor tendons. It innervates the muscles in the hypothenar compartment and the interossei as it crosses the palm and then terminates in the adductor pollicis. The ulnar nerve also innervates the dorsal skin of the medial one and a half digits. This is accomplished by the dorsal cutaneous branch, which arises proximal to the wrist.

## Radial Nerve

The radial nerve descends in the posterior compartment of the arm by curving obliquely around the posterior aspect of the midshaft of the humerus in the **spiral groove**. Its branches to the triceps brachii muscle arise both in the axilla and in the spiral groove. At about the midarm level, it enters the anterior compartment by piercing the lateral intermuscular septum and is positioned between the brachioradialis and the brachialis muscles. Just proximal to the elbow and deep to the brachioradialis, it provides branches to the superficial muscles in the posterior compartment of the forearm and divides into **superficial** and **deep branches**. The superficial branch is cutaneous and descends in the forearm deep to the brachioradialis. It enters the subcutaneous tissue in the distal forearm and is cutaneous to the dorsal surface of the lateral three and a half digits and corresponding part of the dorsal hand. The deep branch is muscular (supplying the deep posterior muscles); it enters the posterior compartment of the forearm by wrapping around the neck of the radius in the substance of the supinator muscle. It then branches into muscular branches and continues through the forearm as the **posterior interosseous nerve**, which terminates at the level of the wrist.

**TABLE 2-10. Peripheral Nerve Supply of the Upper Limb (Major Nerves)**

| Nerve | Cutaneous Distribution | Muscular Distribution (Functions of Muscles) |
|---|---|---|
| Musculocutaneous | Lateral aspect of the forearm | Anterior compartment of the arm (flexion and supination of the forearm) |
| Axillary | Lateral aspect of the midarm | Deltoid and teres minor (abduction and external rotation of the arm) |
| Radial | Posterolateral arm and hand, posterior forearm | Posterior compartments of the arm and forearm (extension of the forearm, wrist, proximal phalanges and thumb, abduction of the thumb, supination of the forearm) |
| Median | Anterior aspect of the lateral three and a half digits, dorsal distal aspects of same digits, anterior lateral hand | Anterior compartment of arm (not flexor carpi ulnaris and ulnar half of flexor digitorum profundus (flexion of wrist and digits, pronation of forearm); thenar compartment (opposition, flexion, and abduction of thumb); second lateral lumbricals (metacarpophalangeal flexion and intraphalangeal extension of index and middle fingers) |
| Ulnar | Medial digit and half corresponding part of palm | Flexor carpi ulnaris and ulnar half of flexor digitorum profundus (flexion of hand and little and ring fingers); hypothenar compartment (flexion and abduction of little finger); all interossei (abduction and adduction of digits); adductor pollicis (adduction of thumb); second medial lumbricals (metacarpophalangeal flexion and interphalangeal extension of little and ring fingers) |

### Axillary Nerve

The axillary nerve passes anteroinferior to the shoulder joint (where it is occasionally stretched in an anterior shoulder dislocation) and then horizontally around the posterior aspect of the surgical neck of the humerus. It then enters the deep surface of the deltoid muscle. This nerve supplies both the teres minor and deltoid muscles.

### Summary of the Innervation of the Upper Limb

The segmental innervation of the upper limb is summarized in Table 2-9; the peripheral innervation is summarized in Table 2-10.

## VESSELS OF THE UPPER LIMB

### Arteries

The blood supply of the upper limb is provided by the **subclavian artery**, which becomes the **axillary artery** as it crosses the first rib (Fig. 2-5). The subclavian artery branches from the aorta on the left and the brachiocephalic trunk on the right. Based on its relationship with the anterior scalene muscle, the artery is divided into three parts. The *first part* is medial to the muscle and usually has three branches. The first branch is the **vertebral artery**, which ascends and passes through the transverse foram-

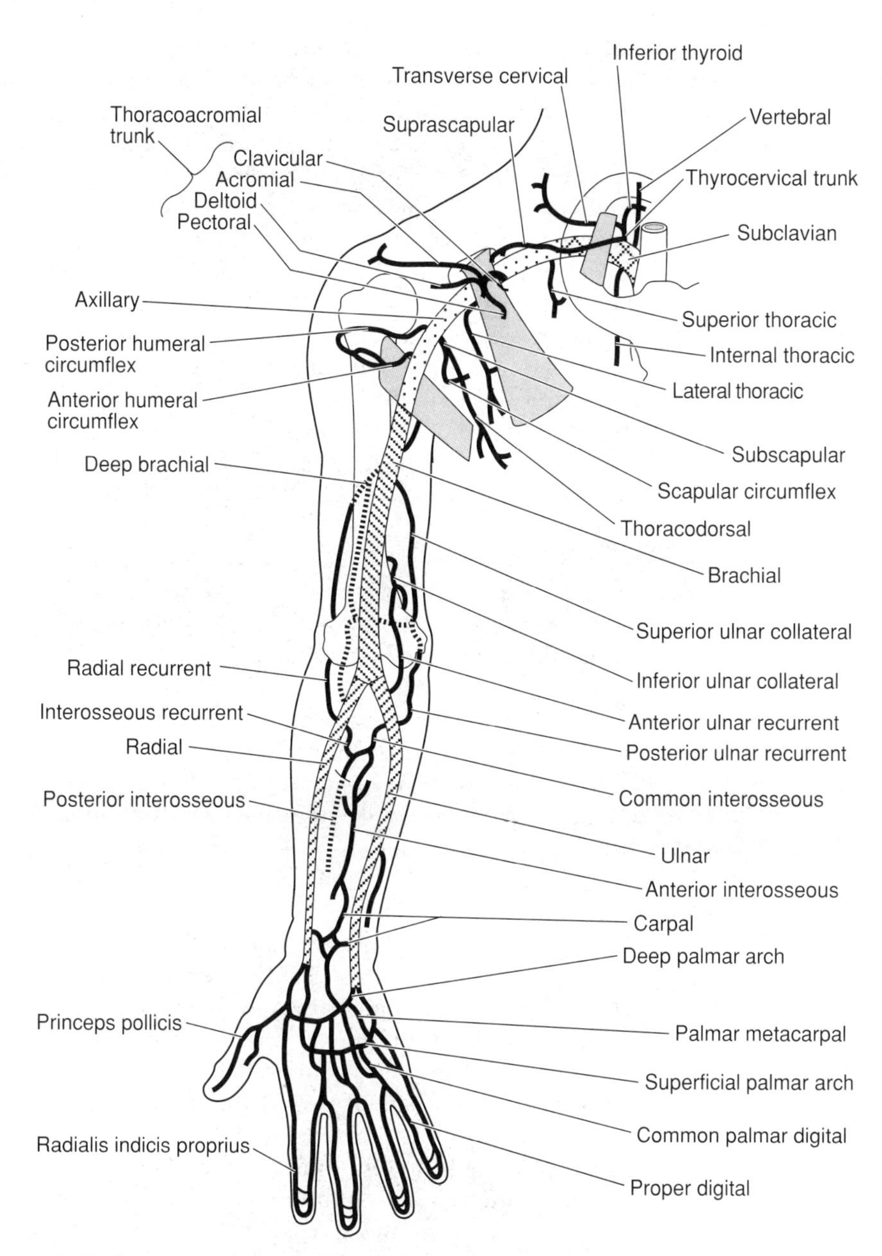


**Figure 2-5.** Summary of the blood supply of the upper limb.

ina of the upper six cervical vertebrae on its way to the cranial cavity. The next branch is the **thyrocervical trunk**, which has branches to the thyroid gland (inferior thyroid artery) and suprascapular region (superficial cervical and suprascapular arteries). The internal thoracic (mammary) artery descends into the thorax on the deep surface of the thoracic wall, just lateral to the sternum. The *second part* of the subclavian artery is posterior to the anterior scalene muscle. This part of the artery usually has a single branch, the **costocervical trunk**. This trunk branches into the **deep cervical artery** to the deep posterior muscles of the neck, and the **superior intercostal artery** to the first and second intercostal spaces. The *third part* of the subclavian artery is lateral to the anterior scalene muscle and typically has no branches.

The **axillary artery** extends from the first rib to the inferior border of the teres major muscle and, based on its relationship to the pectoralis minor muscle, is divided into three parts. The *first part* of the axillary artery extends from the first rib to the pectoralis minor muscle; the *second part* is deep to the muscle; and the *third part* is lateral to the muscle. The **superior thoracic artery** to the upper chest wall branches from the first part of the artery. There are two branches of the second part: (1) the **thoracoacromial trunk**, which supplies the acromial, deltoid, pectoral, and clavicular regions, and (2) the **lateral thoracic artery**, which descends along the anterolateral chest wall. The branches of the third part of the artery are the **subscapular artery**, which bifurcates into the circumflex scapular and thoracodorsal arteries, and the anterior and posterior humeral circumflex arteries, which arise at the level of the surgical neck of the humerus. In the region of the scapula, the circumflex scapular artery forms potential anastomoses with branches of the subclavian artery. The posterior humeral circumflex artery accompanies the axillary nerve as it passes around the posterior aspect of the proximal humerus.

The **brachial artery** is the continuation of the axillary artery, passing through the proximal part of the arm in the medial neurovascular bundle. In the distal half of the arm, it inclines laterally and crosses the elbow by passing through the cubital fossa. In the cubital fossa, it is medial to the tendon of the biceps brachii muscle and lateral to the median nerve (between these two structures), and it is separated from the more superficial median cubital vein by the bicipital aponeurosis. It divides into the radial and ulnar arteries just opposite the radial head. Its largest branch is the deep brachial artery, which arises in the axilla and spirals around the humerus with the radial nerve.

The **ulnar artery** passes through the medial aspect of the forearm deep to the flexor carpi ulnaris muscle, and at the wrist it is just lateral to that muscle's tendon. It enters the hand by passing superficial to the deep part of the flexor retinaculum and lateral to the pisiform. Its largest branch is the common interosseous artery, which arises high in the forearm and immediately divides into the anterior and posterior interosseous arteries.

The course of the **ulnar artery in the hand** is similar to that of the ulnar nerve in that it has superficial and deep branches. The superficial branch gives rise to the superficial palmar arch. This arch is at the level of the distal border of the extended thumb between the palmar aponeurosis and the long flexor tendons. The arch is completed by the superficial palmar branch of the radial artery, which branches proximal to the wrist and passes superficially through the thenar muscles. The arch has common digital branches to the fingers and a proper digital branch to the thumb.

The **radial artery** descends through the lateral part of the forearm deep to the brachioradialis muscle. Just proximal to the wrist, it is lateral to the tendon of the flexor carpi radialis and readily palpable. It then inclines dorsally and enters the dorsum of the hand by passing through the anatomical snuff box. It then passes between (passing dorsal to ventral) the first and second metacarpals. This course brings it into the deep part of the lateral palm, where it gives rise to the deep palmar arterial arch. This arch, completed by the deep branch of the ulnar artery, accompanies the deep ulnar nerve. The position of this arch is proximal to the superficial arch and deep to the long flexor tendons. The palmar metacarpal branches of the deep arch communicate with branches of the superficial arch and the dorsal carpal arterial network.

## Veins

The veins of the upper extremity consist of two sets: the deep and the superficial. The **deep veins** accompany the arteries and communicate frequently with the superficial veins, which are in the subcutaneous tissue.

The **superficial veins** are the basilic, cephalic, and median cubital veins. The **basilic vein** arises on the ulnar side of the dorsum of the hand and extends along the ulnar side of the forearm to the elbow. It crosses the anteromedial aspect of the elbow, pierces the brachial fascia, and empties into the brachial vein. The **cephalic vein** begins at the radial side of the dorsal venous network and continues proximally

**TABLE 2-11. Lymphatics of the Upper Limb**

| Afferents From | Lymph Nodes | Efferents To |
|---|---|---|
| Medial superficial lymphatics of hand and forearm | Supratrochlear (cubital) | Lateral axillary nodes |
| Lateral superficial lymphatics of hand and forearm | Deltopectoral | Apical axillary nodes |
| Deep lymphatics of hand, forearm, and arm; supratrochlear nodes | Lateral axillary nodes | Central axillary nodes |
| Lateral axillary nodes | Central axillary nodes | Apical axillary nodes |
| All other axillary nodes; breast via pectoral nodes | Apical axillary nodes | Subclavian trunks → deep cervical lymph nodes → right lymphatic trunk or thoracic duct |

through the lateral forearm and arm. At the shoulder, it passes through the deltopectoral groove (separating the deltoid and pectoralis major muscles), after which it empties into the axillary vein. The **median cubital vein** interconnects the cephalic and basilic systems superficial to the cubital fossa and the bicipital aponeurosis and is commonly used for venipuncture.

## LYMPHATICS OF THE UPPER LIMB

The lymphatic drainage of the upper limb is summarized in Table 2-11.

Chapter 3

# Lower Limb

This chapter is organized much like Chapter 2. Initial discussion is of the osteology and articulations of the entire limb, followed by each region, proximal to distal. Finally, the neurovascular structures are considered.

## OSTEOLOGY

### Bones of the Pelvis

The adult pelvis is composed of the two hip bones (os coxae) and the sacrum. These three bones are united at the two sacroiliac joints and the single symphysis pubis, thus forming the **pelvic ring**.

**Os coxae**. The os coxae is composed of three bones that fuse early in life: the pubis, ischium, and ilium (Fig. 3-1). The **pubis** is located anteromedially and consists of a body (which forms the symphysis pubis with the body of the opposite side) and posterolaterally directed superior and inferior rami. The superior pubic ramus unites with the ilium; the inferior unites with the ischium (forming the ischiopubic ramus). The palpable pubic tubercle projects from the anterosuperior aspect of the body, and the pectin pubis is a ridge that extends superolaterally along the superior pubic ramus from the tubercle.

The **ischium** is the posteroinferior portion of the hip bone. It has a heavy body from which the ramus of the ischium projects anteromedially, a palpable ischial tuberosity, and an ischial spine. The tuberosity is the inferior-most aspect of the hip bone and the point of origin of the hamstring muscles as well as the weight-bearing bone during sitting. The ischial spine projects posteromedially above the tuberosity and is palpable via rectal or vaginal examination.

The **ilium** is the most superior part of the os coxae, and its body is united with the ischium and the superior ramus of the pubis. The flattened iliac wing flares superiorly and has a thickened crest superiorly that terminates anteriorly and posteriorly in the anterior and posterior superior iliac spines. The lateral aspect of the wing is referred to as the **gluteal surface** because of the attachment of the gluteal muscles. The medial iliac fossa is the origin of the iliacus muscle. The posteromedial aspect of the iliac body forms the auricular surface that articulates with the sacrum to form the sacroiliac joint. The arcuate line is the prominent elevation that extends anteriorly and inferiorly from the auricular surface toward the pectin pubis.

Inferiorly, portions of the pubis and ischium surround the **obturator foramen**. Superior to the obturator foramen and on the lateral aspect of the os coxae, the three component bones form the socket of the hip joint, the **acetabulum**. Posteriorly and inferiorly, the indentation between the ischial tuberosity and ischial spine is the lesser sciatic notch; the area between the ischial spine and the posterior inferior iliac spine is the greater sciatic notch.

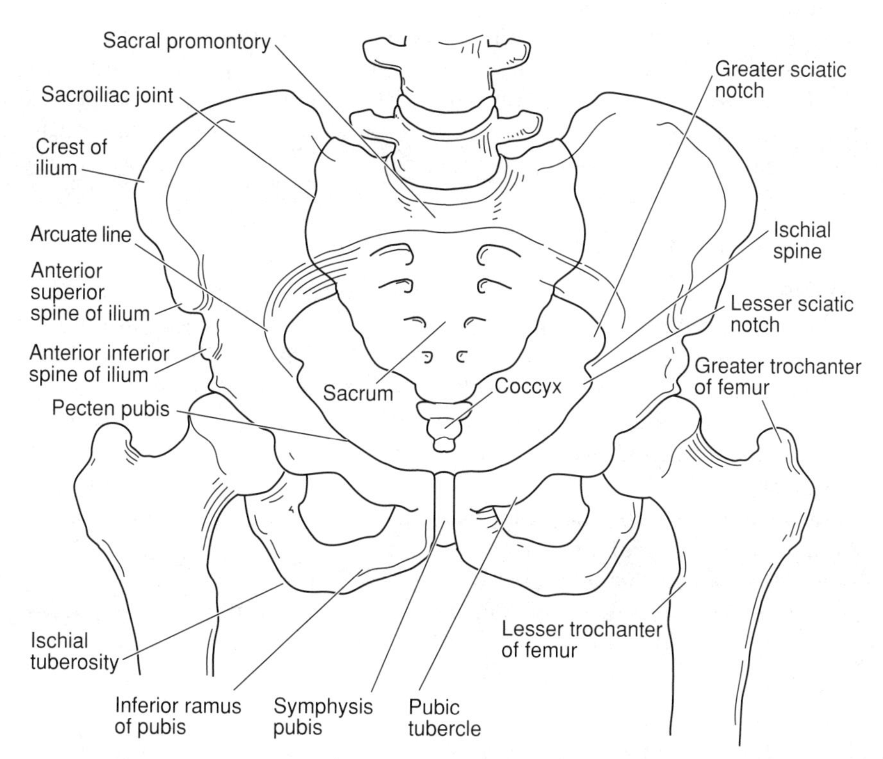


**Figure 3-1.** Anterior view of the pelvis, lower lumbar vertebrae, and proximal portions of the femurs.

**Sacrum.** The sacrum has anterior and posterior surfaces and, when viewed from the front, is triangular in shape. Its flat base is positioned superiorly, formed largely by the superior surface of the body of vertebral segment S1, and oriented obliquely to form an anteriorly inclined plane. The anterior edge of this surface, the **sacral promontory**, is concave from superior to inferior, has four pairs of anterior **sacral foramina**, and is marked by transverse lines that mark the separation of the sacral vertebral segments. The posterior surface is convex and marked by a longitudinal ridge called the **median crest**. In addition to four pairs of dorsal sacral foramina, the posterior surface has an inferior opening, which is the sacral hiatus that marks the inferior termination of the vertebral canal. The **sacral cornua** are prominences on either side of this opening. The auricular surfaces are boomerang-shaped articular surfaces on either side of the bone.

## Femur

The femur is the largest and longest bone in the body. Proximally, the superomedially directed head is separated from the shaft by the neck, which joins the shaft at the trochanteric region. The head is slightly more than half of a sphere and is covered with articular cartilage, except in the region of the fovea. The angle between the neck and shaft is approximately 90 degrees in the female and greater than that in the male. The palpable greater trochanter is located superolaterally at this junctional area; the lesser trochanter is inferomedial and directed somewhat posteriorly. The **trochanters** are interconnected anteriorly by the intertrochanteric line and posteriorly by the intertrochanteric crest, a very prominent ridge that creates a fossa (the intertrochanteric fossa) between it and the neck. The shaft of the femur is long and inclined medially from proximal to distal. The linea aspera, a ridge along the posterior aspect of the shaft, ends inferiorly in the diverging medial and lateral supracondylar lines. The distal end of the femur is greatly enlarged and consists of the medial and lateral **condyles**. The condyles consist entirely of articular surface, are curved from anterior to posterior, and are separated inferiorly and posteriorly by an **intercondylar fossa**. Anteriorly, the condyles join to form the **patellar groove**, which is the articular surface for the patella. The medial and lateral femoral epicondyles are prominences above each of the condyles. The medial epicondyle ends superiorly as the prominent adductor tubercle.

### Tibia and Fibula

**Tibia**. The tibia is the weight-bearing bone of the leg and, therefore, the major bone forming both the ankle and knee joints. Superiorly, it is expanded to form the flat tibial plateau, which is composed of the medial and lateral **tibial condyles**. These condyles are entirely covered by articular surface and are separated by the intercondylar area, which contains the **intercondylar eminence**. The anterior border of the shaft is subcutaneous throughout its length with a prominent superior protrusion, the **tibial tuberosity**. The distal end of the tibia is expanded and has a medial extension, the **medial malleolus**. Both the distal aspect of the tibia and medial aspect of the medial malleolus are articular surfaces.

**Fibula**. The fibula is the lateral bone of the leg. This thin bone articulates with the tibia both proximally and distally and is held firmly to the tibia by the **interosseous membrane**. Although the fibula does not participate in the formation of the knee joint, its inferior aspect (the **lateral malleolus**) helps form the ankle joint.

### Bones of the Foot

The bones of the foot are the **tarsals**, **metatarsals**, and **phalanges**. The seven tarsal bones are arranged in two rows with one bone between the rows. Proximally, the talus sits on the calcaneus; distally, the bones (medial to lateral) are the medial, intermediate, and lateral cuneiforms and the cuboid. The navicular is positioned between the cuneiforms and the talus. The **talus** is the most superior bone in the foot and receives all of the superincumbent weight from the leg. The metatarsals and phalanges are similar in number and position to the metacarpals and phalanges of the hand.

The bones of the foot are arranged to provide flexibility and stability because they accommodate weight bearing while providing for a soft landing and forceful takeoff. These requirements are met by the presence of several **arches**, which are arranged such that weight hits the floor at the calcaneal tuberosity posteriorly and at the heads of the metatarsals anteriorly. The most important arch is the medial longitudinal arch, which consists of the calcaneus, talus, navicular, three cuneiforms, and the three medial metatarsals. The lateral longitudinal arch is made up of the calcaneus, cuboid, and two lateral metatarsals. A transverse arch is found at the level of the distal row of tarsals and bases of the metatarsals. The major static support of these arches (primarily the medial longitudinal arch) is provided by **ligaments**, which are the very important plantar calcaneonavicular (spring) ligament, the long plantar ligament, and the ligament-like plantar aponeurosis. Dynamic support is added by the intrinsic muscles of the foot. Three extrinsic **muscles** of the foot, the tibialis anterior and posterior and the peroneus longus, are thought to provide additional dynamic support.

## ARTICULATIONS OF THE LOWER LIMB

### Pelvis

**Sacroiliac joint**. The sacroiliac joint is an unusual synovial joint in that it permits very little, if any, motion. It is formed by the highly irregular but congruent articular surfaces (auricular surfaces) of the sacrum and ilium. These surfaces are shaped like a boomerang, with the angle directed anteriorly. The two bones are bound tightly together by anterior and posterior **sacroiliac ligaments** as well as a large mass of **interosseous ligaments**, which occupies the area between the two limbs of the boomerang. This joint commonly fuses in the fourth or fifth decade, particularly in males.

**Symphysis pubis**. The pubic symphysis is a cartilaginous joint formed between the pubic bodies. Each articular surface is covered by a hyaline cartilage plate; the two plates are usually separated by a narrow slit. The bones are bound together by strong superior pubic and arcuate pubic ligaments, and thus only very limited motion is available.

The bones of the pelvis are involved in the formation of the walls of both the pelvic and abdominal cavities. The plane of the **pelvic inlet** separates the true pelvis below from the false pelvis (part of the abdominal cavity) above. The inlet is formed posteriorly by the sacral promontory, laterally by the arcuate line of the ilium and the iliopectineal line, and anteriorly by the superior aspects of the pubic bodies and the symphysis.

## Hip Joint

This joint is formed by the acetabulum of the os coxae and the head of the femur. The **acetabulum** is a deep socket with the horseshoe-shaped articular surface oriented so that its open end is directed inferiorly (Fig. 3-2). The articular surface is therefore incomplete centrally (**acetabular fossa**) and inferiorly (**acetabular notch**). The head of the femur is slightly larger than a hemisphere and also has a nonarticular area, the centrally located **fovea**. The two bony surfaces are quite congruent. The fibrocartilaginous acetabular labrum attaches to the periphery of the acetabulum and bridges the acetabular notch as the transverse acetabular ligament. In addition to deepening the socket, the labrum also reduces its diameter and thereby holds the femur in place. The fovea of the head of the femur and the nonarticular portion of the acetabulum

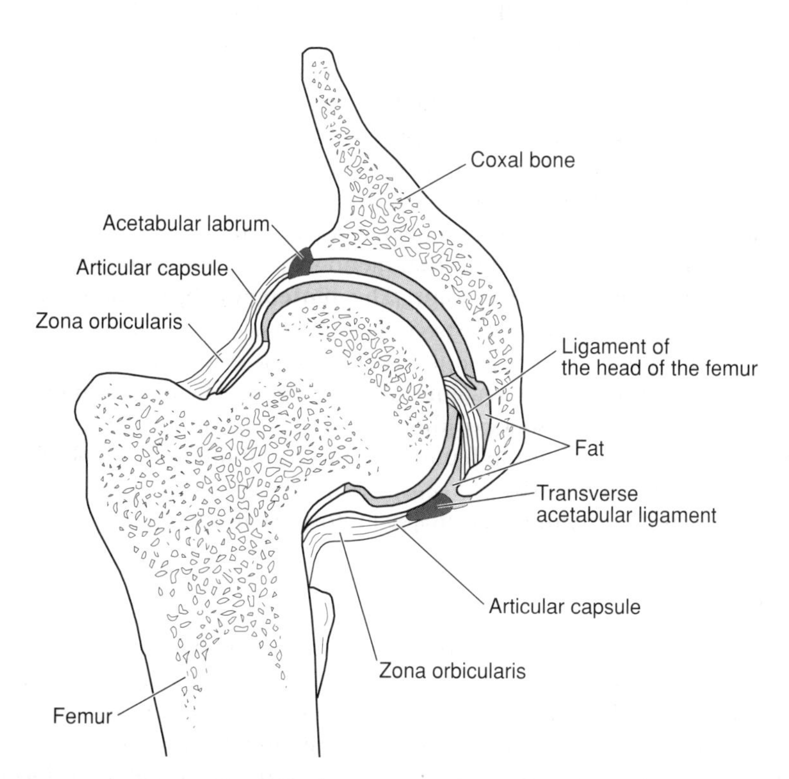


**Figure 3-2.** Coronal section through the hip joint showing the relationship of the joint capsule to the neck of the femur and the extent of the joint cavity.

**TABLE 3-2. Primary Muscles Producing Motion at the Knee**

| Motion | Muscles Producing Motion |
|---|---|
| Extension | Quadriceps femoris (rectus femoris, vastus medialis, vastus lateralis, vastus intermedius) |
| Flexion | Hamstrings (biceps femoris, semitendinosus, semimembranosus), gastrocnemius |
| Medial (internal) rotation | Semimembranosus, semitendinosus, sartorius, popliteus |
| Lateral (external) rotation | Biceps femoris |

between the medial and lateral malleoli. The anterior aspect of the trochlea is wider than the posterior aspect. As a result, in dorsiflexion the widest portion of the trochlea is wedged between the malleoli, producing good bony stability. This support is lost in plantar flexion. Thus, most **sprains** occur when the ankle is plantar flexed.

The distal tibia and fibula are lashed together by anterior and posterior **tibiofibular ligaments**. The major ligaments supporting the ankle joint are the medial (deltoid) and lateral collateral ligaments. The deltoid ligament is a broad band connecting the medial malleolus with the talus, navicular, and calcaneus. The lateral collateral ligament is composed of three distinct bands: the anterior and posterior talofibular ligaments and the calcaneofibular ligament. The **anterior talofibular ligament is the most commonly injured ligament**, accompanying the frequent plantar flexion-inversion sprain of the ankle. The motions of the ankle and their major motors are listed in Table 3-3.

## Subtalar and Transverse Tarsal Joints

Although the motions of the foot (other than toe motion) are the sum totals of the individual amounts of motion that occur at each intertarsal joint, most of this motion occurs at two articulations: the subtalar and transverse tarsal (midtarsal) joints. The

**TABLE 3-3. Primary Muscles Producing Motion of the Foot**

| Motion | Muscles Producing Motion |
|---|---|
| Plantar flexion | Gastrocnemius, soleus, tibialis posterior |
| Dorsiflexion | Tibialis anterior |
| Inversion | Tibialis anterior, tibialis posterior, flexor hallucis longus, flexor digitorum longus |
| Eversion | Peroneus longus, peroneus brevis |
| Flexion of the toes | Flexor digitorum longus, flexor digitorum brevis, flexor hallucis longus |
| Extension of the toes | Extensor digitorum longus, extensor digitorum brevis, extensor hallucis longus, extensor hallucis brevis |

motions of the foot are inversion, which is a combination of adduction and supination, and eversion, which is a combination of abduction and pronation. The **subtalar joint** is inferior to the talus, between the talus and the calcaneus. The two sets of articular surfaces that form this joint are separated by a strong interosseous ligament. These joint spaces are aligned so they form an inclined plane that is directed anteriorly, medially, and inferiorly. The **transverse tarsal joint** extends transversely across the foot. The medial articulation is between the talus and the navicular; the lateral between the calcaneus and the cuboid. The motors of these foot motions are listed in Table 3-3.

# JUNCTIONAL REGIONS BETWEEN THE TRUNK AND LOWER LIMB

## Gluteal Region

The gluteal region includes the anterior and lateral aspects of the hip. The gluteal muscles are large and more superficial in position; the short external rotators are

**TABLE 3-4. Muscles of the Hip Region**

| Muscle | Origin | Insertion | Action | Innervation |
|---|---|---|---|---|
| Gluteus maximus | Posterolateral ilium, posterior sacrum | Iliotibial tract, gluteal tuberosity | Thigh extension and external rotation | Inferior gluteal nerve |
| Gluteus medius | Upper lateral iliac wing | Greater trochanter of femur | Thigh abduction | Superior gluteal nerve |
| Gluteus minimus | Inferior lateral iliac wing | Greater trochanter of femur | Thigh abduction | Superior gluteal nerve |
| Tensor fascia latae | Anterior iliac crest | Iliotibial tract | Flexion and internal rotation of thigh | Superior gluteal nerve |
| Piriformis | Anterior surface of sacrum | Medial greater trochanter | External rotation of thigh | Direct branches of sacral plexus |
| Obturator internus | Deep rim of obturator foramen | Medial greater trochanter | External rotation of thigh | Direct branches of sacral plexus |
| Superior gemellus | Ischial spine | Medial greater trochanter | External rotation of thigh | Direct branches of sacral plexus |
| Inferior gemellus | Ischial tuberosity | Medial greater trochanter | External rotation of thigh | Direct branches of sacral plexus |
| Quadratus femoris | Lateral body of ischium | Intertrochanteric crest of femur | External rotation of thigh | Direct branches of sacral plexus |
| Iliopsoas | | | | |
| Psoas major | Anterolateral lumbar vertebral bodies and transverse processes | Lesser trochanter of femur | Flexion of thigh | Direct branches of lumbar plexus |
| Iliacus | Medial aspect of iliac wing | Lesser trochanter of femur | Flexion of thigh | Femoral nerve |

**TABLE 3-5. Muscles of the Thigh**

| Muscle | Origin | Insertion | Action | Innervation |
|---|---|---|---|---|
| **Anterior compartment** | | | | |
| Rectus femoris | Anterior inferior iliac spine | Tibial tuberosity via patella | Thigh flexion, extension of leg | Femoral nerve |
| Vastus lateralis | Linea aspera | Tibial tuberosity via patella | Extension of leg | Femoral nerve |
| Vastus medialis | Linea aspera | Tibial tuberosity via patella | Extension of leg | Femoral nerve |
| Vastus intermedius | Proximal anterior two-thirds of femur | Tibial tuberosity via patella | Extension of leg | Femoral nerve |
| Sartorius | Anterior superior iliac spine | Anteromedial proximal tibia | Thigh external rotation and flexion; leg flexion and internal rotation | Femoral nerve |
| **Medial compartment** | | | | |
| Adductor longus | Body of pubis | Middle third of linea aspera | Thigh adduction | Obturator nerve |
| Pectineus | Superior pubic ramus | Posteromedial proximal femur | Thigh flexion and adduction | Femoral or obturator nerve |
| Adductor magnus | Ischiopubic ramus | Entire length of posteromedial femur | Thigh adduction and extension | Obturator and tibial nerves |
| Adductor brevis | Body and inferior ramus of pubis | Proximal linea aspera | Thigh adduction | Obturator nerve |
| Obturator externus | Superficial rim of obturator foramen | Trochanteric fossa | Thigh external rotation | Obturator nerve |
| Gracilis | Body and inferior ramus of pubis | Anteromedial proximal tibia | Thigh adduction, leg flexion | Obturator nerve |
| **Posterior compartment** | | | | |
| Semitendinosus | Ischial tuberosity | Proximal anteromedial tibia | Thigh extension, leg flexion and medial rotation | Tibial nerve |
| Semimembranosus | Ischial tuberosity | Proximal posteromedial tibia | Thigh extension, leg flexion and medial rotation | Tibial nerve |
| Biceps femoris | Ischial tuberosity, distal half of linea aspera | Head of fibula | Thigh extension, leg flexion and lateral rotation | Tibial and common peroneal nerves |

small and short and located deep to the gluteus maximus muscle. These muscles are described in Table 3-4. The superior and inferior gluteal arteries, branches of the internal iliac artery, supply most of the region. The major nerves of the region are the superior and inferior gluteal nerves that supply the gluteal muscles, and the **sciatic nerve**, which supplies the posterior thigh and the entire leg and foot.

## Femoral Triangle

The femoral triangle is the anterior junctional region between the abdomen and thigh, and contains important structures that supply the lower limb. The triangle is defined by the inguinal ligament above, the sartorius muscle laterally, and the medial border of the adductor longus muscle medially. The floor is formed by the adductor longus, pectineus, and iliopsoas muscles. The **iliopsoas** is described in Table 3-4. The superficial and deep **inguinal lymph nodes** are found, respectively, superficially and deep to the investing fascia in this triangle. In addition to receiving superficial lymphatics from the lower extremity, the superficial nodes drain the lower abdominal wall, buttock, perineum, and lower portions of the anal canal and vagina. From lateral to medial, the femoral nerve, femoral artery, and femoral vein pass under the inguinal ligament and descend through the triangle. The femoral artery is midway between the anterior superior spine of the ilium and the pubic tubercle. The **femoral sheath** is an extension of transversalis fascia that forms a sleeve around the femoral vessels as they enter the thigh. The area just medial to the vein, the **femoral canal**, is within the sheath and the usual path of a femoral hernia. After leaving the triangle, the femoral vessels pass through the thigh just deep to the sartorius muscle in the **adductor (subsartorial) canal**.

# THIGH

The investing fascia of the thigh is the **fascia lata**. It is dramatically thickened laterally as the iliotibial tract or band. From its proximal attachment to the iliac crest, this band extends distally and crosses the anterolateral aspect of the knee before attaching to the anterolateral aspect of the lateral tibial condyle. Two septa, the **medial and lateral intermuscular septa**, extend from the investing fascia to the femur. These septa separate the thigh into anterior and posterior compartments, which are actually anterolateral and posteromedial in position. The muscles in the **anterior compartment** are innervated by the femoral nerve and consist of the sartorius and the four components of the quadriceps femoris: the rectus femoris, vastus medialis, vastus intermedius, and vastus lateralis. The **posterior compartment** has two groups of muscles: the **medial femoral muscles** or adductors and the posterior femoral muscles or hamstrings. The **medial femoral muscles** are primarily innervated by the obturator nerve and consist of the adductor longus, adductor brevis, adductor magnus, pectineus, gracilis, and obturator externus. The **hamstrings** are primarily innervated by the tibial portion of the sciatic nerve and consist of the biceps femoris, semitendinosus, and semimembranosus. The muscles of the thigh are described in Table 3-5.

# POPLITEAL FOSSA

The popliteal fossa is a deep, diamond-shaped area behind the knee. It is bounded superomedially by the tendons of the semitendinosus and semimembranosus muscles, superolaterally by the tendon of the biceps femoris, and inferiorly by the heads of the gastrocnemius muscle. Its floor is the posterior (supracondylar or popliteal) portion of the distal femur. The **popliteal vessels** pass vertically through this fossa with the artery closest to the bone and thereby are **vulnerable to laceration** when this part of the femur is fractured. The tibial nerve passes through the center of this fossa superficially, and the common peroneal nerve follows the tendon of the biceps femoris muscle. This space is packed with loose connective tissue.

**TABLE 3-6. Muscles of the Leg**

| Muscle | Origin | Insertion | Action | Innervation |
| --- | --- | --- | --- | --- |
| **Anterior compartment** | | | | |
| Tibialis anterior | Proximal two-thirds of lateral tibia | Inferomedial base of first metatarsal and medial cuneiform | Dorsiflexion and inversion of foot | Deep peroneal nerve |
| Extensor digitorum longus | Proximal three-fourths of anterior fibula | Extensor aponeurosis of 4 lateral toes | Extension of proximal phalanges–4 lateral toes | Deep peroneal nerve |
| Peroneus tertius | Separate head of extensor digitorum longus | Dorsal base of fifth metatarsal | Dorsiflexion and eversion of foot | Deep peroneal nerve |
| Extensor hallucis longus | Middle anteromedial fibula | Dorsal base of distal phalanx of great toe | Extension of great toe | Deep peroneal nerve |
| **Lateral compartment** | | | | |
| Peroneus longus | Proximal half of lateral fibula | Inferolateral base of first metatarsal and medial cuneiform | Eversion and plantar flexion | Superficial peroneal nerve |
| Peroneus brevis | Distal two-thirds of lateral fibula | Superolateral base of fifth metatarsal | Eversion and plantar flexion | Superficial peroneal nerve |
| **Posterior compartment** | | | | |
| Gastrocnemius | Medial and lateral aspects of supracondylar femur | Posterior calcaneus | Plantar flexion and leg flexion | Tibial nerve |
| Soleus | Proximal posterior tibia and fibula | Posterior calcaneus | Plantar flexion | Tibial nerve |
| Plantaris | Lateral aspect of supracondylar femur | Posterior calcaneus | Weak plantar flexion and leg flexion | Tibial nerve |
| Flexor hallucis longus | Distal two-thirds of posterior fibula | Plantar base of great toe distal phalanx | Flexion of great toe | Tibial nerve |
| Flexor digitorum longus | Middle half of posterior tibia | Plantar bases of 4 lateral toes | Flexion of 4 lateral toes | Tibial nerve |
| Tibialis posterior | Proximal posterior tibia and fibula | Tubercle of navicular, plantar aspects of cuboid, cuneiforms, 3 medial metatarsal bases | Plantar flexion and inversion of foot | Tibial nerve |
| Popliteus | Lateral epicondyle | Proximal posterior femoral tibia | Medial rotation of leg | Tibial nerve |

## LEG

The leg has anterior, lateral, and posterior compartments. The major partition consists of the subcutaneous **tibia**, the **interosseous membrane**, the **fibula**, and the **posterior intermuscular septum**, which separates the posteromedially situated posterior compartment from an anterolateral region. This anterolateral area is subdivided into anterior and lateral compartments by the anterior intermuscular septum. The anterior compartment is just lateral to the tibia and contains the tibialis anterior, extensor hallucis longus, extensor digitorum longus, and peroneus tertius muscles, all of which are innervated by the deep peroneal nerve. These muscles dorsiflex the ankle, invert the foot, and extend the toes. The **lateral compartment** is superficial to the fibula, and its muscles are primarily everters of the foot. The two muscles in this compartment, the peroneus longus and brevis, are innervated by the superficial peroneal nerve. All posterior compartment muscles are innervated by the tibial nerve and function to plantar flex and invert the foot and flex the toes. The **posterior compartment** muscles are the gastrocnemius, soleus, and plantaris superficially; the flexor hallucis longus, flexor digitorum longus, tibialis posterior, and popliteus form the deep group. The muscles of the leg are summarized in Table 3-6.

# FOOT

The organization of the foot is similar to that of the hand, but the compartmentation is less complete in the foot. There is a definitive compartment associated with the small toe and a deep one between the metatarsals. However, fascial separations complete the formation of neither a central compartment nor one associated with the great toe.

For the most part, the muscles of the foot are similar to those of the hand but are usually described in layers rather than in compartments. All of the **plantar muscles** are innervated by either the medial or lateral plantar nerves, both branches of the tibial nerve. The *superficial layer* consists of the abductor hallucis (medial plantar), flexor digitorum brevis (medial plantar), and abductor digiti minimi (lateral plantar). The *intermediate layer* is limited to the central area of the foot and is formed by the tendons of the flexor digitorum longus and related muscles: the quadratus plantae (lateral plantar) and lumbricals (medial and lateral plantar nerves). The *deep layer* consists of the flexor hallucis brevis (medial plantar), adductor hallucis (lateral plantar), and flexor digiti minimi (lateral plantar). A *fourth layer* consists of both the plantar and dorsal interossei, all of which are supplied by the lateral plantar nerve. The foot differs from the hand also in that it has dorsal intrinsic muscles. These muscles, the extensor hallucis brevis and extensor digitorum brevis, extend the toes and are innervated by the deep peroneal nerve. The muscles of the foot are summarized in Table 3-7.

# NERVES OF THE LOWER LIMB

## Lumbosacral Plexus

Branches of the lumbosacral plexus supply the lower abdominal wall and lower extremity (Fig. 3-3). The **lumbar plexus** is formed in the substance of the psoas major muscle from the ventral rami of L1 through L4. All of L1 and part of L2 give rise to cutaneous nerves that supply the skin of the lower abdominal wall, the anterior part of the perineum, and the proximal portion of the lower limb. The rest of the plexus forms major nerves of the lower extremity. The **sacral plexus** (L4–S4) is formed in the pelvis on the anterior surface of the piriformis muscle. The contribution of L4 and L5 to the sacral plexus is provided by the **lumbosacral trunk**, which enters the pelvis by crossing the arcuate line just lateral to the sacral promontory.

**TABLE 3-7. Muscles of the Foot**

| Muscle | Origin | Insertion | Action | Innervation |
|---|---|---|---|---|
| **Superficial layer** | | | | |
| Flexor digitorum brevis | Calcaneal tuberosity–anterior aspect | Plantar aspects of 4 lateral middle phalanges | Middle phalangeal flexion–4 lateral toes | Medial plantar nerve |
| Abductor hallucis brevis | Calcaneal tuberosity–medial aspect | Plantar base–great toe proximal phalanx | Flexion and abduction of great toe | |
| Abductor digiti minimi | Calcaneal tuberosity–lateral aspect | Lateral base–little toe proximal phalanx | Abduction of little toe | Lateral plantar nerve |
| **Intermediate layer** | | | | |
| Quadratus plantae | Calcaneus | Flexor digitorum longus tendon | Assist FDL and straighten its pull | Lateral plantar nerve |
| Lumbricals | FDL tendons | Extensor aponeurosis | MTP flexion; interphalangeal extension | Medial and lateral plantar nerves |
| **Deep layer** | | | | |
| Adductor hallucis | Heads and bases of metatarsals 2, 3, and 4 | Lateral base of great toe proximal | Adduction of great toe | Lateral plantar nerve |
| Flexor hallucis brevis | Plantar cuboid and lateral cuneiform | Medial and lateral great toe proximal phalanx | Flexion of proximal phalanx of great toe | Medial plantar nerve |
| Flexor digiti minimi brevis | Plantar base of fifth metatarsal | Base of little toe proximal phalanx | Flexion of little toe proximal phalanx | Lateral plantar nerve |
| **Fourth layer** | | | | |
| Plantar interossei | Plantar shafts–3 lateral metatarsals | Medial plantar bases–3 lateral proximal phalanges | Adduction of 3 lateral toes | Lateral plantar nerve |
| Dorsal interossei | Sides of adjacent metatarsal shafts | Bases of proximal phalanges 2, 3, and 4 | Abduction of toes 2, 3, and 4 | Lateral plantar nerve |
| **Dorsal intrinsic muscles** | | | | |
| Extensor hallucis brevis | Superior aspect of calcaneus | Dorsal base–proximal phalanx of great toe | Extension of great toe proximal phalanx | Deep peroneal nerve |
| Extensor digitorum brevis | Superior aspect of calcaneus | Dorsal bases–proximal phalanges of toes 2, 3, and 4 | Extension of proximal phalanges–toes 2, 3, and 4 | Deep peroneal nerve |

*FDL, flexor digitorum longus; MTP, metatarsophalangeal joint.*

**Cutaneous branches.** The **iliohypogastric** (T12, L1) and **ilioinguinal** (L1) **nerves** supply the muscles and skin of the lower abdominal wall. In addition, the ilioinguinal nerve has an anterior labial (scrotal) branch that supplies the anterior perineum. The **genitofemoral nerve** (L1, L2) descends on the anterior surface of the psoas major muscle and supplies the cremaster muscle and skin on the proximal anterior thigh.

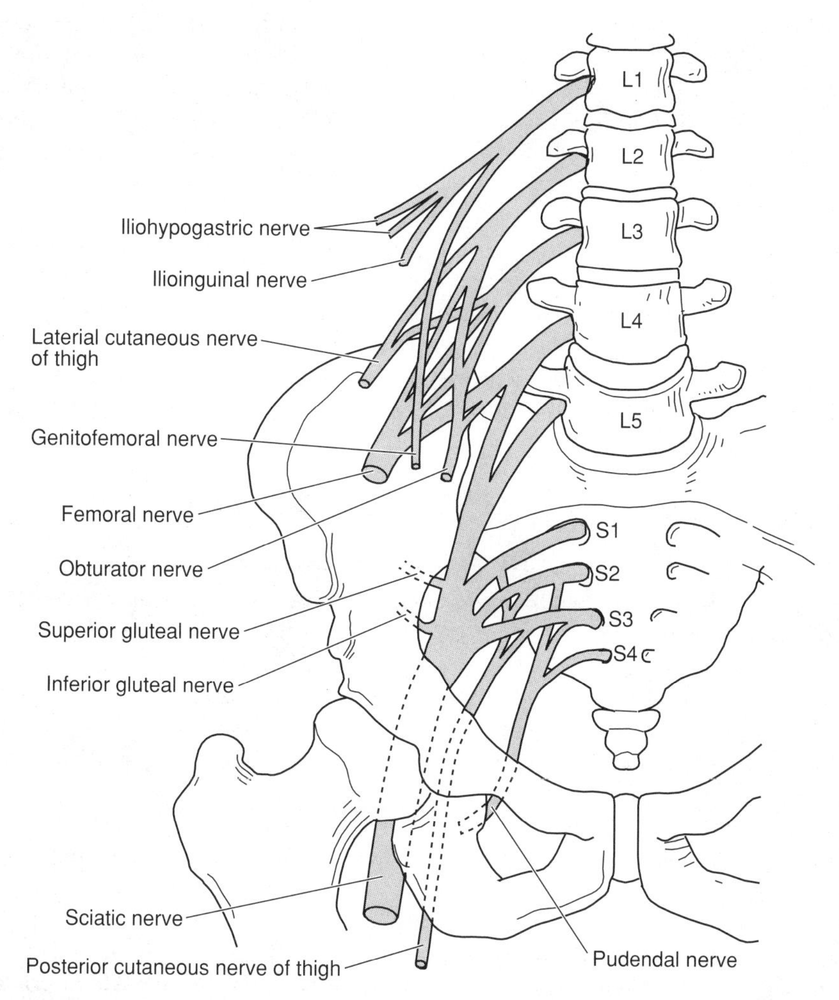


**Figure 3-3.** Diagram of the lumbosacral plexus and its relationship to the lumbar spine and pelvis.

The **lateral femoral cutaneous nerve** (L1–L3), usually the largest of these branches, passes laterally across the iliac fossa, passes through the fascia lata just below the anterior superior iliac spine, and supplies the skin of the anterolateral thigh. The **subcostal nerve** (T12) also supplies the muscles and skin of the lower abdominal wall.

## Femoral Nerve

The femoral nerve (L2–L4) emerges from the lateral aspect of the psoas major muscle and enters the thigh as the most lateral structure passing deep to the inguinal ligament. On entering the femoral triangle, it immediately branches into many muscular branches and the **saphenous nerve**. This terminal branch is cutaneous. It continues through the thigh in the adductor canal, enters the subcutaneous tissue in the distal medial thigh, and supplies the skin on the medial aspects of the knee, leg, and foot.

## Obturator Nerve

The obturator nerve (L2–L4) emerges from the medial aspect of the psoas major muscle just above the pelvic brim. It then enters the pelvic cavity and passes anteriorly and inferiorly toward the obturator canal, through which it enters the medial aspect of the thigh. At the adductor brevis muscle, the nerve separates into anterior and posterior divisions, both of which have multiple muscular branches to the adductor muscles and cutaneous branches to the medial aspects of the thigh and knee.

## Gluteal Nerves

The **superior gluteal nerve** (L4–S1) arises in the pelvis, but immediately exits by passing above the piriformis muscle and through the greater sciatic notch. It passes anteriorly between the gluteus medius and minimus muscles, supplying both, and terminates by supplying the tensor fasciae latae muscle. The **inferior gluteal nerve** (L5–S2) also arises in the pelvis and exits immediately by passing below the piriformis and through the greater sciatic notch, directly into the substance of the gluteus maximus muscle.

## Sciatic Nerve

The sciatic nerve consists of the tibial and common peroneal nerves enclosed in a common connective tissue sheath. This nerve leaves the pelvis by passing below the piriformis (usually) and through the greater sciatic notch. Its course through the gluteal region is curved, first passing midway between the ischial tuberosity and the posterior inferior iliac spine and then between the greater trochanter and the ischial tuberosity. It descends through the center of the posterior thigh between the medial and lateral hamstring muscles, and typically divides into the common peroneal and tibial nerves in the distal thigh as it enters the popliteal fossa.

**Common peroneal nerve**. The common peroneal nerve passes superficially through the lateral aspect of the popliteal fossa, just medial to the biceps femoris muscle and its tendon. It passes superficial to the posterior aspect of the lateral femoral condyle and then wraps around the lateral aspect of the neck of the fibula. The nerve divides into its terminal branches, the superficial and deep peroneal nerves, as it passes the neck of the fibula. The **superficial peroneal nerve** enters the lateral compartment of the leg, supplies the muscles in that compartment, and then descends to innervate most of the skin on the dorsum of the foot. The **deep peroneal nerve** passes through the lateral compartment into the anterior compartment. It descends through the compartment, supplies the muscles in the compartment, and then enters the dorsum of the foot, where it supplies the extensor digitorum brevis and the extensor hallucis brevis muscles. It also has a very small cutaneous distribution to the web space between the great and second toes.

**Tibial nerve**. The tibial nerve descends superficially through the center of the popliteal fossa. It enters the posterior compartment of the leg by passing between, and then deep to, the two heads of the gastrocnemius muscle. It descends through the leg between the superficial and deep groups of muscles (both of which it supplies) and enters the foot by passing posterior to the medial malleolus. As it enters the foot, it divides into the medial and lateral plantar nerves. The **medial plantar nerve** is similar in course and distribution to the median nerve of the hand. It passes into the medial aspect of the foot and divides into muscular branches as well as the cutaneous branches (plantar digital nerves) that supply the skin on the plantar surface of the medial three and a half toes and corresponding part of the ball of the foot. The **lateral plantar nerve** is similar to the ulnar nerve. It passes diagonally across the plantar aspect of the foot by going through the abductor hallucis and then between the flexor digitorum brevis and quadratus plantae muscles. It then terminates by dividing into superficial and deep branches. The superficial branch has muscular branches, and is cutaneous to the plantar aspect of the lateral one and a half toes and corresponding part of the sole in that area. The deep

**TABLE 3-8. Segmental Innervation of the Lower Limb**

| Spinal Cord Segment | Cutaneous Distribution | Muscular Distribution (Functions of Muscles) |
|---|---|---|
| L1 | Proximal medial thigh | — |
| L2 | Proximal anterior thigh | L2 (L1, L3): iliopsoas (flexion of the thigh) |
| L3 | Distal anteromedial thigh and knee | L3 (L2, L4): anterior compartment of the thigh (extension of leg); medial compartment of thigh (adduction of thigh) |
| L4 | Anteromedial leg | L4 (L5): tibialis anterior (dorsiflexion of the foot) |
| L5 | Anterolateral leg, dorsomedial foot and plantar aspect of great toe | L5 (L4, S1): extensor hallucis longus (extension of great toe) |
| S1 | Heel, most of plantar aspect of the foot, dorsal aspect of the lateral foot | S1 (L5, S2): gluteus maximus and hamstrings (extension of thigh and flexion of leg); posterior compartment of leg (plantar flexion of foot and flexion of toes) |
| S2 | Posterior aspects of the thigh and proximal leg | — |

**TABLE 3-9. Peripheral Nerve Supply of the Lower Limb (Major Nerves)**

| Nerve | Cutaneous Distribution | Muscular Distribution (Functions of Muscles) |
|---|---|---|
| Superior gluteal | — | Gluteus medius and minimus, tensor fascia latae (abduction of thigh) |
| Inferior gluteal | — | Gluteus maximus (extension of thigh) |
| Femoral | Anterior thigh and medial knee, leg and foot | Anterior compartment of thigh (extension of leg) |
| Obturator | Inferior medial aspect of thigh | Medial compartment of thigh (adduction of thigh) |
| Tibial | Most of plantar foot | Posterior thigh muscles (extension at hip and flexion at knee); posterior compartment of leg (plantar flexion of foot and flexion of toes); plantar muscles of foot (flexion of toes) |
| Superficial peroneal | Distal anterior aspect of leg, most of dorsum of foot | Lateral compartment of leg (eversion and plantar flexion of foot) |
| Deep peroneal | Dorsal aspect of first web space of foot | Anterior compartment of leg (dorsiflexion and inversion of foot, extension of toes); dorsal intrinsic muscles of foot (extension of toes) |

branch passes medially across the foot deep to long flexor tendons and terminates by supplying the adductor hallucis.

## Summary of the Innervation of the Lower Limb

The segmental innervation of the lower limb is summarized in Table 3-8; the peripheral innervation is summarized in Table 3-9.

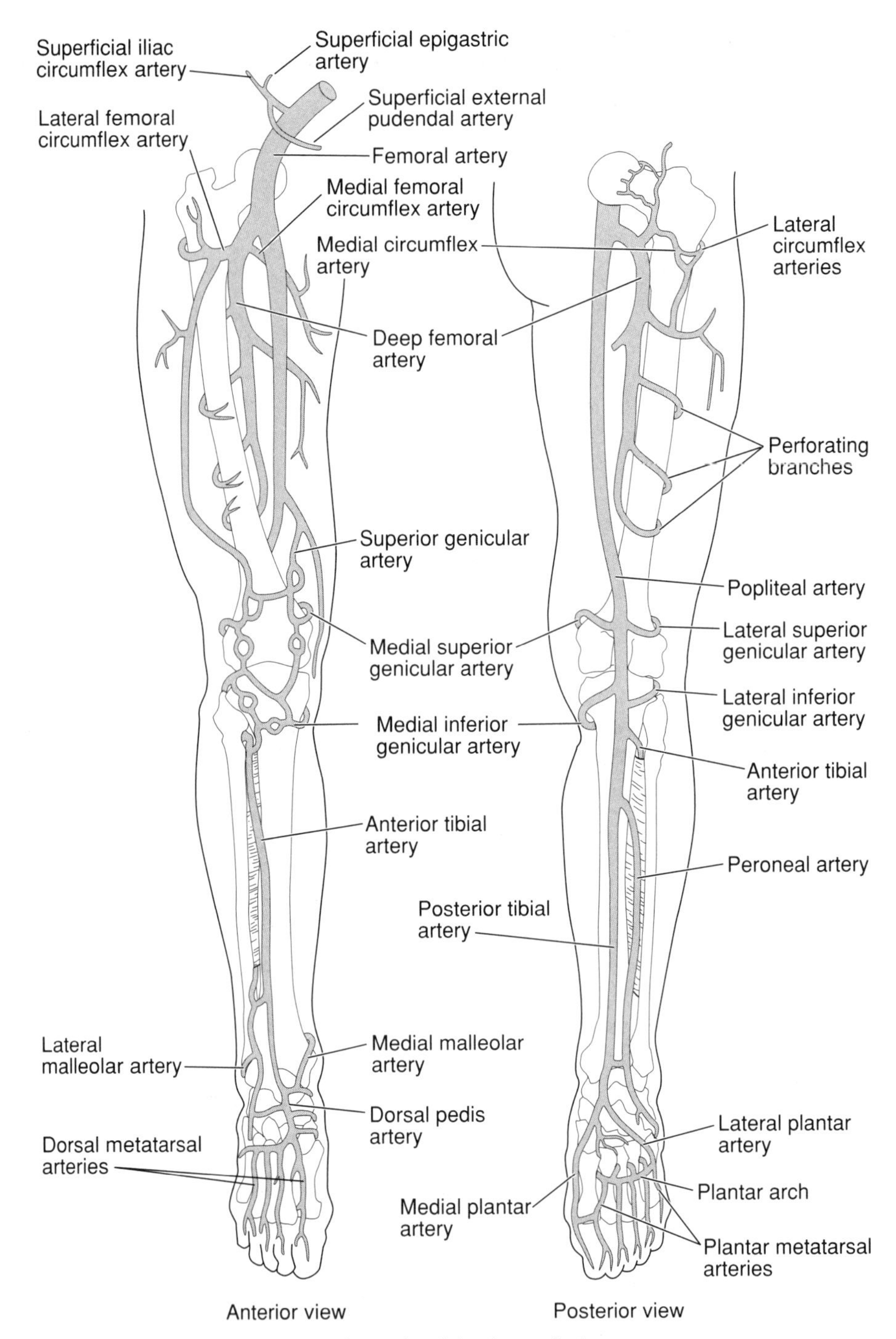


**Figure 3-4.** Summary of the blood supply of the lower limb.

# VESSELS OF THE LOWER LIMB

## Arteries

The arteries of the lower limb are shown in Fig. 3-4.

**Gluteal arteries**. The superior and inferior gluteal arteries are branches of the **internal iliac** artery. The **superior gluteal** artery exits from the pelvis through the greater sciatic foramen and above the piriformis muscle. It then passes anteriorly between the gluteus medius and minimus muscles toward the tensor fascia lata muscle. The **inferior gluteal artery** passes through the greater sciatic foramen, inferior to the piriformis, and directly into the gluteus maximus muscle.

**Femoral artery.** The femoral artery is the continuation of the external iliac artery, which enters the thigh by passing deep to the inguinal ligament about midway between the anterior superior iliac spine and the pubic tubercle. It descends through the femoral triangle and adductor canal and enters the popliteal fossa by passing through the adductor hiatus, where it becomes the popliteal artery. The largest branch of the femoral artery is the **deep femoral artery**, which arises in the femoral triangle and descends just medial to the femur. The **medial and lateral circumflex arteries** arise from the deep femoral artery. The two arteries encircle the trochanteric region of the femur and form potential anastomoses with the gluteal arteries and the first perforating branch of the deep femoral. The deep femoral artery also has perforating branches that pass into the posterior compartment of the thigh.

The **popliteal artery** passes through the deepest portion of the popliteal fossa, directly on the supracondylar or popliteal surface of the femur. It has paired **superior** and **inferior genicular branches** as well as a single **middle genicular artery**. As the artery enters the posterior compartment of the leg, it ends by dividing into anterior and posterior tibial arteries.

The **anterior tibial artery** enters the anterior compartment of the leg by passing above the superior margin of the interosseous membrane. It descends through the anterior compartment and enters the foot by crossing the anterior aspect of the ankle; there, it becomes the **dorsalis pedis artery**. On the dorsum of the foot, the artery is between the tendons of the extensor hallucis longus and the extensor digitorum longus. The dorsalis pedis artery ends in several branches, most of which supply the dorsum of the foot. One of these branches, the **deep plantar branch**, enters the plantar aspect of the foot by passing between the first and second metatarsals and there helps form the **plantar arterial arch**.

The **posterior tibial artery** descends through the deep portion of the posterior compartment of the leg, inclining medially as it descends and entering the foot by passing posterior to the medial malleolus. It ends as it passes around the medial malleolus by dividing into medial and lateral plantar arteries. It has a large branch high in the posterior compartment, the **peroneal artery**, which descends in the lateral part of the posterior compartment and terminates around the ankle.

The **medial and lateral plantar arteries** correspond in course and distribution to the medial and lateral plantar nerves. There is only one arterial arch in the foot, which corresponds to the deep arch of the hand, and is formed by the deep branch of the lateral plantar artery and the deep plantar branch of the dorsalis pedis artery.

## Veins

The veins of the lower extremity consist of deep and superficial veins. The pattern of the **deep veins** corresponds to that of the arteries. The two saphenous veins are the main **superficial veins**. The **greater saphenous vein** begins on the dorsomedial aspect of the foot, passes just anterior to the medial malleolus, and ascends along the anteromedial aspect of the leg and thigh. It terminates by passing through the saphenous opening of the fascia lata and emptying into the femoral vein just distal to the inguinal ligament. The **lesser saphenous vein** begins as a network on the dorsolateral aspect of the foot. It ascends behind the lateral malleolus and through the middle of the calf, and ends by emptying into the popliteal vein in the popliteal fossa.

**TABLE 3-10. Lymphatics of the Lower Limb**

| Afferents From | Lymph Nodes | Efferents To |
|---|---|---|
| Deep and lateral superficial lymphatics of the leg and foot | Popliteal | Deep inguinal nodes |
| Superficial lymphatics of the lower limb, lower anterolateral abdominal wall, gluteal region, and superficial perineum | Superficial inguinal | Deep inguinal nodes → external iliac nodes |

## LYMPHATICS OF THE LOWER LIMB

The lymphatic drainage of the lower limb is summarized in Table 3-10.

Chapter 4

# Head and Neck

## ORGANIZATION OF THE HEAD AND NECK

### Major Surface Landmarks and Superficial Regions

The anterolateral aspect of the neck is divided into anterior and posterior triangles by the prominent sternocleidomastoid muscle. The **anterior triangle** is in front of this muscle and extends to the inferior margin of the mandible and to the midline. Posterior to the sternocleidomastoid, the **posterior triangle** is also delimited by the superior border of the trapezius muscle and the middle third of the clavicle. The face can be divided into several areas, which include the orbit, nose, forehead, temporal region, maxillary region, and mandibular region. It is convenient to start with the well-defined **orbit**, which is protected and delimited by the prominent supraorbital and infraorbital margins, which meet laterally. Extending posteriorly at the level of the infraorbital margin, the zygomatic arch separates the temporal region above from the mandibular or lower jaw region below. The **mandibular region** extends inferiorly and then anteromedially to join the same region of the opposite side. The maxillary or **upper jaw region** is inferior to the infraorbital margin and lateral to the nose. The posterior part of the head is the occipital region.

**Structure location versus vertebral levels**. Several anterior midline structures are easily located and can be used to approximate the locations of a variety of other structures. The hyoid bone is at vertebral level C3, and the thyroid prominence at C4. The bifurcation of the common carotid artery, and hence the carotid sinus, is found laterally between these two levels. The cricoid cartilage is at vertebral level C6. Also found at this level are the carotid tubercle (anterior tubercle of C6 transverse process), the vertebral artery entering the transverse process of C6, the inferior thyroid artery entering the gland, the larynx-trachea and pharynx-esophagus junctions, and the middle cervical sympathetic ganglion. The tracheal ring and overlying isthmus of the thyroid gland are palpable inferior to the cricoid cartilage.

### Organization of the Blood Vessels of the Head and Neck

The common carotid and vertebral arteries are the major arteries to the head and neck. The external carotid supplies most of the head and neck structures outside of the cranial cavity; the internal carotid supplies the orbit and, along with the vertebral artery, the cranial cavity. On the right, the **common carotid artery** is one of the main terminal branches of the brachiocephalic trunk, whereas on the left, it is a direct branch from the arch of the aorta. The **vertebral artery** is a branch of the first part of the subclavian artery. It has no branches in the neck, and enters the cranial cavity through the foramen magnum.

The **jugular system of veins** drains most of the head and neck. The main vessel is the internal jugular vein, which begins at the base of the skull where it receives most

**TABLE 4-1. Lymphatics of the Head and Neck**

| Afferents From | Lymph Nodes | Efferents To |
|---|---|---|
| Superficial and deep structures of head and upper neck (buccal, mandibular, parotid, retroauricular, occipital, submental, submandibular, retropharyngeal nodes) | Superior deep cervical | Inferior deep cervical |
| Posterior scalp and neck, superior deep cervical nodes, axillary nodes, tracheal nodes | Inferior deep cervical (scalene) | Thoracic duct on left; right lymphatic duct on right |

of the venous blood from the cranial cavity. It descends through the neck in company with the carotid arteries, receiving multiple tributaries from outside the cranial cavity, and ends by joining the subclavian vein to form the brachiocephalic vein.

## Lymphatic System of the Head and Neck

The general pattern of the lymphatic drainage of the head and neck is summarized in Table 4-1.

## Cutaneous Innervation of the Head and Neck

The cutaneous innervation is easily defined if the head and neck are divided into three areas (Fig. 4-1). The *first area* includes the entire face and the anterior part of the scalp, extending posteriorly to a line across the top of the head that connects the

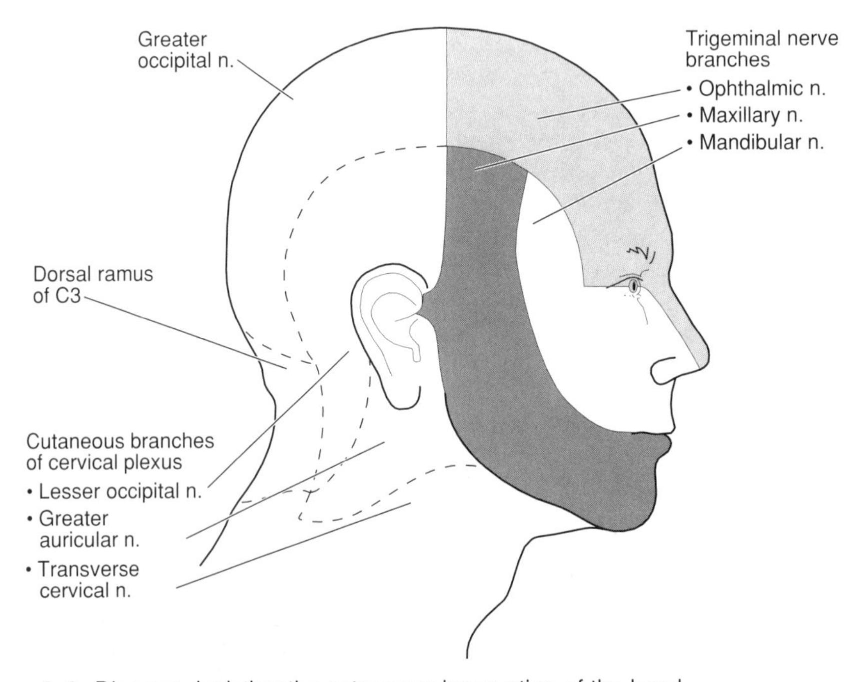


**Figure 4-1.** Diagram depicting the cutaneous innervation of the head.

two external auditory meatuses (interauricular line). The skin of this area is innervated by the three divisions of the **cranial (trigeminal) nerve V (CN V)**. The **ophthalmic nerve** innervates the bridge of the nose, upper eyelid and cornea, forehead, and scalp. The **maxillary division** covers the lateral aspect of the nose, cheek, and anterior temporal region. The **mandibular nerve** supplies the area overlying the mandible and the posterior temporal region. The *second area* includes the posterior neck and the corresponding part of the occipital region, which extends anteriorly to the interauricular line. This area is supplied by **cutaneous branches of the dorsal rami of cervical spinal nerves**. The *third area* includes the anterolateral neck, posterior triangle and shoulder pad region, and the skin surrounding the ear posteriorly. This third area is innervated by the **cutaneous branches of the cervical plexuses** (lesser occipital, great auricular, transverse cervical, and supraclavicular nerves).

## Cranial Nerves

This section provides a comprehensive, concise account of the cranial nerves. The text traces the course and branches of each nerve from the brainstem distally. Table 4-2 lists the functional types of fibers (functional components) found in each nerve and the specific structures each supplies. Table 4-3 lists the cranial parasympathetics. In this table, the parasympathetic fibers are traced from their origins from the brainstem, through the regions and nerves or branches they pass, to the structures they supply.

The cranial nerves are also discussed in Part IV, Neuroanatomy. The level and nuclei of origin or termination of the functional components of each nerve are summarized in Table 36-1. The internal (within the brainstem) courses of the fibers of each nerve are discussed in Chapter 36.

## Courses of the Cranial Nerves

**Olfactory nerve (CN I)**. This nerve consists of the rhinencephalon, olfactory tracts, and bulbs. The olfactory bulbs sit on the cribriform plate of the ethmoid in the anterior cranial fossa; the olfactory nerve consists of many small rootlets that pass through the foramina in the ethmoid into the most superior part of the nasal cavity. These rootlets supply only the uppermost of the nasal cavity and represent the entire sensory area of the nerve.

**Optic nerve (CN II)**. This nerve is also part of the brain, consisting of the occipital part of the cerebral cortex and the optic tracts, chiasm, and nerve. The **optic tracts** pass around the cerebral peduncles and along the base of the brain and join at the optic chiasm, which occupies the **chiasmatic groove**. The optic nerves result from the chiasm, passing anterolaterally through the **optic canals** into the orbit. Each optic nerve passes anteriorly through the center of the orbit and enters the central aspect of the eyeball posteriorly. The fibers of the nerve arise in the retina where their cell bodies are found. This nerve is unique because the meningeal layer of dura mater as well as the arachnoid and pia mater extend along it to the point at which it attaches to the sclera of the eyeball. As a result, the **subarachnoid space** extends along this nerve. In addition, the central artery and vein of the retina enter the nerve in the posterior third of the orbit. Thus, both the artery and vein are also surrounded by the subarachnoid space.

**Oculomotor nerve (CN III)**. This nerve exits from the brainstem in the interpeduncular fossa of the midbrain, between the superior cerebellar and posterior cerebral arteries. It passes anteriorly, crosses above the free edge of the tentorium cerebelli, and enters the cavernous sinus. It passes through the sinus as the most superior nerve in the wall and enters the orbit through the superior orbital fissure. In the orbit, it separates into a superior division to the superior rectus and levator palpebrae superioris muscles and an inferior division to the medial and inferior rectus muscles as well as the inferior oblique. It also has the motor root to the ciliary ganglion.

**TABLE 4-2. Functional Components of the Cranial Nerves (Contained in the Nerves as They Exit the Brainstem)**

| Cranial Nerve (CN) | Functional Components | Structure(s) Supplied |
|---|---|---|
| CN I: olfactory | Special visceral afferent (SVA) | Smell: olfactory epithelium |
| CN II: optic | Special somatic afferent (SSA) | Vision: retina |
| CN III: oculomotor | General visceral afferent (GVE) | Parasympathetic: sphincter pupillae and ciliary muscles |
| | General somatic efferent (GSE) | Extraocular eye muscles: levator palpebrae superioris, superior, inferior, and medial recti, inferior oblique |
| CN IV: trochlear | GSE | Extraocular eye muscle: superior oblique |
| CN V: trigeminal | General somatic afferent (GSA) | General sensation: skin of face, oral and nasal cavities, teeth, anterior tongue, eye |
| | Special visceral efferent (SVE) | Muscles of mastication and tensor tympani, tensor veli palatine, anterior digastric, mylohyoid |
| CN VI: abducens | GSE | Extraocular eye muscle: lateral rectus |
| CN VII: facial | GSA | General sensation: posterior ear, external auditory meatus |
| | SVA | Taste: anterior tongue |
| | GVE | Parasympathetic: submandibular, sublingual, and lacrimal glands; palatal, nasal, and paranasal mucosa |
| | General visceral afferent (GVA) | Visceral sensation: salivary and lacrimal glands, nasal mucosa |
| | SVE | Muscles of facial expression and stapedius, stylohyoid and posterior digastric |
| CN VIII: vestibulocochlear | SSA | Hearing: organ of Corti<br>Proprioception: utriculus, saccule, semicircular ducts |
| CN IX: glossopharyngeal | GSA | General sensation: posterior ear and external auditory meatus |
| | GVA | Visceral sensation: middle ear, pharynx and posterior tongue, carotid sinus, parotid gland |
| | SVA | Taste: posterior tongue |
| | GVE | Parasympathetic: parotid gland |
| | SVE | Stylopharyngeus |
| CN X: vagus | GSA | General sensation: posterior ear and external auditory meatus |
| | GVA | Visceral sensation: gastrointestinal tract and organs of thorax |
| | SVA | Taste: epiglottis and valleculae |
| | GVE | Parasympathetic: muscle and glands of gastrointestinal tract to splenic flexure, heart, pulmonary system |
| | SVE | Skeletal muscle of larynx, pharynx, palate (not tensor veli palatine and stylopharyngeus): these fibers from bulbar IX |
| CN XI: accessory | SVE (in bulbar portion) | See SVE of vagus nerve |
| | GSE (in spinal portion) | Sternocleidomastoid and trapezius muscles |
| CN XII: hypoglossal | GSE | Muscles of tongue (intrinsic and extrinsic except palatoglossus) |

**TABLE 4-3. Cranial Parasympathetics**

| Cranial Nerve of Origin | Cranial Nerve Branches to Ganglion | Parasympathetic Ganglion | Branches from Ganglion | Target Structures |
|---|---|---|---|---|
| CN III: oculomotor | Motor root of inferior division of CN III | Ciliary | Short ciliary nerves | Sphincter pupillae and ciliary muscles |
| CN VII: facial | Chorda tympani → lingual | Submandibular | Lingual | Submandibular, sublingual, and lingual glands |
| CN VII: facial | Greater petrosal → nerve of the pterygoid canal | Pterygopalatine | Zygomatic → lacrimal<br>Branches of maxillary CN V | Lacrimal gland<br>Nasal and palatal glands |
| CN IX: glossopharyngeal | Tympanic nerve → tympanic plexus → lesser petrosal nerve | Otic | Auriculotemporal | Parotid gland |

*Note: The focus of this table is the four cranial parasympathetic ganglia in the third column. The first and second columns contain the preganglionic input to the ganglia; the third and fourth columns show the postganglionic routes and structures supplied.*

**Trochlear nerve (CN IV).** The trochlear nerve is the smallest of the cranial nerves and supplies only one extraocular eye muscle. It leaves the brainstem from the dorsal aspect of the midbrain, passes around the midbrain, and enters the cavernous sinus by piercing the dura in the posterior cranial fossa. It passes anteriorly in the wall of the sinus, just inferior to the oculomotor nerve, and enters the orbit through the **superior orbital fissure**. It then passes medially to the superior oblique muscle, which it supplies.

**Trigeminal nerve (CN V).** This nerve is the largest of the cranial nerves and, in addition to supplying general sensation to the entire head anterior to the interarticular line, it provides pathways for fibers from other nerves to reach their destinations through its three major divisions. The trigeminal nerve exits from the anterolateral aspect of the mid pons in the posterior cranial fossa. It passes anteriorly and pierces the dura just inferior to the free edge of the tentorium cerebelli. Its large, sensory ganglion, the **semilunar ganglion**, occupies a depression on the anterior face of the petrous apex. The *three branches* of the nerve—the ophthalmic, maxillary and mandibular divisions—emerge from this ganglion.

The **ophthalmic division** continues anteriorly in the wall of the cavernous sinus and enters the orbit through the superior orbital fissure. In the orbit, it branches into the frontal, lacrimal, and nasociliary nerves, which supply the forehead and eyelid, lacrimal gland, and eyeball and nasal cavity, respectively.

The **maxillary division** passes through the most inferior aspect of the wall of the cavernous sinus and passes through the foramen rotundum en route to the pterygopalatine fossa. In this fossa, it communicates with the **pterygopalatine ganglion** (via pterygopalatine nerves) and has branches to the nasal cavity (i.e., nasopalatine, lateral nasal), maxillary sinus and dentition (i.e., posterior superior alveolar), skin of the anterior temporal and zygomatic regions (i.e., zygomaticotemporal, zygomaticofacial), and palate (i.e., palatine). The nerve continues anteriorly as the infratemporal nerve. It passes along the floor of the orbit in the infraorbital canal, has middle and anterior superior alveolar branches to the maxillary sinus and teeth, and ends by passing onto the face through the infraorbital foramen, where it supplies the skin of the nose, cheek, and upper lip.

The **mandibular division** does not pass through the cavernous sinus, but rather descends through the foramen ovale into the infratemporal fossa. In this fossa, the nerve communicates with the otic ganglion and has both cutaneous and muscular branches. The cutaneous nerves are the auriculotemporal to the posterior temporal region, the buccal to the cheek, and the inferior alveolar (which also supplies the mandibular teeth) to the lower jaw and lip (via the mental nerve). The muscular branches supply the muscles of mastication as well as the mylohyoid, anterior belly of the digastric, tensor veli palatini, and tensor tympani. The lingual nerve provides general sensation and taste to the anterior two-thirds of the tongue and communicates with the submandibular ganglion.

**Abducens nerve (CN VI)**. This nerve exits from the brainstem at the pons-medulla junction. It passes anteriorly and pierces the dura on the basilar portion of the occipital bone, ascends, crosses the apex of the petrous temporal bone, and enters the cavernous sinus, where it passes anteriorly as the most medial nerve in the sinus. After entering the orbit via the superior orbital fissure, it passes laterally toward the lateral rectus muscle.

**Facial nerve (CN VII)**. This nerve exits from the lateral aspect of the brainstem at the pons-medulla junction (cerebellopontine angle). It then passes laterally and into the internal auditory meatus, which is continuous with the facial canal. While in the facial canal, the nerve first passes laterally, then posteriorly, and finally descends and exits from the temporal bone through the stylomastoid foramen. While in the facial canal, the facial nerve has *several branches*: the greater petrosal nerve (from the geniculate ganglion), the nerve to the stapedius, and chorda tympani. After exiting from the facial canal, the nerve passes into the parotid gland, where it separates into multiple branches that supply the muscles of facial expression.

**Vestibulocochlear nerve (CN VIII)**. This nerve joins the brainstem at the most lateral aspect of the pons-medulla junction (cerebellopontine angle). It has a short, lateral course in the posterior cranial fossa and then enters the internal auditory meatus, where it splits into its vestibular and cochlear portions.

**Glossopharyngeal nerve (CN IX)**. This nerve exits from the anterolateral aspect of the medulla as the most superior nerve in the posterior olivary sulcus, then passes laterally and through the jugular foramen. As it exits, it has the tympanic branch, which passes into the middle ear. After exiting from the skull, the nerve descends along the stylopharyngeus, has branches to the pharyngeal plexus and carotid sinus, and finally passes across the tonsillar fossa.

**Vagus nerve (CN X)**. This "vagrant" nerve has an extensive course as it "wanders" through the neck, thorax, and abdomen. It begins in the posterior cranial fossa, where it exits from the lateral aspect of the medulla in the postolivary sulcus. After passing through the jugular foramen, it descends through the neck in the carotid sheath. Along this course, there are pharyngeal, superior laryngeal, and inferior (recurrent) laryngeal branches. The inferior laryngeal nerve, recurring around the aortic arch on the left and the subclavian artery on the right, ascends to the larynx in the groove between the esophagus and trachea. The details of its course through the thorax and abdomen are described with those regions in Chapter 3.

**Accessory nerve (CN XI)**. This nerve arises from the upper part of the cervical spinal cord, ascends to enter the posterior cranial fossa via the foramen magnum, and then exits through the jugular foramen. It then descends deep to the sternocleidomastoid muscle and enters the posterior cervical triangle. The nerve crosses this triangle superficially and passes deep to the superior border of the trapezius, which it supplies.

**Hypoglossal nerve (CN XII)**. This nerve arises from the preolivary sulcus of the medulla and passes anterolaterally through the posterior cranial fossa toward the

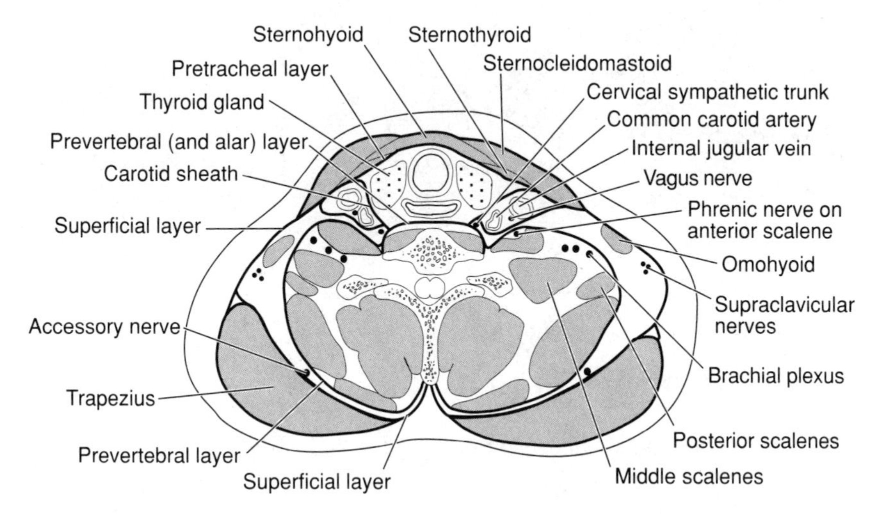


**Figure 4-2.** Cross section of the neck, inferior to the level of the hyoid bone, depicting the major layers of cervical fascia.

hypoglossal canal, through which it exits the cranial vault. It continues anteriorly, passing between the internal jugular vein and internal carotid artery. Where it reaches the floor of the mouth, it is positioned between the mylohyoid and hyoglossus muscles.

# NECK

## Fascia and Triangles of the Neck

**Fascial planes.** The investing layer of **cervical fascia** encircles the neck and encloses the sternocleidomastoid and trapezius muscles (Fig. 4-2). The **prevertebral fascia** surrounds the vertebral column and its associated muscles: the longus capitis and colli, the scalenes, and the deep muscles of the back in the cervical region. The visceral structures of the neck are enclosed in a sleeve of fascia called the **pretracheal fascia** anteriorly and laterally and the **buccopharyngeal fascia** (between the pharynx or esophagus and the vertebral column) posteriorly. The infrahyoid muscles have their own fascia, which is between the investing and pretracheal layers of cervical fascia. In the lateral part of the neck and deep to the plane of the sternocleidomastoid muscle, the several layers of cervical fascia meet and contribute to the formation of the vertically oriented **carotid sheath**.

**Anterior triangle.** This triangle provides access to most of the visceral and neurovascular structures of the neck. It is bounded by the midline, inferior margin of the mandible and the sternocleidomastoid muscle. The part of the anterior triangle above the digastric muscle is the **submandibular** triangle, and the area below consists of the **carotid** triangle above the omohyoid muscle and **muscular** triangle below. The carotid sheath structures—the vagus nerve, the carotid artery, and the internal jugular vein—pass vertically through this area, located deep to the sternocleidomastoid muscle. The cervical sympathetic chain is deep in the triangle on the anterolateral aspects of the cervical vertebrae. The lateral lobes of the thyroid gland are immediately adjacent to the lateral aspects of the trachea and lower larynx, while the isthmus of the thyroid crosses the midline in front of the upper rings of the trachea. The parathyroids are related to the posterior surface of the upper and lower aspects of the lateral lobes of the thyroid gland. The esophagus lies posterior to the trachea and larynx. The sub-

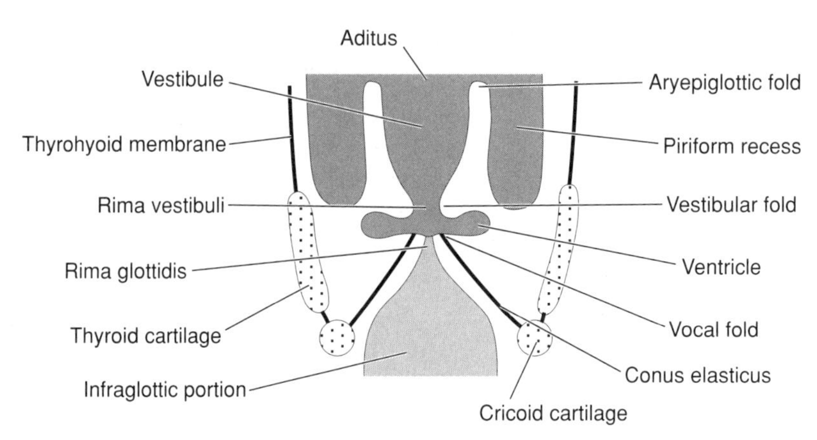


**Figure 4-3.** Diagram of a frontal section through the larynx depicting its subdivisions. The lumen of the larynx is shaded.

mandibular triangle contains the submandibular gland, the facial artery and vein, the mylohyoid vessels and nerves, and the hypoglossal nerve.

**Posterior triangle**. The posterior cervical triangle is described with the upper limb because the major neurovascular structures of the upper limb pass through this triangle (see Chapter 2).

## Visceral Structures of the Neck

**Larynx**. The larynx is that portion of the airway between the pharynx and the trachea. It is anterior to cervical vertebrae C4 through C6, and is related anteriorly to the infrahyoid muscles and laterally to the inferior constrictor muscle of the pharynx and the lobes of the thyroid gland. It is a tubular organ composed of nine cartilages that are connected by elastic membranes and synovial joints (Fig. 4-3). It is lined by a mucosa that also covers the vocal cords and is innervated by branches of the vagus nerve.

The **cartilages** of the larynx consist of three single cartilages (i.e., epiglottis, thyroid, and cricoid) and three pairs (i.e., arytenoids, corniculates, and cuneiforms). The **thyroid cartilage** consists of a pair of lamina that are connected anteriorly but separated posteriorly; superior and inferior cornua project, respectively, superiorly and inferiorly from the posterior aspects of the laminae. The anterosuperior aspect of the cartilage is especially prominent and called the **laryngeal prominence** or **Adam's apple**. The **cricoid cartilage** is ring-shaped and formed by a pair of posterior lamina and an anterior arch. The two cartilages articulate at the cricothyroid joints, synovial joints that are formed between the inferior cornua and the lateral aspects of the cricoid lamina.

The **arytenoid cartilages** are small and triangular. They are positioned on the superior aspects of the cricoid lamina (cricoarytenoid joints), pivot and slide on the lamina, and each has vocal and muscular processes. The vocal ligament attaches to the vocal process; movement of the arytenoid cartilages is important in the movement and tenseness of the vocal ligaments. When the vocal processes move medially, the vocal ligaments are approximated or adducted; lateral movement produces abduction. The small **corniculate cartilages** sit on the apices of the arytenoid cartilages. The **epiglottis** (composed of elastic cartilage) is expanded and rounded superiorly and tapers to a point inferiorly. It is connected to both the hyoid bone and thyroid cartilage anteriorly, and its superior and lateral edges form part of the laryngeal aditus. The **cuneiform cartilages** also form part of the aditus.

The vocal apparatus (**glottis**) consists of the true vocal folds and the opening (rima glottidis) between the folds. The area above the true folds, extending to the

## TABLE 4-4. Muscles of Larynx

| Muscle | Origin | Insertion | Action | Innervation |
|---|---|---|---|---|
| Cricothyroid | Anterior superior cricoid | Inferior thyroid lamina | Tense vocal cords | Superior laryngeal |
| Posterior cricoarytenoid | Posterior aspect of cricoid | Muscular process of arytenoid | Abduction of vocal cords | Recurrent laryngeal |
| Lateral cricoarytenoid | Superior posterior arch of cricoid | Muscular process of arytenoid | Adduction of vocal cords | Recurrent laryngeal |
| Transverse arytenoid (single muscle) | Interconnects posterior aspects of arytenoid cartilages | | Adduction of vocal cords | Recurrent laryngeal |
| Oblique arytenoid and aryepiglotticus | Muscular process of arytenoid | Aryepiglottic fold | Sphincter of laryngeal aditus | Recurrent laryngeal |
| Vocalis | Deep aspect of thyroid cartilage (with vocal ligament) | Vocal process of arytenoid | Relax vocal cord | Recurrent laryngeal |
| Thyroarytenoid | Deep surface of thyroid lamina | Lateral aspect of arytenoid | Sphincter of vestibule | Recurrent laryngeal |

laryngeal aditus (entrance), is the vestibule or supraglottic portion. The false vocal folds are above the true folds, and the area extending laterally between the true and false folds is the ventricle. The vocal fold (true vocal cord) contains the thin cranial edge of the conus elasticus, which is the vocal cord. The vocal fold is divided into anterior intramembranous and posterior intracartilaginous portions. The intramembranous portion stretches between the thyroid and arytenoid cartilages and is capable of tension change and vibration. The intracartilaginous portion is formed by the arytenoid cartilage. With the exception of the cricothyroid, the numerous intrinsic muscles moving the laryngeal cartilages lie deep to the thyroid cartilage. These muscles are described in Table 4-4. The internal branch of the superior laryngeal nerve is sensory to the supraglottic portion of the larynx and the adjacent part of the pharynx; the recurrent (inferior) laryngeal nerve innervates the infraglottic mucosa of the larynx. The blood supply is provided by the superior and inferior thyroid arteries.

**Pharynx**. The pharynx extends from the base of the skull to the beginning of the larynx and esophagus. Posteriorly, it is in contact with the upper six cervical vertebrae; laterally, it is related to the internal and the common carotid arteries, the internal jugular vein, the sympathetic trunk, and the last four cranial nerves. Anteriorly, it communicates with the nasal cavity and the oral cavity; inferiorly, it is continuous with the larynx and esophagus.

That portion of the pharynx above the level of the soft palate is the **nasopharynx**. The **auditory (eustachian) tube** opens on the lateral wall of the nasopharynx. Its projecting cartilage of the auditory tube produces a marked elevation (the torus tubarius) around the opening, and the **pharyngeal recess** is the fossa behind the posterior lip of the torus. The pharyngeal tonsil (**adenoid**) is found on the posterior wall of the nasopharynx. The **oropharynx** is posterior to the oral cavity and is limited above by the soft palate and below by the superior aspect of the epiglottis. The oral pharynx communicates with the oral cavity through the fauces (throat), which is below the soft palate and above the root of the tongue. Laterally, the **fauces** are bounded by two mucosa-covered muscular columns (**pillars of the fauces**): the anterior palatoglossal muscle and fold and the posterior palatopharyngeal muscle and fold. The area between the folds is the **tonsillar fossa**, which contains the palatine tonsil. The **laryngopharynx** extends from the superior edge of the epiglottis inferiorly to the lower border of the cricoid cartilage and surrounds the larynx laterally and posteriorly. The vertical groove between the lateral aspect of the larynx and the pharyngeal wall is the **piriform recess**.

The **muscles of the pharynx** are the three constrictors and the stylopharyngeus, which passes downward between the superior and the middle constrictors. The constrictors overlap each other from below upward and surround the pharynx. The **pharyngobasilar fascia** is found between the muscles and mucous membrane of the pharynx, and it is especially strong superiorly, where it attaches to the basilar process of the occipital bone and the petrous portion of the temporal bone. CN IX (glossopharyngeal) and CN X (vagus) are the primary nerves of the pharynx, with CN X supplying the constrictor muscles and CN IX supplying the stylopharyngeus and pharyngeal mucosa. The **gag reflex**, then, is mediated by CN IX (sensory limb) and CN X (motor).

**Thyroid and parathyroid glands**. The **thyroid gland** consists of a pair of lobes that are interconnected by an isthmus. The lobes are located on either side of the larynx and extend from the oblique line of the thyroid cartilage inferiorly to the level of the fourth or fifth costal cartilages. The isthmus crosses the midline at about the second costal cartilage. There are usually two pairs of **parathyroid glands**, the superior and inferior, which are located on the superior and inferior aspects of the thyroid lobes, respectively. The blood supply to both the thyroid and parathyroid glands is very rich and provided by superior and inferior thyroid arteries.

## Neurovascular Structures of the Neck

**Common carotid artery**. In a plane deep to the sternocleidomastoid muscle, this artery ascends along a line that passes posterior to the sternoclavicular joint and through the midpoint between the angle of the mandible and the mastoid process. It ends by dividing into the internal and external carotid arteries between the levels of the hyoid bone (C3) and prominence of the thyroid cartilage (laryngeal prominence or Adam's apple; C4). The carotid body and sinus are located at this bifurcation. The **internal carotid** has no branches in the neck and passes directly toward the carotid canal, through which it enters the cranial cavity.

The **external carotid** artery ascends to the neck of the mandible, where it divides into its terminal branches: the **maxillary artery**, which passes deep to the neck of the mandible into the infratemporal fossa, and the **superficial temporal artery**, which ascends into the temporal region to supply that region and the anterior part of the scalp. *Six other branches* commonly arise from the external carotid artery. The superior thyroid artery supplies primarily the thyroid gland and larynx. The lingual artery passes deeply toward the tongue. The facial artery crosses the inferior margin of the mandible just anterior to the angle and then follows a tortuous course obliquely across the face toward the angle between the nose and medial aspect of the eye. The ascending pharyngeal artery supplies the pharyngeal and palatal regions. The occipital and posterior auricular branches supply primarily the superficial regions designated by their names.

**Vertebral artery**. This artery branches from the first part of the subclavian artery and is located on the anterolateral aspect of vertebral body C7, usually just anterior to the inferior cervical sympathetic ganglion. The artery then ascends and passes through the transverse foramina of the upper six cervical vertebrae before entering the cranial cavity via the foramen magnum. As it ascends, it has segmental radicular branches that enter the vertebral canal to supply the meninges and spinal cord.

# CRANIAL CAVITY

## Osteology

The cranial cavity is formed by a roof (calvaria) and a floor (Fig. 4-4). The calvaria is formed by the single frontal and occipital bones, the paired parietal bones, portions of the greater wings of the sphenoid, and the squamous parts of the temporal bones. The parietal bones are united by the sagittal suture and with the frontal bone at the coronal suture. Posteriorly, the occipital and parietal bones are united at the lambdoidal suture.

**Anterior cranial fossa**. The floor of the cranial cavity is divisible into the anterior, middle, and posterior cranial fossae. The anterior cranial fossa is formed medially by the cribriform plate of the ethmoid and the crista galli and laterally by the orbital plate of the frontal bone. Posteriorly, both the body of the sphenoid and its lesser wing participate in the formation of this fossa, which is related anteromedially to the frontal sinus and inferiorly to the nasal cavity, ethmoid sinuses, and the orbit. The multiple openings in the **cribriform plate** transmit the rootlets of the olfactory nerve into the superior aspect of the nasal cavity. The foramen cecum is anterior to the crista galli and may transmit an emissary vein between the nasal cavity and the superior sagittal sinus. The anterior cranial fossa houses the frontal lobes of the brain and the olfactory bulbs and tracts.

**Middle cranial fossa**. The middle cranial fossa has a central and two lateral parts. The central part is formed by the body of the sphenoid and consists of the **sella turcica** posteriorly and the **chiasmatic groove** anteriorly. It houses the hypophysis (pitu-

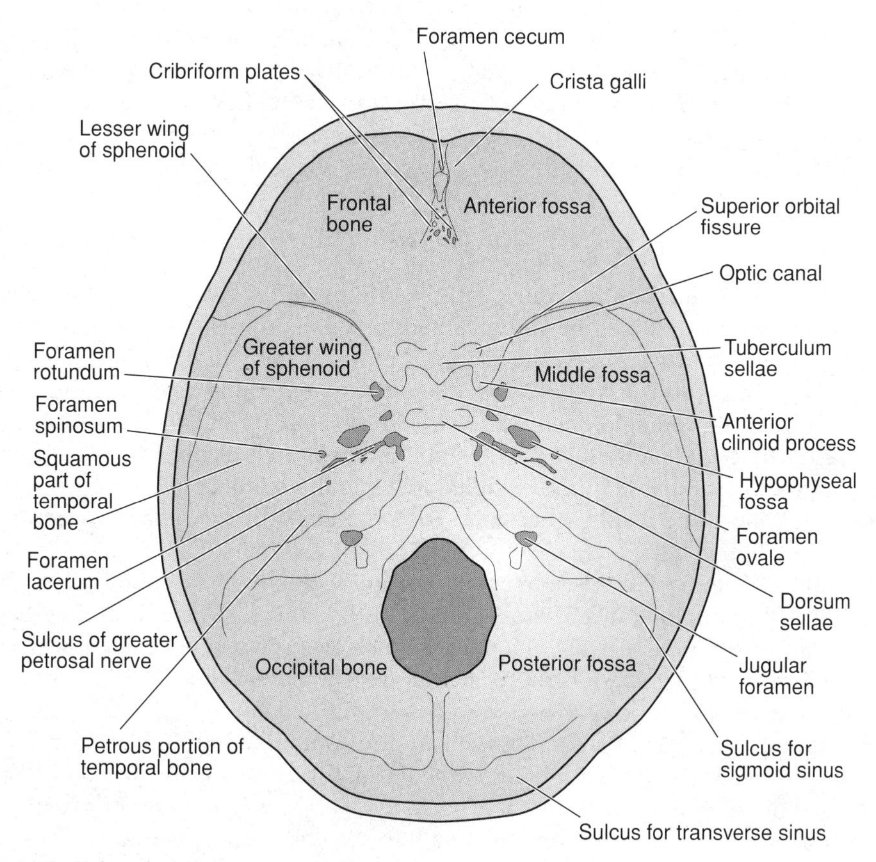


**Figure 4-4.** Interior of the base of the skull.

itary) and optic chiasm. This part of the fossa is related inferiorly (and anteriorly) to the sphenoid sinus and laterally to the cavernous sinus. The **optic canals** convey the optic nerves into the orbits.

The **lateral part of the middle cranial fossa** houses the temporal lobe of the brain and is formed by the greater wing of the sphenoid and parts of both the squamous and petrous portions of the temporal bone. It is bounded anteriorly by the sphenoid ridge and posteriorly by the petrous ridge. This part of the fossa is related inferiorly to the infratemporal fossa and middle ear cavity, laterally to the temporal fossa, anteriorly to the orbit, and medially to the sphenoid sinus. There are a number of openings in this fossa: The **superior orbital fissure** (containing the ophthalmic division of the trigeminal, oculomotor, trochlear, and abducens nerves, and the ophthalmic veins) opens into the orbit; the **foramen rotundum** (containing the maxillary division of the trigeminal nerve) opens into the pterygopalatine fossa; the **foramen ovale** (containing the mandibular division of the trigeminal nerve) and the **foramen spinosum** transmitting the middle meningeal artery) open into the infratemporal fossa. The **foramen lacerum** is an irregularly shaped opening at the apex of the petrous pyramid. This opening is filled by fibrous tissue and is the floor of the carotid canal, thus nothing passes through the opening.

**Posterior cranial fossa**. The posterior cranial fossa is formed by the posterior aspect of the body of the sphenoid, a portion of the petrous portion of the temporal bone, and the occipital bone. Anteriorly in the midline, the inclined plane formed by the sphenoid and the basilar portion of the occipital bone is occupied by the brainstem. The midbrain is surrounded by the notch of the tentorium cerebelli. The cerebellum occupies the rest of the posterior fossa. The **foramen magnum**, the large single opening in this fossa, transmits the spinal cord–brainstem junction and its meningeal coverings; the accessory nerve; the vertebral, anterior, and posterior spinal arteries; and the communication between the dural venous sinuses and the internal vertebral venous plexus. The **hypoglossal canal** transmits the hypoglossal nerve; the **jugular foramen** contains the glossopharyngeal, vagus, and accessory nerves as well as the communication between the dural venous sinuses and the internal jugular vein. The **internal acoustic meatus** is an opening on the posterior face of the petrous pyramid that transmits the facial and vestibulocochlear nerves and the labyrinthine artery.

## Meningeal Coverings of the Brain

**Dura mater**. The cranial dura mater consists of outer periosteal and inner meningeal layers, which are fused together and attached to the deep surfaces of the bones that form the skull. The layers separate where septa and the dural venous sinuses are formed. The **falx cerebri** is the sagittal septum that separates the cerebral hemispheres in the longitudinal cerebral fissure (Fig. 4-5). The falx cerebri is continuous posteriorly with another dural reflection, the **tentorium cerebelli**, lying in the transverse cerebral fissure between the occipital lobes of the telencephalon above and the cerebellum below. The **diaphragma sellae** forms a roof over the sella turcica, but has a small central opening through which the pituitary stalk passes.

**Pia mater**. The pia mater is attached to the surface of the brain and closely follows its contours, dipping into all sulci and so on. This layer is vascular, containing the vessels that supply the neural tissue. The pia mater forms the **tela choroidea,** which are part of the **choroid plexuses** of the ventricles. **Cerebrospinal fluid (CSF)** is produced in these plexuses.

**Arachnoid**. The arachnoid is intermediate in position and separated from the pia mater by the **subarachnoid space**. This space is enlarged into subarachnoid cisterna at the cerebellum-medulla junction (cisterna magna), between the cerebral peduncles (interpeduncular cistern), superior and lateral to the midbrain (cisterna ambiens), and at several other areas. Arachnoid granulations project into the superior sagittal sinus and

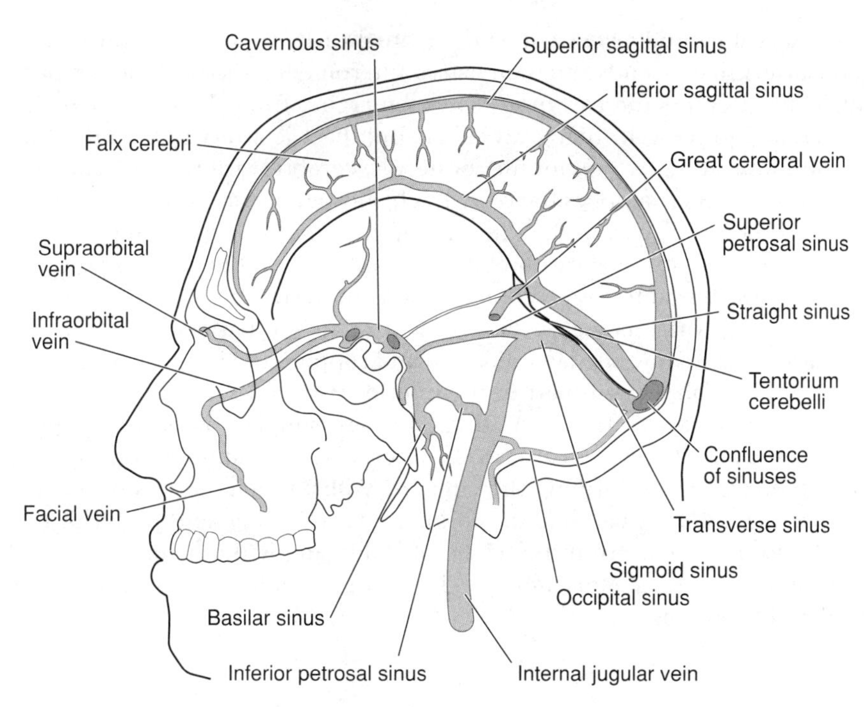


**Figure 4-5.** Diagram depicting the positions of the dural venous sinuses.

serve as the means by which CSF flows from the subarachnoid space into the venous system. Even though this layer is held closely against the dura mater by the CSF, it is not attached to the dura. As a result, the subdural space is normally only a potential space.

**Dural venous sinuses**. The dural venous sinuses are interconnected and drain toward the jugular bulb in the upright position (see Fig. 4-5). The **superior sagittal sinus**, in the attached margin of the falx cerebri, drains toward the **transverse sinuses**, which are in the attached margin of the tentorium cerebelli. These in turn drain to the internal jugular veins via the **sigmoid sinuses**. The **inferior sagittal sinus**, in the free margin of the falx cerebri, and the great vein of Galen from the brain join to form the **straight sinus**, which is in the junction of the falx and tentorium. The straight sinus drains to the transverse sinus. An anterior group of sinuses includes the **cavernous**, **intercavernous**, **superior** and **inferior petrosal**, and the **basilar plexus**. There is no actual epidural space because the dura is attached to the deep surfaces of the bones; however, one of the meningeal arteries (usually the middle) may be lacerated when the skull is fractured. The artery can leak and separate the dura from the bones, thus forming an **epidural hematoma**.

## Blood Supply of the Brain

The brain is supplied by the vertebral and internal **carotid arteries**. Soon after entering the skull through the foramen magnum, the **vertebral artery** has posterior inferior cerebellar and anterior and posterior **spinal arteries**. The vertebral arteries unite at the pons-medulla junction to form the basilar artery. The basilar artery passes along the basilar sulcus of the pons and terminates at the level of the mesencephalon by branching into posterior **cerebral arteries**. It has anterior inferior cerebellar, labyrinthine, paramedian, short and long circumferential and superior cerebellar branches. Branches of the vertebral and basilar arteries supply the cerebellum, mesencephalon, pons, medulla, medial portion of the occipital lobe, and part of the temporal lobe and diencephalon.

The **internal carotid artery** enters the cranial cavity via the carotid canal. In this canal, the artery first ascends and then passes anteromedially toward the petrous apex. At that point, it enters the cavernous sinus, emerging from the sinus medial to the anterior clinoid process. It then gives off the ophthalmic artery to the orbit; the posterior communicating artery to the posterior cerebral artery; and the anterior choroidal artery, which supplies the optic tract, choroid plexus of the lateral ventricle, basal ganglia, posterior part of the internal capsule, and the hippocampus. Lateral to the optic chiasm, the internal carotid bifurcates into the middle cerebral and anterior cerebral arteries. The anterior cerebral arteries, connected by the anterior communicating artery, supply the medial surface of the frontal and parietal lobes. The middle cerebral artery passes into the lateral fissure and supplies the insula, lateral surface of the cerebral hemisphere, and part of the inferior surface of the temporal lobe. The basal ganglia and part of the internal capsule are supplied by the lenticulostriate branches of the middle cerebral artery. The internal carotid and vertebral arterial supplies are interconnected, forming the **circle of Willis**, which consists of the anterior communicating, anterior cerebral, posterior communicating, and posterior cerebral arteries. Central branches from the circle of Willis supply the basilar portion of the diencephalon and basal ganglia and give rise to the hypophyseal portal arterial system of the adenohypophysis.

# FACE

## Facial Muscles

The muscles of facial expression are found in the subcutaneous tissue of the face, neck (platysma), and scalp (epicranius). They function to move the skin and regulate the shapes of the openings on the face. Most of these muscles are associated with the muscle surrounding the mouth (orbicularis oris) and those surrounding the eyes (orbicularis oculi). The buccinator muscle is deep in the cheek and represents the only muscle in the interval between the maxilla and mandible. All of these muscles are innervated by CN VII (facial nerve).

## Parotid Gland

The largest of the salivary glands is located anteroinferior to the ear and extends inferiorly to the level of the angle of the mandible. It has a deep portion that extends posterior and then medial to the ramus of the mandible. The main trunk of the facial nerve enters the posterior aspect of the gland and divides into its main divisions within the substance of the gland. The **parotid (Stensen's) duct** passes anteriorly around the masseter muscle and empties into the oral cavity just opposite the second upper molar.

## Neurovascular Structures of the Face

The branches of the **facial nerve** supply the muscles of facial expression. Within the parotid gland, this nerve divides into multiple branches, which diverge across the face and anterior neck. These *regional branches* are grouped into the temporal, zygomatic, buccal, mandibular, and cervical branches. Branches of the three divisions of the **trigeminal nerve** supply the skin of the face. These branches emerge through various foramina: the **supraorbital nerve** (ophthalmic) to the forehead via the supraorbital foramen, the **infraorbital nerve** (maxillary) to the cheek via the infraorbital foramen, and the **mental nerve** (mandibular) to the mandibular region via the mental foramen. Another branch of the mandibular division, the **auriculotemporal nerve**, supplies the posterior temporal region. The **facial artery** and **vein** cross the face diagonally, from just anterior to the angle of the mandible to the angle formed by the eye and nose. The vein begins superomedially, where it is called the **angular vein** and it communicates with the veins of the orbit. After crossing the inferior border of the mandible, the facial vein usually joins the retromandibular vein before ending at the internal jugular vein. After crossing the inferior border of the mandible where a facial pulse is easily taken, the artery has a tortuous course across the face, passing both superficial and deep to the muscles. It has inferior and superficial labial branches and a lateral nasal branch.

# TEMPORAL AND INFRATEMPORAL FOSSAE

## Temporal Fossa

The temporal fossa is superficial to those areas of the frontal, parietal, and squamous portions of the temporal and greater wing of the sphenoid bones that are bounded superiorly and posteriorly by the temporal lines. It extends inferiorly to the zygomatic arch and anteriorly to the frontal process of the zygomatic bone.

## Infratemporal Fossa

**Osteology**. The infratemporal fossa is deep to the ramus of the mandible and the zygomatic arch. It is limited above by the infratemporal crest of the sphenoid, medially by the lateral pterygoid plate, anteriorly by the maxilla, and inferiorly by the alveolar border of the maxilla. The infratemporal fossa is continuous medially with the pterygopalatine fossa via the **pterygomaxillary fissure**. It communicates with the middle cranial fossa by openings in its roof: the foramen ovale and foramen spinosum. Connections with the orbit are established through the **inferior orbital fissure.**

**Temporomandibular joint**. The temporomandibular articulation is formed between the anterior portion of the mandibular fossa and the articular tubercle of the temporal bone above and the condyle of the mandible below. The articular surfaces are covered by fibrocartilage and are separated by an intraarticular disk. This disk separates the joint space into two compartments. The joint capsule is loose between the temporal bone and articular disk, but tighter and stronger between the disk and mandibular condyle. As a result, different types of movement occur between the temporal bone and disk and between the disk and condyle. The disk and condyle slide as a unit relative to the mandibular fossa, and the condyle rotates as a hinge on the disk. For example, opening the mouth (depression of the mandible) involves anterior translation of the disk and condyle as well as rotation between the condyle and disk. The joint is reinforced by three ligaments: the **lateral temporomandibular ligament** directly reinforces the lateral aspect of the joint capsule, the **stylomandibular ligament**, and the **sphenomandibular ligament**. The sphenomandibular ligament extends from the spine of the sphenoid to the lingula of the mandible, suspending the mandible as a sling and permitting the mandibular condyle to translate anteriorly and posteriorly.

**Muscles of mastication**. The muscles of mastication are the masseter and temporalis and the medial and lateral pterygoids. They are found both in the temporal and infratemporal fossae as well as superficial to the ramus of the mandible. These muscles are listed in Table 4-5.

**Neurovascular structures of the infratemporal fossa**. In addition to the pterygoid muscles, the infratemporal fossa contains the proximal portion of the maxillary artery, the mandibular division of the trigeminal nerve, and the pterygoid plexus of veins. The **maxillary artery** is the larger of the two terminal branches of the external carotid. It arises

**TABLE 4-5. Muscles of Mastication**

| Muscle | Origin | Insertion | Action | Innervation |
|---|---|---|---|---|
| Masseter | Zygomatic arch | Ramus of mandible, lateral aspect | Elevation | Mandibular division of the trigeminal nerve |
| Temporalis | Temporal fossa | Coronoid process | Elevation, retraction | Same as masseter |
| Medial pterygoid | Lateral pterygoid plate, medial aspect | Ramus of mandible, medial aspect | Elevation, deviation to opposite side | Same as masseter |
| Lateral pterygoid | Lateral pterygoid plate, lateral aspect; greater wing of sphenoid, zygomatic surface | Neck of mandible, intraarticular disk | Protrusion, deviation to opposite side | Same as masseter |

in the substance of the parotid gland and enters the infratemporal fossa by passing deep to the ramus of the mandible. It passes obliquely through the fossa (either deep or superficial to the lateral pterygoid muscle) on its course to the pterygopalatine fossa via the pterygomaxillary fissure. While in the infratemporal fossa, it has anterior tympanic, deep auricular, middle meningeal, inferior alveolar, pterygoid, masseteric, buccal, and deep temporal branches. The **middle meningeal artery** enters the middle cranial fossa through the foramen spinosum and is the major arterial supply to the cranial dura mater.

The **mandibular nerve** enters the infratemporal fossa through the foramen ovale. The main trunk of this nerve is short (1 cm) so that it branches high in the fossa. It has muscular branches to the muscles of mastication, along with the mylohyoid, tensor tympani, and tensor veli palatini. Its sensory branches are the buccal nerve to the cheek, the auriculotemporal nerve to the posterior temporal region, the lingual nerve to the oral cavity, and the inferior alveolar nerve to the mandibular dentition and the skin covering the chin. Parasympathetic fibers are also distributed in certain branches of this nerve. The lingual nerve contains taste fibers to the anterior two-thirds of the tongue. These fibers reach the brainstem via the facial nerve, passing from the lingual to the facial in the chorda tympani nerve.

The **pterygoid plexus of veins** surrounds the pterygoid muscles. It receives blood from the face, nasal cavity, orbit, palate, cranial cavity, pharynx, and infratemporal fossa. It drains into the maxillary vein, which joins the superficial temporal vein to form the retromandibular vein.

# ORBIT

## Osteology

The **orbital cavities** are four-sided pyramids. The roof of each is formed by the orbital plate of the frontal bone and the lesser wing of the sphenoid. The floor is formed by the orbital surface of the maxilla, the orbital process of the zygoma, and the orbital process of the palatine bone. The medial wall is formed by the nasal process of the maxilla, the lacrimal, the ethmoid, and the sphenoid bones. The lateral wall is formed by the orbital process of the zygomatic bone and the greater wing of the sphenoid. The orbit is related

TABLE 4-6. Extraocular Eye Muscles

| Muscle | Origin | Insertion | Action | Innervation |
|---|---|---|---|---|
| Superior rectus | Tendinous ring around apex of orbit | Superior aspect of sclera | Elevation, intorsion, adduction | Oculomotor |
| Inferior rectus | Same as superior rectus | Inferior aspect of sclera | Depression, extorsion, adduction | Oculomotor |
| Medial rectus | Same as superior rectus | Medial aspect of sclera | Adduction | Oculomotor |
| Lateral rectus | Same as superior rectus | Lateral aspect of sclera | Abduction | Abducens |
| Superior oblique | Body of sphenoid | Superoposterior sclera via pulley at anteromedial roof of orbit | Abduction, depression, intorsion | Trochlear |
| Inferior oblique | Anterior floor of orbit | Inferoposterior sclera | Abduction, elevation, extorsion | Oculomotor |
| Levator palpebrae superiorus | Lesser wing of sphenoid | Superior tarsal plate | Elevation of upper eyelid | Oculomotor, sympathetics |

superiorly to the frontal sinus and anterior cranial fossa; medially to the nasal cavity, ethmoid air cells, and sphenoid sinus; inferiorly to the maxillary sinus; and laterally to the temporal fossa and the middle cranial fossa. The orbit communicates with the cranial cavity by way of the superior orbital fissure and the optic canal, with the infratemporal and pterygopalatine fossae by way of the inferior orbital fissure, and with the sphenoid sinuses and nasal cavity by way of the anterior and posterior ethmoidal foramina.

## Organization

The contents of the orbit are enclosed in a tough, conically shaped layer of fascia, the **periorbita**. The periorbita is only loosely attached to the walls of the orbit (in reality it is the periosteum of these bones), and it is continuous at the apex of the orbit through the optic canal and the superior orbital fissure with the periosteal layer of cranial dura. The meningeal layer of cranial dura forms a tubular sleeve around the optic nerve. This layer blends with the sclera of the eyeball. As the arachnoid and pia also follow the optic nerve to the eyeball, the subarachnoid space surrounds the optic nerve and extends the same distance. Within the confines of the periorbita, the extraocular structures are embedded in fat.

## Extraocular Muscles

The extraocular eye muscles control the movements of the eyeball and the upper eyelid. These muscles are listed in Table 4-6. Rotation of the eye is based on the 12 o'clock point on the eyeball moving medially (intorsion) or laterally (extorsion).

## Neurovascular Structures

The **optic nerve** enters the orbit through the optic canal and passes through the center of the orbital cone toward the eyeball. The oculomotor, trochlear, ophthalmic, and abducens nerves enter by way of the superior orbital fissure. The **oculomotor nerve** has two divisions: the superior, which supplies the superior rectus and levator palpebrae muscles, and the inferior, which provides the motor root (preganglionic parasympathetics) to the ciliary ganglion and innervates the medial and inferior rectus muscles and the inferior oblique. The **trochlear nerve** is very small and passes superomedially to the superior oblique muscle. The **ophthalmic nerve** trifurcates into the lacrimal nerve, which supplies the lacrimal gland; the frontal nerve, which terminates as the supratrochlear and supraorbital nerves; and the nasociliary nerve, which has ethmoidal and infratrochlear branches. The abducens nerve passes through the lateral part of the orbit to the lateral rectus.

**Ciliary ganglion**. The ciliary ganglion is a parasympathetic ganglion located in the posterior third of the orbit just lateral to the optic nerve. The parasympathetic pathways involving this ganglion are noted in Table 4-2. The nasociliary nerve provides a sensory root to the ciliary ganglion, through which sensory fibers reach the bulb by way of the ganglion and the short ciliary nerves. **Sympathetic fibers** reach the bulb through either long or short ciliary nerves as well as in periarterial plexuses. The sympathetics innervate the **dilator pupillae muscle** and the superior **tarsal muscle**.

**Ophthalmic artery**. The ophthalmic artery arises from the internal carotid artery as it passes the optic nerve. The ophthalmic artery then enters the orbit by passing inferior to the optic nerve and through the optic canal. The **central artery of the retina** enters the optic nerve about halfway along the orbital course of the nerve. It passes to the retina within the substance of the nerve (along with the central vein of the retina) and is therefore surrounded by the subarachnoid space. Other major branches of the ophthalmic artery are the **ciliary branches** to the eyeball, the **lacrimal artery**, and the **ethmoidal branches** to the nasal cavity. The superior and inferior ophthalmic veins drain primarily into the cavernous sinus, although there are communications with the pterygoid plexus and the veins of the face.

### Lacrimal Apparatus

The lacrimal apparatus consists of the lacrimal gland, lacrimal ducts, lacrimal sac, and the nasolacrimal duct. The **lacrimal gland** is situated near the front of the lateral roof of the orbit. Its main ducts open onto the upper lateral half of the conjunctival fornix. Tears from the lacrimal gland move across the eyeball to the medial angle of the eye, where they enter the lacrimal canaliculi. The canaliculi arise on the medial margins of the upper and lower lids at the lacrimal puncta on the lacrimal papillae. The **lacrimal ducts** carry the lacrimal secretion medially to the **lacrimal sac**, which is an upward expansion of the **nasolacrimal duct**.

## NASAL CAVITY AND PARANASAL SINUSES

### Nasal Cavity

**Osteology**. The nasal cavity extends from the base of the anterior cranial fossa to the roof of the mouth (palate) and is divided into right and left sides by the **nasal septum** (Fig. 4-6). The septum is formed by the perpendicular plate of the ethmoid, the vomer, and the septal cartilage. The nasal cavity is related superiorly to the anterior cranial fossa; laterally to the ethmoid air cells, maxillary sinus, and orbit; inferiorly to the oral cavity; and posterosuperiorly to the sphenoid sinus. The nasal cavity opens anteriorly on the face by way of the vestibule and nares and is continuous posteriorly by the choanae with the nasopharynx. The superior, middle, and inferior **conchae** divide the cavity into superior, middle, and inferior **meatus**. The area posterior and superior to the superior concha is the **sphenoethmoidal recess**.

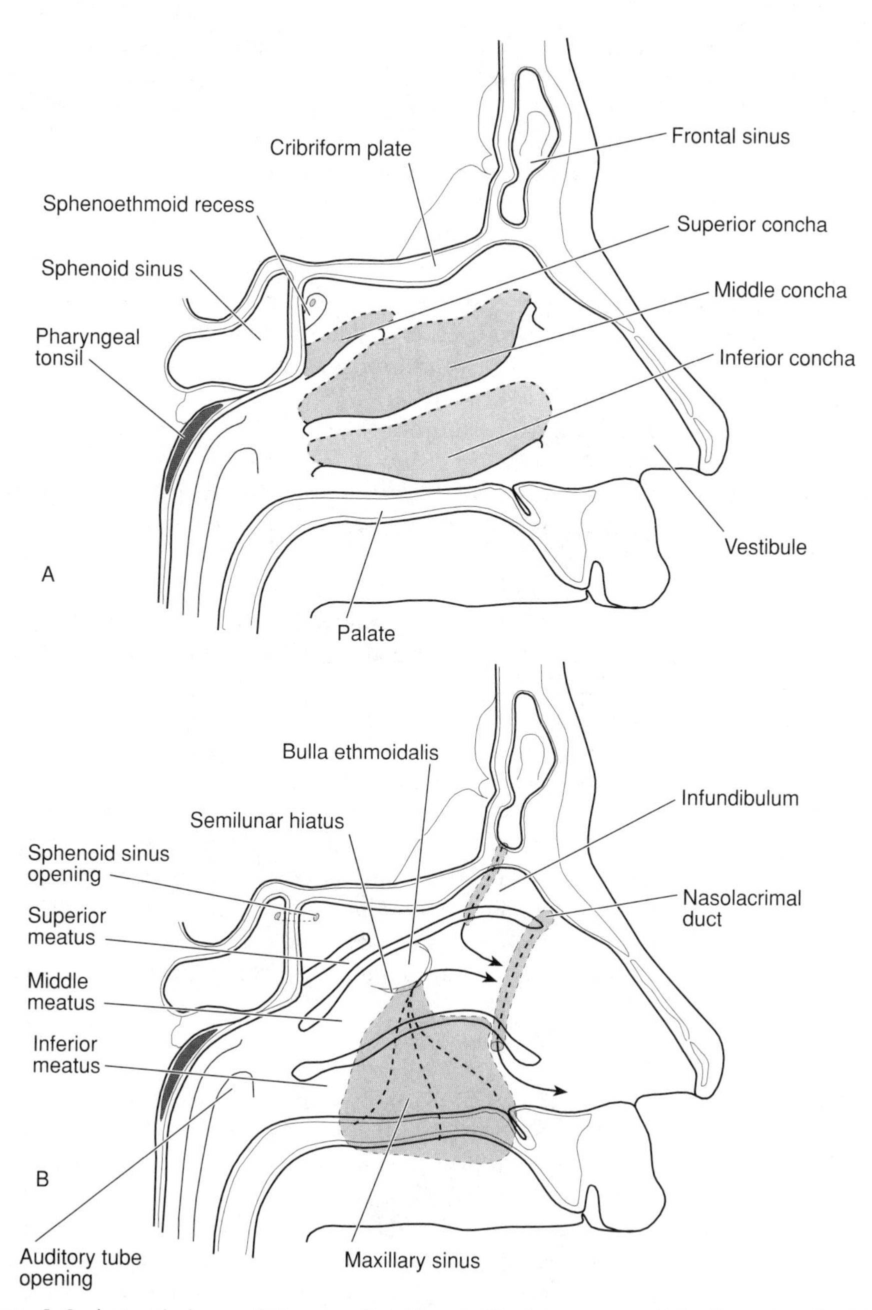


**Figure 4-6.** Lateral views of the nasal cavity. ***A:*** The structures of the lateral wall; ***B:*** The conchae removed to demonstrate the drainage pathways of the paranasal sinuses and the nasolacrimal duct.

**Neurovascular structures**. The **general somatic afferent (GSA) fibers** to the mucosa of the nasal cavity are from the anterior and posterior ethmoidal nerves (ophthalmic, CN V), and the lateral nasal and nasopalatine nerves (maxillary, CN V). The olfactory epithelium in the upper part of the nasal cavity is innervated by **special visceral afferent (SVA) fibers** from the olfactory nerve. All parasympathetic **[general visceral efferent (GVE)] fibers** are from the pterygopalatine ganglion and distributed by branches of the maxillary division of the trigeminal nerve. The blood supply to the nasal cavity is provided by three arteries: the maxillary, ophthalmic, and facial. The anterosuperior portions of the lateral wall and septum are supplied by the anterior and posterior ethmoidal branches of the ophthalmic artery. Most of the lateral wall (posteroinferior portion) is supplied by the lateral nasal branches of the maxillary artery, and the same area of the septum is supplied by the sphenopalatine branch of the maxillary. The area around the external nares is supplied by the superior labial branches of the facial artery.

## Paranasal Sinuses

**Maxillary sinus**. The maxillary sinus is related to the orbit superiorly, the nasal cavity medially, the posterior maxillary teeth inferiorly, the infratemporal fossa posterolaterally, and the cheek anterolaterally. The sinus empties into the middle meatus by way of the hiatus semilunaris and is best drained when lying on the opposite side. Because the opening is well above the inferior extent of the sinus, the top of the head should be lower than the lower jaw for complete drainage to occur.

**Frontal sinus**. The frontal sinus is in the frontal bone deep to the superciliary ridge. It is related anteriorly to the forehead; posteriorly to the anterior cranial fossa; and inferiorly to the orbit, ethmoid air cells, and the nasal cavity. It empties into the middle meatus and is best drained in the upright position.

**Ethmoid air cells**. The ethmoid air cells are interposed between the upper portion of the nasal cavity and the orbit. They are related superiorly to the anterior cranial fossa and inferiorly to the maxillary sinus. These sinuses drain into the superior and middle meatus. The locations of these openings are variable and, hence, the optional drainage position varies between upright and lying on the opposite side.

**Sphenoid sinus**. The sphenoid sinus is in the body of the sphenoid bone. It is inferior to the sella turcica, the hypophysis, and the optic nerve, and bounded laterally by the cavernous sinus. The pterygoid canal is in the floor of the sinus, and the sinus empties into the sphenoethmoidal recess. It drains best with the head flexed more than 90 degrees.

The mucosa of the paranasal sinuses is continuous with that of the nasal cavity. The secretomotor fibers to the mucosa are postganglionic parasympathetics from the pterygopalatine ganglion, which are distributed primarily with branches of the maxillary nerve.

## Pterygopalatine Fossa

**Osteology**. The **pterygopalatine fossa** is between the maxilla anteriorly and the pterygoid portion of the sphenoid posteriorly. Medially, it is bound by the perpendicular plate of the palatine bone. It communicates laterally with the infratemporal fossa by the pterygomaxillary fissure, medially with the nasal cavity by the sphenopalatine foramen, inferiorly with the oral cavity by the palatine canal and the greater and lesser palatine foramina, posterosuperiorly with the middle cranial fossa by the foramen rotundum, anterosuperiorly with the orbit by the inferior orbital fissure, and posteromedially with the pharynx by the pharyngeal canal. The pterygoid canal is a canal through the pterygoid portion of the sphenoid that opens onto the posterior wall of the pterygopalatine fossa.

This fossa contains the terminal portion of the maxillary artery, the maxillary division of the trigeminal nerve, and the pterygopalatine parasympathetic ganglion. The parasympathetic pathways involving the pterygopalatine ganglion are listed in Table 4-3. The sympathetic fibers that pass through the ganglion are derived from the internal carotid plexus and pass sequentially through the deep petrosal nerve and nerve of the pterygoid canal. After passing through the ganglion, the fibers are distributed through branches of the maxillary nerve.

**Maxillary nerve**. The maxillary nerve continues as the infraorbital nerve, which passes into the floor of the orbit and eventually terminates on the face via the infraorbital foramen. The branches of the maxillary nerve and the proximal portion of the infraorbital nerve are the palatine (greater and lesser) to the palate, the nasopalatine and lateral nasal branches to the nasal cavity, the pharyngeal nerve to the nasopharynx, the zygomatic nerve to the skin of the zygomatic region of the face, and the superior alveolar nerves to the maxillary sinus and dentition.

**Pterygopalatine artery**. The branches of the pterygopalatine artery (third part of the maxillary artery) are the posterior superior alveolar to the maxillary sinus and maxillary dentition, the infraorbital artery, the descending palatine artery to the palate, the pharyngeal artery, and the sphenopalatine artery to the nasal cavity.

# ORAL CAVITY

The mouth or oral cavity consists of a vestibule and the mouth proper. The vestibule of the mouth lies between the lips and cheeks externally and the gums and teeth internally. It receives the parotid duct opposite the second upper molar tooth.

The **mouth proper** is bounded laterally and in front by the alveolar arches and the teeth; posteriorly, it communicates with the pharynx through the fauces. It is roofed by the hard and the soft palates. The floor is composed of the tongue and the reflection of its mucous membrane to the gum lining the inner aspect of the mandible. The midline reflection is elevated into a fold called the **frenulum linguae**. On each side of this fold is the caruncula sublingualis, containing the openings of the submandibular (Wharton's) ducts. Behind these are the openings of the ducts of the sublingual glands.

## Lips and Cheeks

The **lips** are muscular folds covered externally by skin and internally by mucosa (mucous membrane). The upper lip extends to the nasolabial sulcus and contains a vertical midline groove, the philtrum. The **mentolabial sulcus** separates the lower lip from the skin of the chin. The lips receive their blood supply from labial branches of the facial artery. Their sensory nerve supply is by infraorbital branches of the trigeminal nerve to the upper lip and mental branches to the lower lip; the facial nerve supplies the orbicularis oris muscle.

## Tongue

The tongue is a muscular organ whose bilateral muscle masses are separated in the midline by a fibrous septum. Extrinsic muscles interconnect the tongue and surrounding structures (Table 4-7). Intrinsic muscles are oriented in vertical, longitudinal, and transverse bundles and function to control the shape of the tongue. These

**TABLE 4-7. Extrinsic Muscles of the Tongue**

| Muscle | Origin | Insertion | Action | Innervation |
|---|---|---|---|---|
| Hyoglossus | Body and greater cornu of hyoid | Inferolateral aspect of tongue | Depression and retraction | Hypoglossal |
| Styloglossus | Styloid process | Lateral aspect of tongue | Elevation and retraction | Hypoglossal |
| Genioglossus | Mental spine of mandible | Body of tongue | Protrusion and deviation to opposite side | Hypoglossal |
| Palatoglossus | Soft palate | Dorsolateral tongue | Elevation and retraction | Vagus |

muscles are supplied by the hypoglossal nerve. At its root, the tongue is connected to the pharynx, palate, and epiglottis; the glossoepiglottic folds attach the root to the epiglottis and bound the vallecula.

## Teeth

Teeth, gums, and alveolar bone provide a wall between the vestibule and the mouth proper. Twenty deciduous teeth and 32 permanent teeth are equally distributed between the upper and lower jaws. Each tooth consists of a free crown, a root buried in an alveolus (socket) of the jaw, and a neck between the crown and root at the gum margin.

The sensory nerves to the maxillary teeth are branches of the maxillary division of the fifth nerve. The posterior superior alveolar nerve supplies the molars, the middle superior alveolar innervates the bicuspids, and the anterior superior alveolar supplies the canine and incisor teeth. These branches and palatine branches of the maxillary nerve also supply the gums. The lower teeth are supplied by the inferior alveolar nerve from the mandibular division of the trigeminal nerve. Blood supply is by way of the superior alveolar branches of the maxillary artery and the inferior alveolar artery.

## Palate, Isthmus of the Fauces, and Palatine Tonsil

The palate forms the roof of the mouth and consists of hard and soft portions. The **hard palate** is formed by the palatine processes of the maxillae and the horizontal portions of the palatine bones. An **incisive canal** penetrates it anteromedially, and greater and lesser palatine foramina lie posterolaterally. The **soft palate** is a muscular organ that extends posteriorly from the hard palate. The palatoglossal muscle, along with its mucosa, forms the anterior pillar of the fauces (**glossopalatine arch**). Posteriorly, the palatopharyngeus muscle and its mucosa constitute the posterior pillar of the fauces (**palatopharyngeal arch**). The uvular muscles extend from the hard palate to the tip of the uvula. The muscles of the palate are listed in Table 4-8. The isthmus of the fauces is the communication between the oral cavity proper and the oral pharynx. It is bounded above by the soft palate, below by the tongue, and laterally by the glossopalatine arch.

The principal artery of the hard palate is the **greater palatine branch** of the maxillary artery. It enters through the greater palatine foramen and runs forward toward

**TABLE 4-8. Muscles of the Palate**

| Muscle | Origin | Insertion | Action | Innervation |
|---|---|---|---|---|
| Levator veli palatine | Inferior petrous temporal bone, auditory tube | Posterior soft palate | Elevation and retraction | Vagus |
| Palatoglossus | Soft palate | Dorsolateral tongue | Depression | Vagus |
| Palatopharyngeus | Posterior hard palate, adjacent soft palate | Pharyngobasilar fascia | Depression of palate, elevation of pharynx | Vagus |
| Musculus uvulae | Posterior hard palate | Two sides blend to form uvula | Elevation of uvula | Vagus |
| Tensor veli palatine | Scaphoid fossa and spine of sphenoid | Anterior soft palate | Tense soft palate | Mandibular |

the incisive canal, where it anastomoses with branches of the sphenopalatine artery. The soft palate is supplied by the **lesser palatine artery**, **ascending palatine branches** of the facial artery, and branches of the **ascending pharyngeal artery**. Palatine veins are tributaries to the pterygoid plexus.

The **palatine tonsil** is located between the anterior and posterior pillars. It bulges into this depression and is covered by a mucosal fold of the anterior pillar; there is a depressed supratonsillar fossa above the tonsil. Efferent lymphatic vessels penetrate the pharyngeal wall and pass to the superior deep cervical nodes, especially the jugulodigastric node. The arterial supply to the palatine tonsil is by the ascending palatine branch of the facial artery, tonsillar branch of the facial artery, palatine branch of the ascending pharyngeal artery, dorsal lingual branch of the lingual artery, and descending palatine branch of the maxillary artery. The nerves innervating the tonsil are branches of the maxillary division of the trigeminal nerve and the glossopharyngeal nerve.

# EAR

## Temporal Bone

The temporal bone houses the middle ear cavity, contains a network of interconnected canals that form the internal ear, and participates in the formation of various cranial and extracranial fossae. It is composed of squamous, mastoid, petrous, and tympanic parts. The **squamous portion** forms part of the mastoid process, external auditory meatus, and the mandibular fossa and has the zygomatic process, which forms part of the zygomatic arch. It helps define the middle cranial fossa. The **mastoid portion** forms most of the mastoid process and part of the wall of the posterior cranial fossa. The **tympanic part** forms most of the external auditory meatus and all of the styloid process. The **petrous portion** projects anteromedially toward the dorsum sellae, where it ends. Its petrous ridge separates the anterior face from the posterior face. The anterior face forms the posterior portion of the floor of the middle cranial cavity; the posterior face is the anterolateral aspect of the posterior cranial fossa and contains the opening of the internal auditory meatus.

## External Ear

The external ear is composed of the external cartilaginous portion, the pinna or auricle, and the external auditory meatus. The **external auditory meatus** is about 3 cm long, connects the auricle and the middle ear cavity, and consists of a lateral cartilaginous and a medial osseous portion. In the infant, the **osseous meatus** is merely a bony ring; in the adult, it is about 2 cm long. It is narrowest at the isthmus, about 0.5 cm from the tympanic membrane. The entire external auditory meatus is S-shaped. The convexity of the outer cartilaginous portion is directed upward and posteriorly, and that of the inner osseous portion is directed downward and anteriorly.

## Middle Ear

The middle ear, or **tympanic cavity**, is generally shaped like a flat cigar box. Its long axis parallels the tympanic membrane so that it is obliquely oriented, sloping medially from above downward and from behind forward. The cavity is divided into three regions: the **middle ear cavity proper** at the level of the tympanic membrane; the **attic**, or **epitympanum**, superior to the membrane; and the **hypotympanum** below the membrane. The cavity communicates posterolaterally with the mastoid air cells by way of the attic and mastoid antrum, and anteromedially with the nasopharynx by way of the auditory tube.

**Lateral wall.** The lateral wall is formed primarily by the tympanic membrane. This membrane is angularly concave with its apex—the **umbo**—directed medially. It is composed of a fibrous stratum covered laterally by skin and medially by mucous membrane. The greater part of the periphery of the membrane is a thickened fibrocartilaginous ring that attaches to the bony tympanic sulcus. Superiorly, the sulcus and fibrous stratum are deficient and, thus, the membrane is lax (**pars flaccida**). The rest of the membrane is called the **pars tensa**.

**Medial wall.** The medial wall of the middle ear cavity separates that cavity from the inner ear. Prominent on that wall are (1) the **promontory**, which corresponds to the first turn of the cochlea; (2) the **oval window**, which contains the foot plate of the stapes and lies above and behind the promontory; (3) the **round window** below the oval window; and (4) the **pyramid** containing the stapedius muscle.

The roof is the **tegmen tympani**, a thin portion of the petrous temporal bone, which forms part of the floor of the middle cranial fossa. The **floor** of the middle ear is formed by the roof of the jugular foramen. The **anterior wall** is formed below by the roof of the carotid canal; superiorly, it is deficient where the auditory tube opens into the tympanic cavity. The **posterior wall** is formed inferiorly by the descending portion of the facial canal and superiorly the attic is in communication with the mastoid antrum.

The **auditory (eustachian) tube** is about 3.8 to 4 cm in length, extending from the tympanum obliquely forward, downward, and inward. Its posterolateral third is bony; the pharyngeal two-thirds is cartilaginous.

The **air cells of the mastoid process** communicate with the middle ear by means of the antrum and the attic. Before 5 years of age, there is usually only one cell, the antrum; after 5 years of age, the mastoid consists of numerous cells communicating with one another and the antrum. This area is lined with a continuation of the mucous membrane of the tympanum.

**Ossicles.** The ossicles of the middle ear are the malleus, incus, and stapes. The manubrium of the **malleus** attaches to the umbo of the tympanic membrane, the foot plate of the **stapes** fits into the oval window, and the **incus** interconnects the stapes and malleus.

**Nerves.** The **chorda tympani** nerve branches from the facial nerve and passes between the incus and malleus as it crosses the medial surface of the tympanic membrane. The chorda tympani conveys taste and preganglionic parasympathetic fibers to the lingual nerve, which it joins in the infratemporal fossa. The **tympanic plexus** is located on the promontory. It contains sympathetics (by way of the caroticotympanic branch of the internal carotid plexus), and sensory and preganglionic parasympathetic fibers (both by way of the tympanic branch of the glossopharyngeal nerve).

## Inner Ear or Labyrinth

The inner ear is contained in the petrous portion of the temporal bone and consists of an osseous labyrinth containing a membranous labyrinth. Between the bony and membranous labyrinth is perilymph. Within the membranous labyrinth is **endolymph**.

The **osseous labyrinth** is a series of cavities in bone consisting of a central vestibule that is continuous with three semicircular canals posterolaterally and the cochlea anteromedially. The vestibule is separated from the laterally situated middle ear cavity by a bony wall containing the fenestra ovalis. This oval window is closed by the foot plate of the stapes. In the posteromedial wall of the vestibule is the opening of the vestibular aqueduct, which extends to the posterior wall of the petrous portion of the temporal bone. The three **semicircular canals** are oriented at right angles to each other. The anterior (superior) and posterior canals are vertically oriented; the lateral (horizontal) canal is horizontally positioned. The anterior semicircular canal is parallel to the posterior canal of the other side. The anterior canal is transverse to the long axis of the petrous bone. Its anterior limb is dilated into an ampulla just before its entrance into the vestibule; the posterior limb joins the anterior limb of the posterior canal to enter the vestibule as a crus commune. The posterior canal is parallel to the posterior wall of the petrous bone, and its posterior limb enters the vestibule just beyond the ampulla. Both limbs of the lateral canal enter the vestibule, and an ampulla is located on the anterior limb. The **cochlea** is conical and has two and a half turns, and its apex is directed forward, outward, and downward.

The blood supply of the labyrinth is by way of the **internal auditory (labyrinthine) and stylomastoid arteries**. The stylomastoid is a branch of the posterior auricular. The internal auditory arises from the basilar artery, or in common with the anterior inferior cerebellar artery, and traverses the internal acoustic meatus before branching into cochlear and vestibular branches. The veins accompany the arteries and drain as internal auditory veins into the superior petrosal or transverse sinuses.

# Chapter 5

# Thorax

The thorax is bound posteriorly by the thoracic vertebrae and the ribs, laterally by the ribs and the intercostal spaces, and anteriorly by the sternum, the costal cartilages, and ribs. The sternal angle (of Louis) is formed by the junction of the manubrium and the body of the sternum. It marks the level of the second costal cartilages, the bifurcation of the trachea, and the lower aspect of the fourth thoracic vertebra.

## THORACIC WALL

### Osteology

The 12 pairs of ribs are similar in that each articulates with the vertebral column, but they differ anteriorly. The upper seven ribs are considered the true ribs because each articulates directly with the sternum. Ribs 8, 9, and 10 articulate indirectly with the sternum because the costal cartilage of each joins the cartilage above. Ribs 11 and 12 do not articulate with the sternum and thus are considered "floating ribs."

**Typical rib**. The typical ribs, 3 through 9, consist of a head, neck, tubercle, and shaft. The head articulates with two vertebral bodies, and the short neck extends laterally to the tubercle, which articulates with the transverse process of vertebra of the same number. The angle is found just lateral to the tubercle where the shaft curves abruptly anteriorly.

**Atypical ribs**. The first rib is short and acutely curved, flattened superoinferiorly, indented superiorly by the subclavian groove, and articulates with only vertebra T1. The second rib is similar to the first but slightly longer. Ribs 10 through 12 have neither tubercles nor angles and articulate with only one vertebra.

### Costovertebral Articulations

The ribs are connected to the vertebrae by costovertebral and costotransverse joints and to the sternum via the costal cartilages. The **costal cartilages** form a flexible interface between the ribs and sternum and thus provide considerable flexibility so the ribs are free to move. Both the **costovertebral** (between vertebral body and head of rib) and **costotransverse** (between tubercle and transverse process) **joints** are synovial joints. Small amounts of gliding motion occur at both of these joints; however, this motion is magnified anteriorly and laterally and thus permits considerable anterior and lateral enlargement of the thorax.

### Muscles

The muscles of the thoracic wall are the external and internal intercostals, the subcostal, and the transversus thoracis. The **intercostal** muscles fill the intercostal spaces, the external sloping medially superolaterally and the internal sloping laterally superolaterally. **Subcostal** muscles are fasciculi of the internal intercostals that extend over two or more intercostal spaces near the angles of the ribs. The **transversus thoracis** muscle arises from the dorsal surface of the lower sternum and xiphoid process and extends superior to inferior to insert on the second through the sixth costal cartilages. In respiration, the intercostals apparently maintain both size and rigidity of the intercostal spaces, and the entire rib cage is elevated by the scalene muscles.

## Neurovascular Structures

**Intercostal nerves**. The intercostal nerves supply the muscles and skin of the thoracic wall and the underlying parietal layer of pleura. The lower intercostal nerves, T7–T12, extend into the abdominal wall, where they also supply the muscles and skin as well as the underlying parietal layer of peritoneum. These nerves are the continuations of the ventral rami of the thoracic spinal nerves. They, along with the intercostal vessels, occupy the **costal grooves** on the inferior aspects of the ribs. The superficial lip of this groove extends more inferiorly than the deep lip; as a result, the vessels and nerves are protected. Each intercostal nerve has two cutaneous branches: the **lateral cutaneous nerve** in the midaxillary line and the **anterior cutaneous nerve** just lateral to the sternum.

**Intercostal arteries and veins**. **Posterior intercostal arteries** arise from the aorta and, in the upper spaces, from the costocervical trunk of the subclavian artery. They supply the deep muscles of the back, contents of the spinal canal, and most of the intercostal space. The **anterior intercostal arteries** branch from the internal thoracic artery, are smaller than the posterior, and supply only the anterior aspects of the intercostal spaces. The **posterior intercostal veins** are tributaries to the azygous system, and the **anterior intercostal veins** empty into the internal thoracic veins. Both systems, the arterial and the venous, form potential anastomoses in the intercostal spaces and thus represent potential collateral vascular routes.

## Mammary Gland

Although the mammary glands are part of the pectoral region that is usually included with the upper limb, they are included here because of their lymphatic drainage. These structures are found within the subcutaneous tissue and extend from the level of the second to sixth or seventh ribs and from the lateral aspect of the sternum into the axilla. The **arterial supply** is provided by branches of the internal thoracic and intercostal arteries along with the thoracoacromial and lateral thoracic branches of the axillary artery. **Efferent lymphatics** drain primarily toward the axilla and internal thoracic lymph nodes, but also can cross the midline and extend inferiorly into the abdominal wall.

## Respiratory Diaphragm

The diaphragm is a thin, dome-shaped muscle consisting of a series of radially oriented muscle fibers that arise from the inner aspect of the thoracic outlet (subcostal margin) and insert into the central tendon. On the right, its dome extends superiorly to the fifth rib, and on the left to the fifth interspace.

The diaphragm arises from three areas: (1) a small sternal part that attaches to the posterior aspect of the xiphoid process, (2) an extensive costal portion that attaches to the subcostal margin, and (3) a lumbar portion. The lumbar portion consists of the right and left **crura**, which arise from the anterior aspects of the lumbar vertebra and surround the **aortic hiatus** (T12), through which the aorta and thoracic duct pass. The **esophageal hiatus** (T10) is anterior to the aortic hiatus; the opening for the **inferior vena cava** (T8) is more anterior and to the right. The diaphragmatic muscle is supplied by the phrenic nerves; the peripheral part of the diaphragm receives sensory fibers from the intercostal nerves. The diaphragm flattens as it contracts, and thus draws the central tendon downward. This movement increases the thoracic volume and decreases the pressure within the thoracic cavity.

The **phrenic nerve** arises from the ventral rami of spinal nerves C3–C5. It descends over the cupula of the pleura, in front of the root of the lung and between the pericardium and pleura, to reach the diaphragm. **Referred pain** from the diaphragm commonly occurs in the **shoulder region** because both structures are supplied by sensory fibers that terminate in the fourth cervical spinal cord segment.

# ORGANIZATION OF THE THORACIC CAVITY

The thorax is separated into **two lateral regions** that contain the lungs and pleura, and a central region called the **mediastinum**. The mediastinum contains all of the thoracic viscera except the lungs and is located between the pleural cavities and the sternum and vertebral column. A line through the sternal angle separates the mediastinum into superior and inferior regions; the inferior region is further subdivided into anterior, posterior, and middle parts.

The **superior mediastinum** is continuous with the neck through the superior thoracic aperture. It contains the aortic arch and its branches, the superior vena cava and its tributaries, the vagus, recurrent laryngeal and phrenic nerves, trachea, esophagus, thoracic duct, left highest intercostal vein, remains of the thymus gland, and some lymph nodes.

The **anterior mediastinum** is small and anterior to the pericardial cavity. It contains the thymus gland, internal thoracic vessels, lymph nodes, and surrounding connective tissue.

The **posterior mediastinum** is posterior to the pericardium. It contains the thoracic aorta, esophagus, vagus nerves, thoracic duct, azygos and hemizygos veins, lymph nodes, and sympathetic trunks.

The **middle mediastinum** contains the heart and pericardium.

# TRACHEA AND BRONCHI

## Gross Structure

**Trachea**. The trachea extends from the cricoid cartilage (vertebral level C6) to the level of the upper border of T5, where it bifurcates into left and right bronchi. It is related posteriorly to the esophagus and anteriorly to the thyroid gland and vessels, the sternohyoid and sternothyroid muscles, the thymus, the manubrium of the sternum, the arch of the aorta and left brachiocephalic vein, and the deep cardiac plexus.

**Main stem bronchi**. The right bronchus is both shorter and wider and diverges from the midline less than the left bronchus. Each bronchus enters the hilus of the lung at the mediastinal surface of the lung. The pulmonary and bronchial arteries and veins, lymphatic vessels and lymph nodes, and autonomic nerve fibers also pass through the hilus. The **right bronchus** divides into three lobar bronchi: the superior, middle, and inferior. The **left bronchus** divides into the superior and inferior lobe bronchi. Each of the lobar bronchi in turn subdivides to supply **bronchopulmonary segments**. In the right lung, these segments consist of the apical, posterior, and anterior segments of the superior lobe; the medial and lateral segments of the middle lobe; and the superior, medial basal, lateral basal, anterior basal, and posterior basal segments of the inferior lobe. In the left lung, these bronchopulmonary segments are (1) the apical-posterior, anterior, superior, and inferior of the superior lobe, and (2) the superior, anteromedial basal, lateral basal, and posterior basal of the inferior lobe. The segmental bronchi further subdivide to smaller bronchi and bronchioles in the substance of the lung.

# LUNGS AND PLEURA

## Lungs

Each lung is conical in shape and has an apex and base. The sides of the cone are formed by the concave **mediastinal surface**, which bears impressions of mediastinal structures, and the convex **costal surface**. The apex extends approximately 2.5 cm into the root of the neck (superior to the clavicle), and the **base** is concave and rests on the convex surface of the diaphragm. The **posterior border** is in the vertical concavity on either side of the vertebral column. The **anterior border** is sharp and extends medially in front of the mediastinum except on the left where the **cardiac notch** is found. The **inferior border** is sharp and formed where the peripheral part of the base extends inferiorly between the diaphragm and thoracic wall. The **right lung** is divided into superior, middle, and inferior lobes by the transverse and oblique fissures; the **left lung** is divided into superior and inferior lobes by an oblique fissure. The surface projection of the lungs is summarized in Table 5-1.

Both the efferent and afferent innervation of the lung is derived from the anterior and the **posterior pulmonary plexuses**, which receive branches from the **vagus** and the thoracic sympathetic trunk. The **visceral afferents** transmit pain and reflex activity and return to the central nervous system by way of both the vagal and sympathetic pathways.

## Pleura

With the exception of the hilus, each lung is surrounded by a **pleural cavity**. The pleura consists of two layers. The **outer parietal layer** is attached to the deep surface

**TABLE 5-1. Surface Projection of the Lungs and Pleura**

| Reference Points | Lungs | Pleura | Oblique Fissure | Horizontal Fissure |
|---|---|---|---|---|
| Medial third of clavicle | 2.5 cm above | 2.5 cm above | | |
| Sternoclavicular joint | Posterior | Posterior | | |
| Sternal angle | Lungs do not meet | Pleura meet in midline | | |
| Level at which border passes laterally | Sixth costal cartilage | Left: fourth costal cartilage<br>Right: seventh costal cartilage | | |
| Midclavicular line | Sixth rib | Eighth rib | Sixth rib | Fourth costal cartilage |
| Midaxillary line | Eighth rib | Tenth rib | Fifth rib | Fifth rib |
| Scapular line (vertebral border) | Tenth rib | Twelfth rib | Fourth rib (base of scapular spine) | |

of the thoracic wall, diaphragm, and mediastinal structures. The **deeper visceral layer** is firmly attached to the lung and dips into the interlobar fissures. These two layers are continuous around the hilar region, and thus the hilus is open so the structure may enter and leave the lung. The area between the two layers is the pleural cavity; in reality, this is only a potential space because the parietal and visceral layers of the pleura are separated by only a thin layer of lubricating fluid.

At certain points (i.e., where the lung extends medially between the thoracic wall and mediastinal structures and inferiorly between the thoracic wall and diaphragm), the pleura extends even beyond the lung. As a result, there are areas where the parietal pleura is in contact with itself rather than visceral pleura; these areas are called **recesses**. Specifically, there are the **costodiaphragmatic** and **costomediastinal recesses**. These recesses exist during only quiet breathing. With deep inhalation, the lung expands and fills the recesses. The surface projection of the pleura is summarized in Table 5-1.

# HEART AND PERICARDIUM

## Gross Structure of the Heart

The **apex** of the heart is the part of the left ventricle that is directed inferiorly and to the left. It is located deep to the left fifth intercostal space about 4 cm below and 2 cm medial to the left nipple, and is overlapped by an extension of the pleura and lungs. The **base** of the heart faces superiorly, to the right and posteriorly. It consists primarily of the left atrium, part of the right atrium, and proximal parts of the great vessels. It is bounded superiorly by the bifurcation of the pulmonary artery, inferiorly by the coronary sulcus, on the left by the oblique vein of the left atrium, and on the right by the sulcus terminalis. The base is separated from the bodies of T5–T8 vertebrae by the thoracic aorta, esophagus, and thoracic duct. The **sternocostal surface** of the heart is formed by the right atrium, right ventricle, and a small part of the left ventricle. The **diaphragmatic surface** is comprised of the two ventricles.

**Right atrium**. The right atrium is larger than the left and comprised of a primary cavity (**sinus venarum**) and an **auricle**. The superior vena cava opens into the upper and posterior part of the sinus venarum, and the inferior vena cava into its lowest part near the interatrial septum. The coronary sinus opens between the ostium of the inferior vena cava and the atrioventricular (A-V) foramen. Rudimentary valves guard the openings of the inferior vena cava and coronary sinus. The auricle has a rough surface because of muscular ridges (musculi pectinati). It is demarcated externally from the sinus venarum by the sulcus terminalis and internally by the crista terminalis.

The dorsal wall of the right atrium is formed by the interatrial septum. The fossa ovalis, representing the embryonic foramen ovale, is an oval depression in the septal wall located above the openings of the coronary sinus and inferior vena cava. It is bounded above and at its sides by the limbus fossa ovalis, representing the embryonic free margin of septum secundum.

**Right ventricle**. The right ventricle is bounded on the right by the coronary sulcus and on the left by the anterior interventricular sulcus. Its superior part, the conus arteriosus (infundibulum), is continuous with the pulmonary trunk. Inferiorly, the ventricular wall forms the acute margin of the heart. The wall of this ventricle is about one-third the thickness of the left ventricle, but the capacity of both ventricles is similar. The internal surface is irregular because of the muscular bundles called **trabeculae carneae**. Some of these project from the wall of the ventricle and insert via chordae tendineae on the apices, margins, and ventricular surfaces of cusps of the right A-V (tricuspid) valve.

The **right A-V valve** has anterior (infundibular), posterior (marginal), and medial (septal) cusps. The anterior cusp is the largest; the posterior is the smallest. These leaflets are composed of strong, fibrous tissue that is continuous at their bases, with the annuli fibrosi of the fibrous skeleton separating the atria from the ventricles. The anterior papillary muscle arises from the anterior and septal walls and is attached to the anterior and posterior cusps by chordae tendineae. The **septomarginal trabecula (moderator band)** extends from the interventricular septum to the base of the anterior papillary muscle. The posterior papillary muscle arises from the posterior wall, and its chordae tendineae insert on the posterior and septal cusps.

The **conus arteriosus** has a smooth inner surface. It is limited from the rest of the ventricle by a ridge of muscular tissue called the **crista supraventricularis**. At the summit of the conus is the orifice of the pulmonary trunk. The **pulmonary valve** is formed by three cusps, anterior, right, and left, and each has a sinus between its distal surface and the wall of the pulmonary trunk. Adjacent cusps attach at a common commissure, and a thin marginal lunula portion of each cusp runs from each commissure to a thickened nodule in the central free margin of the cusp. When the valve is closed, the lunulae and nodules of the cusps are in contact.

**Left atrium**. Like the right atrium, the left consists of a sinus venarum and an auricle. The smooth-walled sinus venarum receives the four pulmonary veins. The part of the interatrial septum covering the fossa ovalis of the right atrium constitutes a valve of the foramen ovale. The left auricle is longer than that of the right atrium. It curves ventrally around the base of the pulmonary trunk, and lies over the proximal portion of the left coronary artery.

**Left ventricle**. The left ventricle is longer, more conical, and thicker than the right. It forms the apex of the heart and is separated from the right ventricle by the muscular and membranous parts of the interventricular septum. It has two openings, the left A-V (mitral), which is regulated by the mitral valve, and the aortic, which has the aortic valve. The **left A-V valve** consists of a large anterior (aortic) cusp and a smaller posterior cusp. Each cusp receives chordae tendineae from both the anterior and posterior papillary muscles. The aortic opening is anterior and to the right of the mitral valve. The portion of the ventricle below the aortic orifice is called the aortic vestibule. The aortic valve consists of three cusps, the posterior, right, and left. The cusps are similar in structure to those of the pulmonary valve, but they are larger and stronger. The right and left **coronary arteries** originate from the right and left **aortic sinuses** (of Valsalva), respectively.

**Cardiac skeleton**. The fibrous skeleton of the heart consists of a series of fibrous rings (**annuli fibrosi**) and **fibrous trigones**, and serves as the base of support to which the valves and cardiac muscle attach. Fibrous rings surround each A-V orifice and the aortic and pulmonary orifices and provide attachment for the cusps of each of the associated valves. The membranous part of the interventricular septum also is continuous with the fibrous tissue of these annuli. The right and left fibrous trigones are found at the junctions of the right and left A-V rings with the aortic ring, respectively.

## Surface Projections of the Heart

The projection of the heart to the anterior surface of the thorax is based on four points, which represent the locations of the angles that are formed at the intersections of the four borders of the heart. These four points, along with the borders of the heart and the locations of the cardiac valves, are summarized in Tables 5-2 and 5-3.

## Neural Structures of the Heart

**Conduction system**. This system is composed of specialized cardiac muscle found in the sinoatrial (S-A) node and in the A-V node and bundle. The heartbeat is initiated in the **S-A node (pacemaker of the heart)**, located in the right atrium in the upper part of the crista terminalis just to the left of the opening of the superior vena cava. From

**TABLE 5-2. Surface Projection of the Heart**

| Part of the Heart | Location of the Chest Wall |
| --- | --- |
| Point I | Second left costal cartilage, 1–2 cm lateral to edge of sternum |
| Point II | Third right costal cartilage, 1–2 cm lateral to edge of sternum |
| Point III | Sixth right costal cartilage, 1–2 cm lateral to edge of sternum |
| Point IV (apex of heart) | Fifth left intercostal space, 7–8 cm lateral to midline |
| Lower (diaphragmatic) border (right atrium, left and right ventricles) | Line extending from apex to point III |
| Right border (right atrium, superior vena cava) | Slightly convex line (laterally) extending from point II to point III |
| Left border (left ventricle, left auricle) | Slightly convex line (laterally) extending from point I to apex |
| Superior border (great vessels) | Line interconnecting points I and II |

the S-A node, the cardiac impulse spreads throughout the atrial musculature to reach the A-V node lying in the subendocardium of the atrial septum, directly above the opening of the coronary sinus. Thereafter, the impulse is conducted to the ventricles by passing through the specialized tissue of the **A-V bundle (of His)**. This bundle consists of a crus commune and right and left **bundle branches**. The common bundle travels from the A-V node into the membranous part of the interventricular septum. It divides into right and left bundle branches that pass in the subendocardium along the muscular part of the septum and distribute to the ventricles as **Purkinje tissue**.

**TABLE 5-3. Surface Projection of Heart Valves and Their Auscultation Points**

| Valve | Surface Projection | Auscultation Point |
| --- | --- | --- |
| Aortic | Lower edge of left third costal cartilage at lateral margin of sternum | Second right intercostal space just lateral to sternum |
| Pulmonary | Upper edge of left third costal cartilage at lateral margin of sternum | Second left intercostal space just lateral to sternum |
| Tricuspid (right atrioventricular) | Right half of sternum at level of fourth intercostal space | Right half of body of sternum just above junction with xiphoid process |
| Mitral (left atrioventricular) | Left half of sternum at level of fourth intercostal space | Apex of heart: left fifth intercostal space 7–8 cm lateral to midline |

**Innervation of the heart**. The innervation of the heart is from both the parasympathetic and sympathetic divisions of the autonomic nervous system via the cardiac plexuses. The **cardiac plexuses** are related to the arch of the aorta; the **superficial plexus** is inferior to the arch, and the **deep plexus** is located between the arch and tracheal bifurcation. Right and left thoracic cardiac branches from the vagus nerves pass to the deep cardiac plexus, where they synapse on postganglionic parasympathetic neurons. The vagus nerves also give rise to superior and inferior cervical cardiac nerves. The left inferior cervical cardiac nerve ends in the superficial cardiac plexus; the rest end in the deep plexus. Superior, middle, and inferior cervical cardiac nerves arising from sympathetic ganglia also descend to cardiac plexuses. The left superior cervical cardiac sympathetic nerve ends in the superficial cardiac plexus; the rest end in the deep plexus. The deep cardiac plexus also receives thoracic cardiac branches from the upper five thoracic sympathetic ganglia. The coronary and pulmonary plexuses and the right and left atria are supplied by branches from the superficial and deep plexuses. The right vagal and sympathetic branches end chiefly in the region of the S-A node; the left branches end chiefly in the region of the A-V node.

**Cardiac pain**. Cardiac pain impulses arise in free nerve endings in the cardiac connective tissue and adventitia of the cardiac blood vessels. They then travel in visceral sensory fibers through the cardiac plexuses, the middle and the inferior cervical cardiac and thoracic cardiac nerves, and the sympathetic chain ganglia of the neck and the upper thorax. All the pain fibers continue through the white rami communicantes of spinal nerves T1 and T5 and traverse the corresponding dorsal roots and their ganglia. Their cell bodies are located in the dorsal root ganglia, and the central fibers pass from these spinal ganglia to the dorsal horns of the upper thoracic cord segments.

Cardiac pain is referred to cutaneous areas that supply sensory impulses to the same segments of the cord that receive the cardiac sensation. Thus, they involve mainly the region of C7 through T5 and lie predominantly on the left side. The C8 and T1 segments are responsible for referred pain along the medial side of the arm and the forearm.

## Blood Vessels of the Heart

**Right coronary artery**. The right coronary artery originates from the right aortic sinus and passes ventrally between the pulmonary trunk and the right atrium (Fig. 5-1). It descends in the right part of the A-V groove, passing onto the posteroinferior aspect of the heart. It terminates by dividing into the **posterior interventricular artery**, which passes toward the apex in the posterior interventricular sulcus where it anastomoses with the anterior interventricular branch of the left coronary, and a short continuation of the main trunk, which anastomoses with the circumflex branch of the left coronary. Its **marginal branch** passes along the lower border of the right ventricle.

**Left coronary artery**. The left coronary artery, which is usually larger than the right, originates from the left aortic sinus. It passes posteriorly and then to the left of the pulmonary trunk, branching into the circumflex and anterior interventricular arteries as it emerges from behind the pulmonary trunk. The **circumflex artery** passes to the left in the left A-V sulcus and anastomoses with the right coronary artery on the posteroinferior aspect of the heart. The **anterior interventricular artery** descends toward the apex in the anterior interventricular sulcus and passes onto the diaphragmatic surface, where it anastomoses with the posterior interventricular branch of the right coronary.

**Cardiac veins**. For the most part, the cardiac veins accompany the coronary arteries and open into the coronary sinus, which in turn empties into the right atrium. The remainder of the drainage occurs by means of small anterior cardiac veins that drain much of the anterior surface of the heart and terminate directly into the right atrium.

The **coronary sinus** is located in the **posterior A-V sulcus** and drains into the right atrium at the left of the mouth of the inferior vena cava. It receives the **great cardiac vein**, which runs in the anterior interventricular sulcus; the **middle cardiac vein**, located in the posterior interventricular sulcus; and the **small cardiac vein**, which accompanies the marginal branch of the right coronary artery.

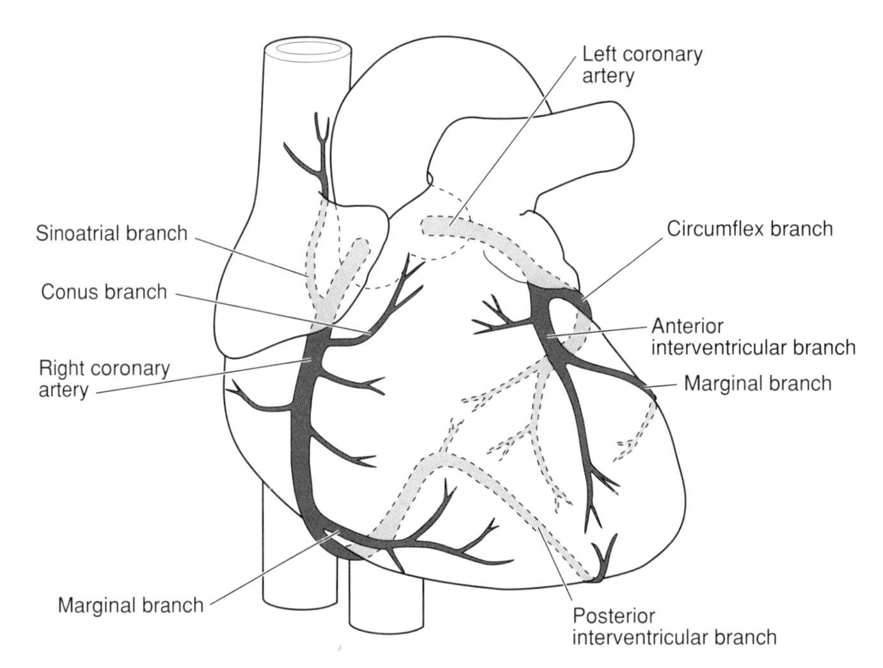


**Figure 5-1.** Anterior view of the heart depicting the coronary arteries and their major branches.

## Pericardium

**Fibrous pericardium**. This tough connective tissue membrane **surrounds the entire heart**. Posteriorly and superiorly, it blends with the adventitia of the great vessels. Inferiorly, it blends with the central tendon of the diaphragm. It is attached to the manubrium by a superior pericardiosternal ligament and to the xiphoid process by an inferior pericardiosternal ligament. In the area between these two ligamentous attachments, most of the anterior surface of the pericardium is separated from the thoracic wall by the lungs and pleural cavities. Only a small portion of the pericardium is intimately related to the lower left portion of the sternum and the medial ends of the fourth through the sixth costal cartilages. Neither lungs nor pleura intervene between the chest wall and fibrous pericardium in this region. This small portion of the pericardium corresponds to the cardiac notch in the left lung and underlies the left fourth and fifth intercostal spaces. This relation permits needle entry to the pericardial cavity without traversing the pleural cavity. The fibrous pericardium is in contact posteriorly with the bronchi, esophagus, and descending thoracic aorta. The lateral outer surfaces of the fibrous pericardium are in close contact with the adjacent parietal pleura. The phrenic nerve and the pericardiacophrenic vessels descend between these two layers.

**Serous pericardium**. This membrane **lines the pericardial cavity** with a smooth, glistening surface that facilitates cardiac movement. The serous pericardium, by its location, is divisible into two layers, parietal and visceral. The **parietal layer** lines the deep surface fibrous pericardium. At the points where the fibrous pericardium blends with the walls of the great vessels entering and leaving the heart, the parietal layer of the serous pericardium is reflected onto the vessels and then the heart muscle to form the **visceral layer (epicardium)** of the serous pericardium. These reflections are in the form of two tubular sheaths, one sheath for the aorta and pulmonary trunk (arterial mesocardium), and the other for the pulmonary veins and venae cavae (venous mesocardium). That portion of the pericardial cavity passing horizontally between the two tubular sheaths remains as the **transverse pericardial sinus**. The reflection of the visceral pleura over the veins forms an inverted, U-shaped cul-de-sac dorsal to the heart that is referred to as the **oblique pericardial sinus**.

## ESOPHAGUS

The esophagus extends from the termination of the pharynx (C6) to the stomach, passing through the diaphragm via the esophageal hiatus (T10). It descends anterior to the vertebral bodies and to the thoracic duct, azygos vein, and right intercostal arteries. It descends initially posterior to the trachea, followed by the left atrium and left main stem bronchus. On the left, the esophagus is related to the aortic arch and mediastinal pleura; on the right, to the arch of the azygos vein and mediastinal pleura. It is supplied by the vagal and sympathetic fibers that form an **esophageal plexus** around the esophagus. In the lower thorax, anterior and posterior vagal trunks accompany the esophagus through the diaphragm to the stomach. Parasympathetic preganglionic fibers penetrate the wall and synapse on postganglionic neurons in the myenteric (Auerbach's) and submucosal (Meissner's) plexuses.

## MAJOR BLOOD VESSELS OF THE THORAX

The great vessels entering and leaving the heart are constituents of either the pulmonary or systemic vascular circuits. The pulmonary circulation is represented by the pulmonary trunk, originating from the right ventricle and branching into left and right pulmonary arteries, and four pulmonary veins returning blood from the lungs to the left atrium. The systemic circulation is represented by the aorta, which originates from the left ventricle, and the superior and inferior venae cavae, which return blood to the right atrium.

**Pulmonary trunk**. This vessel arises from the infundibulum of the right ventricle and ascends obliquely and dorsally to the level of the sternal end of the second left costal cartilage, where it divides into the left and right pulmonary arteries. In its course, it passes in front of, and then to the left of, the ascending aorta. Anteriorly, the pulmonary trunk is separated from the sternal end of the second left intercostal space by the left lung, pleura, and pericardium. At its origin, it is related on the left to the left auricle and the left coronary artery; on the right, to the right auricle and occasionally the right coronary artery.

**Pulmonary arteries**. The **right pulmonary artery** is longer and wider than the left. It passes horizontally to the right and enters the hilus of the lung immediately below the upper lobe (eparterial) bronchus. In its course, it passes dorsal to the ascending aorta, the superior vena cava, and the superior right pulmonary vein. It is ventral to the esophagus, right bronchus, and anterior pulmonary plexus. The **left pulmonary artery** passes laterally and posteriorly toward the root of the left lung, passing anterior to the left bronchus and the descending aorta. The **ligamentum arteriosum** connects the arch of the aorta above with the left pulmonary artery below. The superior left pulmonary vein lies at first ventral and then inferior to the left pulmonary artery.

**Pulmonary veins**. These vessels emerge from each of the five lobes of the lungs. Upon entering the lung root, however, those from the superior and middle lobes of the right lung unite; thus, four terminal pulmonary veins course from the roots of the lungs to the left atrium. The **superior right pulmonary vein** passes dorsal to the superior vena cava; the **inferior right pulmonary vein** passes behind the right atrium before both enter independently through the dorsal and right wall of the left atrium. The **superior** and **inferior left pulmonary veins** pass anterior to the descending aorta and enter separately through the posterior wall of the left atrium near its left border. Fusion of the left pulmonary veins into a common trunk is not uncommon.

**Aorta**. This vessel is the main arterial trunk of the systemic circulation. It ascends from the left ventricle, arches to the left and dorsally over the root of the left lung, descends within the thorax on the left side of the vertebral column, and enters the abdominal cavity through the aortic hiatus of the diaphragm. As a result, the parts of the aorta are the ascending aorta, the arch of the aorta, and the thoracic and abdominal portions of the descending aorta.

The **ascending aorta** arises from the base of the left ventricle at the caudal level of the third left costal cartilage. It passes obliquely upward to the right as far as the level of the second right costal cartilage. Initially, it is related anteriorly to the pulmonary trunk and the right auricle; more superiorly, it is separated from the sternum by the pericardium, variable portions of the right pleura, ventral margin of the right lung, loose areolar tissue, and the remains of the thymus. The coronary arteries arise from the ascending aorta.

The **arch of the aorta** is in the superior mediastinum. It begins at the upper border of the second right sternocostal articulation. It curves cranially and dorsally to the left and then descends along the left side of the vertebral column to the level of the intervertebral disk between the fourth and fifth thoracic vertebrae, where it continues as the descending aorta. In its course, it first passes ventral to the trachea and then to the left of this structure and the esophagus. Arising from the superior aspect of the aortic arch are three large vessels: the brachiocephalic artery (innominate), the left common carotid artery, and the left subclavian artery.

The **brachiocephalic artery** is the first branch from the arch of the aorta and arises posterior to the middle of the manubrium and courses obliquely upward toward the right sternoclavicular joint, where it divides into the right subclavian and common carotid arteries. The **left common carotid artery** arises from the arch of the aorta behind and immediately to the left of the brachiocephalic artery. Its thoracic portion extends up to the level of the left sternoclavicular joint. It passes anterior to the trachea, left recurrent laryngeal nerve, esophagus, thoracic duct, and left subclavian artery. The **left subclavian artery** arises from the arch of the aorta about 2.5 cm distal to the left common carotid. It ascends almost vertically on the left side of the trachea to the root of the neck, where it arches upward and laterally. It passes posterior to the left vagus nerve, left phrenic nerve, left superior cardiac sympathetic nerve, and left brachiocephalic vein. It passes in front of the left lung, pleura, esophagus, and thoracic duct.

The **thoracic portion of the descending aorta** lies in the posterior mediastinum. It extends downward from the upper border of the body of the fifth thoracic vertebra to the aortic opening in the diaphragm (T12). It has branches that supply the walls and viscera of the thorax, and is related posteriorly to the left pleura and lung, the vertebral column, and the hemiazygos veins. Anteriorly, superior to inferior, it is related to the root of the left lung, pericardium, esophagus, and diaphragm. It passes to the left of the azygos vein and thoracic duct, and to the right of the left lung and pleura.

**Superior vena cava**. This vessel returns blood to the right atrium from the upper half of the body. It arises from the junction of the right and left brachiocephalic veins at the level of the lower border of the first right costal cartilage and enters the right atrium at the level of the third right costal cartilage. The superior vena cava is related posteromedially to the trachea and anteromedially to the ascending aorta. The phrenic nerve is positioned between the superior vena cava and the parietal layer of mediastinal pleura on the right. The superior vena cava receives the **azygos vein** on its posterior surface at the level of the second costal cartilage. The azygos vein enters the thorax through the aortic hiatus in the diaphragm, ascends along the right side or anterior aspect of the vertebral column, receives the hemiazygos vein at the T9 level (and possibly the accessory hemiazygos vein at the T8 level), and finally passes posterior to the root of the right lung before arching anteriorly over the root to enter the superior vena cava.

Both **brachiocephalic veins** are formed by the union of the internal jugular and subclavian veins. The right brachiocephalic vein arises dorsal to the sternal end of the clavicle and passes almost vertically downward in front of the trachea and vagus nerve and behind the sternohyoid and sternothyroid muscles and right lung and pleura. The left brachiocephalic vein is longer than the right. From its origin deep to the medial end of the clavicle, it passes obliquely downward and to the right, where it joins the right brachiocephalic vein to form the superior vena cava at the lower border of the right first costal cartilage. The thoracic portion of each brachiocephalic vein receives an internal thoracic and often an inferior thyroid vein. In addition, the left receives the left highest intercostal vein.

**Inferior vena cava**. This vessel returns blood to the right atrium from the caudal half of the body. It enters the thorax by piercing the diaphragm (T8) between the middle and right leaflets of the central tendon. It ascends in a slightly anteromedial direction in the middle mediastinum and pierces the fibrous pericardium. The inferior vena cava is separated in its extrapericardial course from the right pleura and lung by the right phrenicopericardiac ligament. In its short intrapericardial course, it is invested on its right and left sides with a reflection of the serous pericardium.

## LYMPHATIC SYSTEM OF THE THORAX

**Thoracic duct.** The thoracic duct drains the lymphatics of the entire body except those of the right upper quadrant above the diaphragm. It originates in the **cisterna chyli** of the abdomen at the second lumbar vertebra, enters the thorax through the aortic hiatus of the diaphragm, and ascends through the posterior mediastinum between the aorta and azygos vein and posterior to the esophagus. At the level of the T5 vertebra, it crosses the midline to the left side, enters the superior mediastinum, and ascends between the esophagus and pleura to enter the venous system at the junction of the left subclavian and internal jugular veins.

**Right lymphatic duct**. This duct drains the lymphatics from the right upper quadrant. Specifically, it receives lymph from the right side of the head and neck through the right jugular trunk; from the right upper extremity by way of the right subclavian trunk; and from the right side of the thorax, right lung, and part of the convex surface of the liver through the right bronchomediastinal trunk.

**Thymus**. The thymus is larger in the infant than it is in the adult. It consists of two lateral lobes invested by a connective tissue capsule that sends septa into the gland and divides it into lobules. The gland lies in the anterior part of the superior mediastinum and extends from the fourth costal cartilage to the lower border of the thyroid gland. It lies anterior to the great vessels and fibrous pericardium.

**Lymph nodes of the thorax**. The lymphatic nodes of the thorax are divided into the more superficial parietal and deeper visceral groups. The **parietal nodes** include the sternal, intercostal, and diaphragmatic nodes. The **visceral lymph nodes** consist of anterior and posterior mediastinal and tracheobronchial nodes. The afferent and efferent vessels of these nodes are listed in Table 5-4.

**TABLE 5-4. Lymphatics of the Thorax**

| Afferents From | Lymph Nodes | Efferents To |
|---|---|---|
| Breast | Axillary | Deep cervical nodes → right lymphatic duct or thoracic duct |
| Breast, anterior mediastinum, anterior aspects of intercostal spaces | Parasternal | Same as breast |
| Lungs and bronchi | Tracheobronchial | Same as breast |
| Mediastinum | Mediastinal | Same as breast |

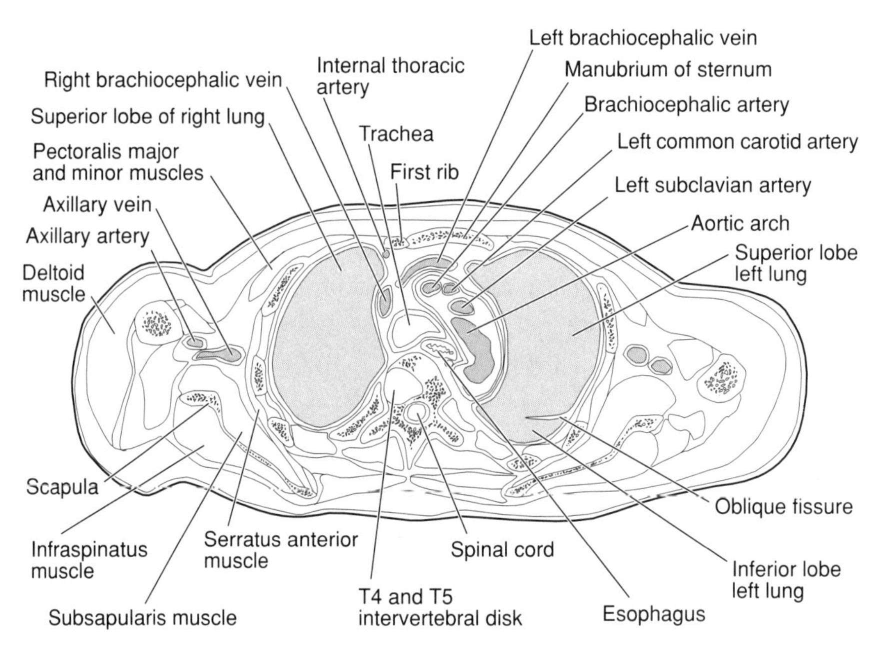


**Figure 5-2.** Cross section through the junction of the superior and inferior parts of the mediastinum.

## CROSS SECTIONS OF THE THORAX

The orientation of all cross sections of the thorax corresponds to that used clinically in displaying the various types of cross-sectional images (e.g., the view is the inferior aspect of the section with the patient's left on the right of the page). In addition, the narrative description of structure location corresponds to the true orientation of the specimen; as a result, something to the left on the illustration is really on the right and is so described.

Figure 5-2 is a section approximately **through the junction of the superior and inferior parts of the mediastinum**. The plane passes through the junction of thoracic vertebrae four and five, the lower aspect of the manubrium of the sternum, and includes ribs one through five. The arch of the aorta passes posterolaterally to the left and is cut so the origins of its three branches are apparent. From proximal to distal, these are the brachiocephalic, left common carotid, and left subclavian arteries. The fourth opening in this section is a slice through a slightly abnormal arching of the aorta. The left brachiocephalic vein is passing to the right in front of the aorta, just superior to the level where it joins the right brachiocephalic vein to form the superior vena cava. The trachea just superior to its bifurcation is separated from the vertebral column by the esophagus, which is somewhat to the left at this level. The thoracic duct is posterolateral to the esophagus. The internal thoracic arteries are posterior to the lateral aspects of the sternum. The lungs are sectioned superior to the hilar regions; thus, the pleural cavities are seen completely surrounding the lungs. The oblique fissure of the left lung separates the anteriorly located superior lobe from the more posteriorly positioned inferior lobe.

The plane of Fig. 5-3 is **through vertebra T8, the junction of the fourth rib with the sternum and the middle mediastinum**. The oblique fissure clearly separates the superior and inferior lobes of the left lung. On the right, both oblique and horizontal fissures are present so the superior, middle, and inferior lobes are partially demarcated.

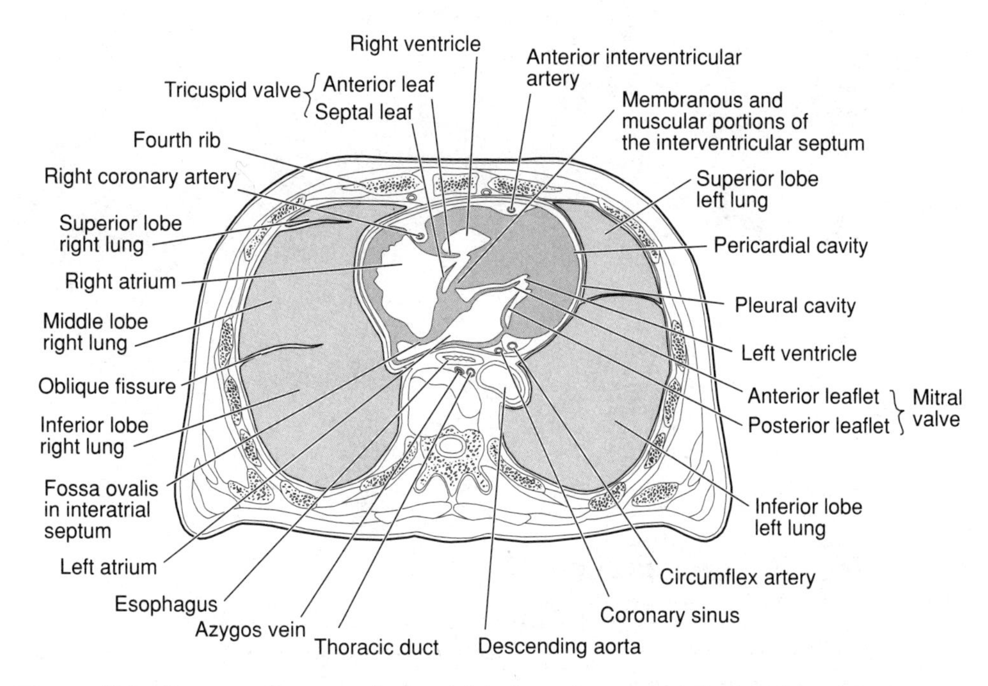


**Figure 5-3.** Cross section through the eighth thoracic vertebra, the junction of the fourth rib with the sternum, and the middle mediastinum.

The **heart** is sectioned so that **all four chambers are visible**. The posterior surface is formed almost entirely by the left atrium, and the right atrium forms the right border. Note that the interatrial septum is essentially in the coronal plane. The right and left ventricles form the anterior and right borders, respectively. The latter two chambers are separated by the obliquely oriented interventricular septum; both its muscular and membranous portions are visible. The tricuspid valve is anterior and somewhat to the right of the mitral valve. The coronary (A-V) sulcus houses the right coronary artery anteriorly and the coronary sinus and circumflex branch of the left coronary posteriorly. The anterior interventricular sulcus is not easily seen, but its location is marked by the anterior interventricular artery.

In the **posterior mediastinum**, the esophagus descends directly posterior to the right atrium. The azygos vein and thoracic duct are essentially between the esophagus and vertebral column, with the azygos vein to the right of the thoracic duct. The descending portion of the thoracic aorta is anterolateral (on the left) to the vertebral bodies.

# Chapter 6

# Abdomen

## SURFACE ANATOMY

### References for Location

**Quadrants**. The abdomen can be divided into four areas by vertical (linea alba) and horizontal lines through the umbilicus. The resulting general areas are the **right** and **left upper** and **lower quadrants**.

**Regions**. The abdomen can be separated into more specific regions by two horizontal and two vertical lines. The **right** and **left semilunar lines** correspond to the lateral edges of the rectus abdominis muscles. These lines are slightly curved and extend from the tips of the ninth costal cartilages to the pubic tubercles. The **transpyloric plane** is a horizontal line through a point halfway between the suprasternal notch and the upper border of the pubic symphysis. It is also midway between the xiphisternal joint and the umbilicus. This plane passes through the pylorus, the body of L1, the tips of the ninth costal cartilages, and, on the right, the fundus of the gallbladder. The **transtubercular plane** is at the level of the top of the iliac crest. These four lines subdivide the abdomen into the following **nine regions** (superior to inferior): In the center are the epigastric, umbilical, and pubic regions; on the sides are the right and left hypochondriac, lumbar, and inguinal regions.

### Surface Projection of Specific Organs

The approximate position of most of the abdominal organs can be projected to the surface. Even though some organs are variable in position, certain parts of these organs are fixed by peritoneal connections and thus have a predictable location. The projection of the abdominal organs is summarized in Table 6-1.

## ABDOMINAL WALL

### Superficial Fascia

This fascia is composed of superficial and deep layers. The superficial or fatty layer (**Camper's fascia**) is continuous with the superficial fascia of adjacent areas. The deep or fibrous layer (**Scarpa's fascia**) attaches to the deep fascia of the thigh, but continues into the perineum as the superficial perineal fascia; thus, fluid collecting in the **superficial perineal space** can extravasate into the abdominal wall (between the fibrous layer of the superficial fascia and the fascia of the external abdominal oblique muscle), but not into the thigh.

**TABLE 6-1. Surface Projection of Abdominal Organs**

| Organ | Surface Location |
|---|---|
| Stomach | Cardiac orifice: seventh costal cartilage, 2 cm to left of xiphisternal joint<br>Pyloric orifice: transpyloric plane, 1.5 cm to left of midline<br>Lesser curvature: downward curving line to right, interconnecting the above two points<br>Greater curvature: extends inferiorly as low as umbilicus (or lower) |
| Small intestine | Duodenum, part I: transverse line from pylorus to junction of transpyloric and right semilunar lines<br>Duodenum, part II: descending line to subcostal margin (L3)<br>Duodenum, part III: transverse line to left about 3 cm to left of midline<br>Duodenum, part IV: ascending line (about 2 vertebral levels) to duodenojejunal junction<br>Jejunum: predominantly left upper quadrant<br>Ileum: predominantly right lower quadrant<br>Ileocolic junction: slightly inferior and medial to junction of the right semilunar and transtubercular lines |
| Large intestine and vermiform appendix | Cecum: right iliac and hypogastric regions<br>Appendix: one-third (5 cm) of distance along a line from right anterior superior iliac spine to umbilicus (McBurney's point)<br>Right colic flexure: just below transpyloric line 2.5 cm lateral to right semilunar line<br>Left colic flexure: just above transpyloric line 2.5 cm lateral to left semilunar line |
| Gallbladder | Tip of the ninth costal cartilage in the midclavicular line |
| Liver | Superior border: horizontal line just below nipples<br>Inferior border: inferior costal margin or right; leaves this margin at tip of ninth costal cartilage, line extends obliquely across subcostal angle to just below left nipple |
| Pancreas | Head: fills C curve formed by duodenum<br>Neck: transpyloric line in midline<br>Body: extends to left from neck and ascends slightly<br>Tail: in contact with spleen |
| Spleen | Deep to left ninth, tenth, and eleventh ribs posteriorly<br>Superior pole: 3 cm lateral (left) to T10 spinous process<br>Inferior pole: midaxillary line at the level of the eleventh rib |
| Kidney | Dorsal body wall: two vertical lines 2.5 and 10.0 cm from midline between T11 and L3 levels<br>Upper pole: 5 cm from midline<br>Lower pole: 7.5 cm from midline<br>Hilum: 5 cm from midline at L1 level<br>Left kidney is 1.0–1.5 cm higher than right kidney |

## Muscles

The interval between the inferior costal margin and the superior aspect of the pelvis (i.e., pubis, inguinal ligament, iliac crest) contains four flat muscles as well as the posteriorly positioned quadratus lumborum, all of which are segmentally innervated by thoracic and lumbar nerves. The external abdominal oblique, internal abdominal oblique, and transversus abdominis are lateral to the semilunar line; the rectus abdominis is medial. The most superficial muscle is the **external oblique**, whose fibers are directed anteriorly, medially, and inferiorly. The fibers of the **internal oblique** are perpendicular to those of the external oblique; those of the **transversus abdominis** are transversely oriented. The aponeuroses of all three muscles extend medially and together form the rectus sheath. The **rectus abdominis** is the only vertically oriented

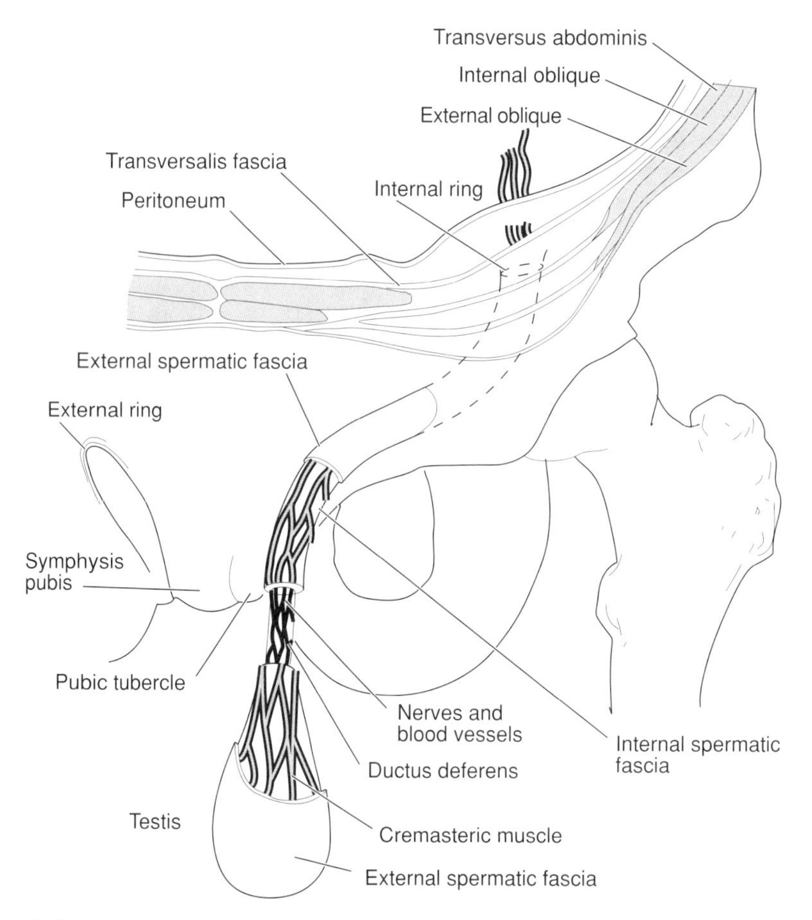


**Figure 6-1.** Inguinal canal and spermatic cord.

muscle, extending between the superomedial aspect of the pubis and the medial aspect of the inferior costal margin and the xiphoid process. Its tendinous intersections account for the "ripples" seen on the surface of the abdomen. The **quadratus lumborum** completes the muscular closure of the interval posteromedially. This muscle extends between the twelfth rib and the iliac crest, and also attaches to the lumbar transverse processes.

## Inguinal Region

The inferior aspect of the abdominal wall is composed primarily of the aponeuroses of the abdominal muscles. The inferior margin of the external oblique aponeurosis forms the **inguinal ligament** as it stretches between the anterior superior iliac spine and the pubic tubercle. This tough band also is an attachment for some of the other abdominal muscles, and it participates in the formation of the inguinal canal.

**Inguinal canal**. The inguinal canal is an obliquely oriented pathway through the abdominal wall (Fig. 6-1). It stretches between the **superficial (external)** and **deep (internal) inguinal rings** and is directed inferomedially just above the inguinal ligament. The general shape of this canal is that of a flattened cylinder; it can be described as having a floor and roof as well as anterior and posterior walls. The structures forming the various parts of the canal are listed in Table 6-2.

**Spermatic cord**. The spermatic cord consists of the vas deferens and its artery, the pampiniform plexus of veins, lymphatics, sympathetic nerve fibers, the ilioinguinal

**TABLE 6-2. Formation of the Inguinal Canal**

| Part of Canal | Structure(s) Forming that Part |
|---|---|
| Superficial ring | Split in external oblique aponeurosis |
| Medial (superior) and lateral (inferior) crura | Medial and lateral edges of superficial ring, formed by external oblique aponeurosis |
| Deep ring | Beginning of sleeve (outpouching) of transversalis fascia |
| Floor | Inguinal ligament |
| Anterior wall | External oblique aponeurosis |
| Posterior wall | Medially: conjoined tendon (falx inguinalis), which is formed by medial arching fibers of internal oblique and transversus abdominis<br>Laterally: transversalis fascia |
| Roof | Conjoined tendon |

nerve, the nerve and artery to the cremasteric muscle, and the testicular artery. Most of these structures are enclosed within the **internal spermatic fascia**, which is derived from the transversalis fascia; the **cremasteric muscle** and **fascia**, which are derived from the internal oblique; and the **external spermatic fascia**, which is derived from the external abdominal oblique aponeurosis.

**Inguinal hernias.** Inguinal hernias are either direct or indirect, based on the path of the herniating sac. The **indirect** hernia follows the course of the testis during development; that is, it passes into the canal through the internal opening and then through the canal. After exiting through the superficial ring, it usually passes into the scrotum. The neck of this type of hernia is found lateral to the inferior epigastric vessels, and the hernial sac is covered with peritoneum and the three fascial coverings of the spermatic cord. A **direct** hernia passes through **Hesselbach's triangle**, the triangular area defined by the lateral border of the rectus abdominis, the inferior epigastric vessels, and the inguinal ligament. The neck of this type of hernial sac is medial to the inferior epigastric vessels and is covered by a layer of peritoneum and transversalis fascia. Because this hernia passes through the superficial ring, it is also covered by the external spermatic fascia.

## PERITONEUM

The peritoneum is similar to the pleura in that it consists of a parietal layer that lines the abdominopelvic cavity and a visceral layer that is reflected over organs. It is also similar in that the peritoneal cavity contains nothing other than a small amount of lubricating fluid and in reality is a potential space. It differs from the pleura in that its visceral layer reflects in varying degrees over multiple organs. Some organs (i.e., the jejunum and ileum) are almost completely covered by peritoneum and attached to the posterior body wall by a mesentery, and thus are said to be **(completely) peritonealized**. Other organs (kidneys) are essentially outside the peritoneum and are covered by peritoneum on only one side. These organs are **extraperitoneal** or, if located on the posterior body wall, **retroperitoneal**. Those organs that are neither peritonealized nor extraperitoneal, but are somewhere in between, are **partially peritonealized**. The peri-

**TABLE 6-3. Peritoneal Spaces, Gutters, Fossae, and Pouches**

| Areas | Location |
|---|---|
| Right subphrenic space | Between the diaphragm and liver, to the right of the falciform ligament and the lesser omentum |
| Left subphrenic space | Between the diaphragm and liver, to the left of the falciform ligament and the lesser omentum |
| Right subhepatic space | Inferior to the right and caudate lobes of the liver |
| Left subhepatic space | Inferior to the quadrate lobe of the liver |
| Right paravertebral gutter | Vertical trough to the right of the lumbar vertebral bodies<br>Right medial paracolic gutter: between lumbar bodies and ascending colon<br>Right lateral paracolic gutter: lateral to ascending colon |
| Left paravertebral gutter | Vertical trough to the left of the lumbar vertebral bodies<br>Left medial paracolic gutter: between lumbar bodies and descending colon<br>Left lateral paracolic gutter: lateral to descending colon |
| Pararectal fossae | On either side of the rectum |
| Rectovesical pouch | Male only: between rectum and bladder |
| Vesicouterine pouch | Female only: between the uterus and bladder |
| Rectouterine pouch (of Douglas) | Female only: between the uterus and rectum |

toneum also differs from the pleura in that in the female the peritoneal cavity is not closed but communicates with the lumen of the reproductive track via the openings in the infundibulum of the uterine tubes.

## Greater and Lesser Omental Bursae

The **peritoneal cavity** is separated into greater and lesser peritoneal sacs. The **lesser sac** (omental bursa) is posterior to the stomach, lesser omentum, and caudate lobe of the liver. The **greater sac** is the rest of the peritoneal cavity. The two areas are connected only through the epiploic foramen (of Winslow), which is small. The anterior border of this foramen is the right free margin of the lesser omentum, which is formed by the hepatoduodenal ligament and contains the common bile duct, portal vein, proper hepatic artery, and lymphatics and nerves.

## Peritoneal Spaces, Gutters, Fossae, and Pouches

These are areas where inflammatory material can accumulate and become sequestered within the peritoneal cavity. These areas are referred to as **spaces**, **gutters**, **fossae**, and **pouches** and are found in both the abdominal and pelvic parts of the abdominopelvic cavity. They are summarized in Table 6-3.

**Folds and fossae** of the peritoneum are also found on the deep aspect of the lower part of the anterior abdominal wall. These consist of the single median umbilical fold along with the medial and lateral umbilical folds. The **median fold** overlies the remains of the urachus; the **medial folds** are formed by the obliterated umbilical arteries and the lateral by the inferior epigastric vessels. The **supravesical fossae** are between the median and medial folds; the **medial** and **lateral inguinal fossae** are medial and lateral, respectively, to the lateral fold. **Indirect hernias** pass through the lateral inguinal fossa, whereas **direct hernias** pass through either the supravesical or medial inguinal fossa.

# ABDOMINAL VISCERA

The relationships of each organ along with its general peritoneal classification (i.e., peritonealized, retroperitoneal) are listed in Table 6-4. The detailed peritoneal relationships are discussed here.

## Stomach

The stomach extends from the cardiac opening of the esophagus to the pylorus. It consists of a **fundus**, **body**, and pyloric portion, which is made up of the **pyloric antrum**, **canal**, and **sphincter**. Its right or upper border forms the lesser curvature, to which the lesser omentum attaches. The lower and left border forms the greater curvature and gives attachment to the greater omentum and gastrolienal ligament. The **pregastric space (Traube)** is anterior to the stomach, semilunar in outline, and bounded by the lower edge of the left lung, anterior border of the spleen, left costal margin, and lower edge of the left lobe of the liver. The blood supply to the stomach is provided by the three main branches of the celiac trunk: the common hepatic, left gastric, and splenic arteries. Each of these has branches that pass along the greater or lesser curvatures of the stomach. Venous and lymphatic drainage parallels the arterial supply. The veins are tributaries of the portal vein, and the lymphatics all eventually pass to the celiac group of aortic lymph nodes.

## Small Intestine

**Duodenum**. The duodenum is a C-shaped tube surrounding the head of the pancreas (Fig. 6-2). It is divided into *four parts*, most of which are retroperitoneal. The first part is short, passes to the right, and ascends slightly from the gastroduodenal junction. The second part descends in the right medial paracolic gutter. The common bile duct and main pancreatic duct empty into it posteromedially. The third part passes from right to left across the midline anterior to the body of vertebra L3. The fourth part is short and curves superiorly and then anteriorly at the duodenojejunal junction. It is suspended from the area of the right crus of the diaphragm by a peritoneal fold called the **suspensory ligament of Treitz**.

**Jejunum and ileum**. The mesenteric portion of the small intestine is divided into the proximal jejunum and distal ileum. The **mesentery** is a fan-shaped, double layer of peritoneum continuous with the serosa of the jejunum and ileum and enclosing their vessels. The root of the mesentery is about 15 cm long, and its attachment to the posterior body wall extends from the duodenojejunal junction obliquely down and to the right, crossing the third part of the duodenum, the inferior vena cava, and the right ureter. The mesentery contains the intestinal and ileocolic branches of the superior mesenteric vessels as well as lymphatics and autonomics. The transition from jejunum to ileum is not abrupt, but certain differences exist. The amount of mesenteric fat tends to increase from jejunum to ileum as does the number of arcades formed by the vessels in the mesentery; however, the blood supply to the jejunum is greater, so its pink color is more intense than that of the ileum. The jejunum has a thicker wall because the **plicae circulares** are higher and more numerous.

**TABLE 6-4. Organ Relationships and Peritoneal Characterization of Abdominal Viscera**

| Organ | Peritoneal Characterization | Anterior Relationships | Posterior Relationships |
|---|---|---|---|
| Stomach | Peritonealized | Liver, diaphragm, abdominal wall, transverse colon | Spleen, splenic artery, diaphragm, left kidney, suprarenal, pancreas, transverse mesocolon |
| Small intestine | | | |
| Duodenum, part I | Retroperitoneal | Gallbladder, liver | Portal vein, common bile duct, gastroduodenal artery, inferior vena cava |
| Duodenum, part II | Retroperitoneal | Transverse colon | Hilus of right kidney, suprarenal, renal vessels, ureter |
| Duodenum, part III | Retroperitoneal | Superior mesenteric vessels | Aorta, inferior vena cava |
| Duodenum, part IV | Retroperitoneal | Root of mesentery, jejunum | Left psoas muscle, aorta |
| Jejunum | Peritonealized | Anterior abdominal wall | Posterior body wall |
| Ileum | Peritonealized | Anterior abdominal wall | Posterior body wall |
| Large intestine | | | |
| Cecum | Peritonealized (usually) | Anterior abdominal wall | Posterior body wall |
| Vermiform appendix | Peritonealized | Cecum (usually) | Posterior body wall |
| Ascending colon | Partially peritonealized | Anterior body wall, ileum | Posterior body wall, right kidney |
| Transverse colon | Peritonealized | Liver, gallbladder, stomach | Spleen |
| Descending colon | Partially peritonealized | Jejunum and ileum | Posterior body wall, left kidney |
| Liver | Partially peritonealized | Anterior body wall, diaphragm | Visceral surface: hepatic flexure of colon, right kidney and suprarenal gland, gallbladder, duodenum, esophagus, stomach |
| Gallbladder | Partially peritonealized | Liver, ninth costal cartilage | Duodenum, pyloric stomach |
| Pancreas | Retroperitoneal | Stomach | Posterior body wall |
| Spleen | Peritonealized | Tail of pancreas | Ninth, tenth, and eleventh ribs, diaphragm |
| Kidney | Retroperitoneal | Left: stomach, spleen, left colic flexure, pancreas, jejunum<br>Right: liver, right colic flexure, second part of duodenum | Diaphragm, twelfth rib, psoas major and quadratus lumborum muscles |
| Suprarenal gland | Retroperitoneal | Right: liver, inferior vena cava<br>Left: splenic artery, pancreas, stomach (?) | Diaphragm, posterior body wall |
| Ureter | Retroperitoneal | Right: duodenum, gonadal, and right colic vessels, root of mesentery<br>Left: gonadal and left colic vessels, sigmoid colon | Psoas major muscle, common iliac vessels |

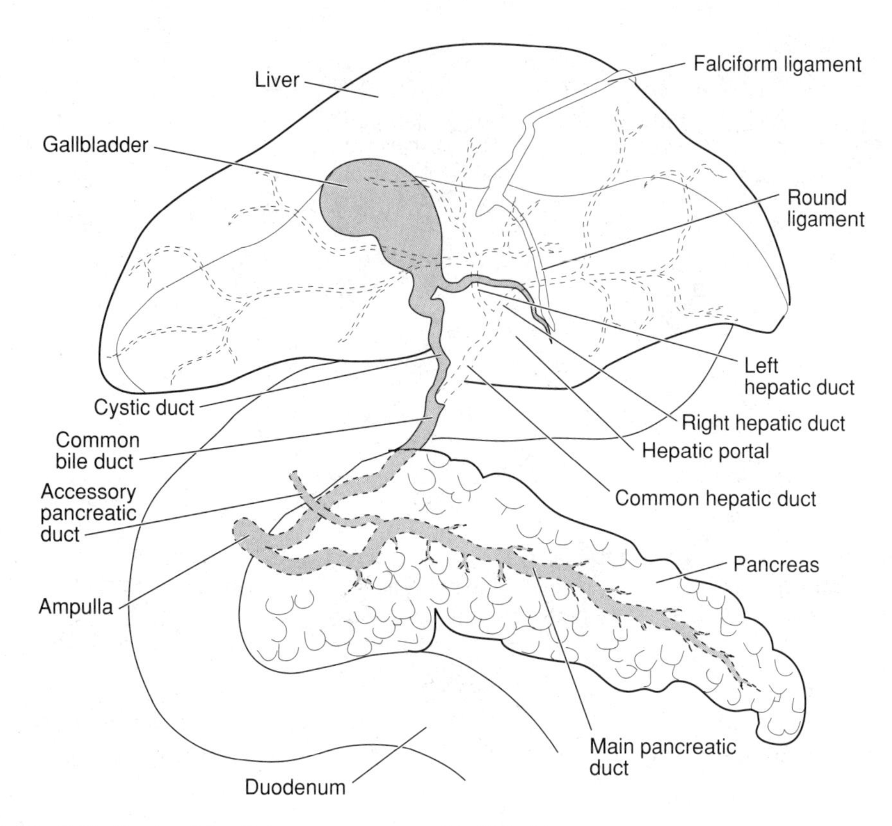


**Figure 6-2.** Ducts of the liver, pancreas, and gallbladder.

## Large Intestine

This portion, the **colon**, extends from the ileocecal valve to the anus. It is characterized by sacculations (**haustra**), which are produced by three longitudinal muscle bands (**taenia**) that converge at the appendix. Between the sacculations are the **semilunar folds**, and along the free surface of the colon are pouches that contain fat (**appendices epiploicae**).

**Cecum**. The cecum is a cul-de-sac of the colon below the entrance of the ileum. It is found in the right iliac fossa and is most often almost entirely enveloped in peritoneum. The **ileocecal valve** is formed by two lip-like folds projecting into the medial aspect of the cecum.

**Vermiform appendix**. The vermiform appendix is a blind tube that attaches to the posteromedial aspect of the cecum about 2.5 cm below the ileocecal valve. It is most commonly found behind the cecum (retrocecal), although it may be found in a variety of positions including extending into the pelvis. It has a small valve and, although variable, is usually about 10 cm in length. It has no true mesentery, but it is covered with a peritoneal fold and is supplied by an appendicular branch of the ileocolic artery.

**Ascending colon**. The ascending colon is the continuation of the cecum that passes superiorly against the posterior body wall and the right kidney. Just below the right lobe of the liver, it makes a sharp bend to the left. This bend is the **right colic** or **hepatic flexure**.

**Transverse colon**. The transverse colon extends between the hepatic and the splenic flexures and is suspended from the posterior abdominal wall by the **transverse mesocolon**. The attachment of the mesocolon crosses the second part of the duodenum and the pancreas, then attaches to the greater curvature of the stomach.

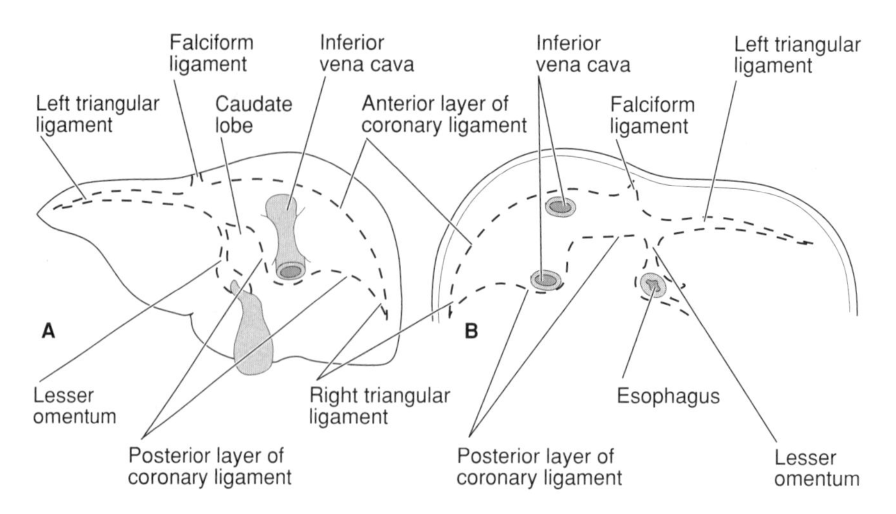


**Figure 6-3.** Posterior view of the liver and its attachments. ***A:*** The posterior diaphragmatic and visceral surfaces of the liver, with the lines of reflection of the peritoneum and associated ligaments. ***B:*** The peritoneal attachments to the diaphragm.

**Descending colon**. The descending colon extends from the **splenic flexure** to the **left iliac fossa**, descending in the left paravertebral gutter before becoming the **sigmoid colon**.

## Liver

The anterior, superior, and posterior surfaces of the liver form a dome that is related to the respiratory diaphragm. Its **visceral surface** faces posteriorly and inferiorly. The visceral surface has an **H-shaped configuration** in which the center bar of the H is the **porta hepatis**, where the portal vein, autonomics, and hepatic artery enter, and where the common hepatic duct and lymphatics exit. The limb of the H extending anteriorly on the right contains the **gallbladder**; the anterior limb on the left contains the **ligamentum teres**; posteriorly on the right is the **inferior vena cava**; and posteriorly on the left is the **ligamentum venosum**. That part of the liver between the two anterior limbs is the **quadrate lobe**; the portion between the two posterior limbs is the **caudate lobe**. Functionally, the liver is separated into a large right lobe and a smaller left lobe, which includes the quadrate and caudate lobes.

The liver is partially enclosed within a complicated system of peritoneal folds (Fig. 6-3). The **falciform ligament** reflects from the anterior belly wall to the liver and contains the ligamentum teres in its inferior border. Anteriorly, the falciform ligament reflects over the right and left lobes of the liver. On the superior and posterior aspects of the dome of the liver, the two layers of this ligament initially diverge toward the sides of the liver. After diverging, the two layers turn sharply medially and posteriorly and converge in the region of the ligamentum venosum. As the two layers of the falciform ligament diverge and then converge in this pattern, they reflect onto the diaphragm. This crown-shaped pattern of peritoneal reflections between the posterosuperior surface of the liver and the inferior surface of the diaphragm is termed the **coronary ligament**. It surrounds and defines the bare area of the liver, an area in which no peritoneum separates the liver and the diaphragm. The right and left parts of the coronary ligament are called the right and left **triangular ligaments**, respectively. The lesser omentum extends from the visceral surface of the liver to the lesser curvature of the stomach. The free edge of the lesser omentum is formed as the peritoneum surrounds the portal vein, hepatic artery, and common bile duct. This free edge extends between the first part of the duodenum and the porta hepatis and forms the anterior boundary of the **epiploic foramen** (of Winslow), which is the entrance to the lesser omental sac.

## Gallbladder and Ducts

The system that stores and transports bile includes the gallbladder, cystic duct, hepatic duct, and common bile duct (see Fig. 6-2). The **gallbladder** is a pear-shaped viscus lying below the liver between the right and the quadrate lobes. It has a broad fundus, a body, and a neck that narrows into the cystic duct. The **cystic duct** unites with the **hepatic duct** to form the **common bile duct**. The common bile duct and the main pancreatic duct unite just proximal to the point at which they empty into the second part of the duodenum at the **major duodenal papilla**. This common opening is protected by the **sphincter of Oddi**.

## Pancreas

The pancreas lies between the stomach and the posterior abdominal wall and extends from the duodenum to the spleen (see Fig. 6-2). It is divided into a head, occupying the concavity of the duodenum; a neck, which is constricted posteriorly by the portal vein; a triangular body; and a tail that rests upon the spleen. The secretions of the neck, body, and tail portions of the pancreas are carried through the **main pancreatic duct**, through which they enter the duodenum. The **accessory pancreatic duct of Santorini** traverses the head of the pancreas and enters the duodenum 2 cm above the major duodenal papilla. The main arteries supplying the pancreas are the superior and the inferior pancreaticoduodenal and the splenic, which runs behind the upper part of the body and the tail of the pancreas.

## Spleen

The spleen occupies the left hypochondrium deep to the ribs posteriorly. Its parietal surface is convex and related to the diaphragm; its visceral surface is concave and related to several organs. It is connected by peritoneal reflections to the posterior body wall in the region of the left kidney by the **lienorenal ligament** and to the stomach by the **gastrolienal ligament**. Its blood supply is provided by the splenic artery.

## Kidneys

The kidneys are surrounded by perirenal fat and supported by the renal fascia. They extend from the last thoracic to the third lumbar vertebrae, the right kidney lower than the left, and occupy the most dorsal position of all the abdominal organs.

The **renal hilus** is directed anteromedially and is the concave aspect of the kidney. The renal artery enters the kidney through this area, and the renal vein and ureter exit at this site. The final collecting cistern for urine is the **renal pelvis**. The pelvis occupies a large percentage of the hilus and is usually at the level of the body of L1. The pelvis narrows rapidly to form the **ureter**, which conducts the urine to the bladder.

The ureters descend through the abdomen; at roentgenographic examination in which they are filled with contrast medium, they are seen to cross the transverse processes of the lumbar vertebrae 3 to 4 cm lateral to the midline. As the ureters enter the pelvis, they incline medially and usually cross the termination of the common iliac arteries (or their branches). Their pelvic courses are inferomedial toward the posterior aspect of the bladder. In the female, the ureter crosses the uterine artery near the lateral fornix. The ureters have three natural constrictions: at the junction of the renal pelvis and the ureter, where they cross the common iliac arteries, and as they pass through the wall of the bladder.

Generally, each kidney is supplied by a single large renal artery, which is a direct branch from the abdominal aorta. A common variation is two or three renal arteries on one or both sides.

### Suprarenal Glands

The suprarenal (**adrenal**) glands sit on the superior poles of each kidney. Each gland receives a rich blood supply via branches from the renal and inferior phrenic arteries and from the aorta.

## BLOOD SUPPLY OF THE ABDOMEN

The arterial supply to the abdomen is provided by branches of the abdominal aorta, and the venous drainage occurs through the inferior vena cava and azygos vein; however, a large portion of the gastrointestinal tract drains initially into the portal vein, which, in turn, passes to the liver. After flowing through the liver, the portal blood is collected into the hepatic veins, which join the inferior vena cava. In the abdomen, therefore, there are **systemic** and **portal venous systems**.

### Arterial Supply of the Abdomen

The abdominal aorta (Fig. 6-4) enters the abdomen through the aortic hiatus (T12) and extends to the lower portion of the fourth lumbar vertebral body, where it terminates by dividing into the large **common iliac arteries** and into its true continuation, the rudimentary **median sacral artery**. The aorta descends vertically along the anterior (or slightly to the left) aspect of the lumbar vertebral bodies. The areas supplied by its branches are listed in Table 6-5; the locations and courses of its branches, from superior to inferior, are as follows:

**Inferior phrenic**. This pair of arteries arises just below the diaphragm or from the celiac trunk and distributes to the inferior surface of the diaphragm.

**Celiac trunk**. The highest of the unpaired vessels, it arises at vertebral level T12. This artery is very short; it passes anteriorly behind the peritoneum and above the pancreas, where it divides into the left gastric, splenic, and common hepatic arteries. The left gastric artery passes to the left and then anteriorly and to the right along the lesser curvature of the stomach. The **splenic artery** passes to the left along the upper surface of the pancreas. It forms part of the floor of the lesser omental sac and has short gastric and left gastroepiploic branches to the greater curvature of the stomach. The **common hepatic artery** runs anterolaterally to the right toward the upper aspect of the first part of the duodenum. Its gastroduodenal branch descends behind the duodenum and has anterior and posterior superior pancreaticoduodenal and right gastroepiploic branches. The continuation of the common hepatic artery after the gastroduodenal branch is the **proper hepatic artery**. This artery passes to the right and ascends to the porta hepatis through the lesser omentum. The **right gastric artery** usually branches from the proper hepatic artery soon after its beginning; the **cystic artery** branches where the proper hepatic artery passes the cystic duct.

**Middle suprarenal**. These vessels are usually single, one passing directly to each suprarenal gland.

**Superior mesenteric**. This single artery branches from the anterior aspect of the aorta (vertebral level L1), just inferior to the celiac trunk, and passes between the pancreas and the third part of the duodenum. Its **inferior pancreaticoduodenal arteries** ascend on either side of the head of the pancreas and anastomose with similar branches of the gastroduodenal artery. The **middle colic artery** enters the transverse mesocolon:

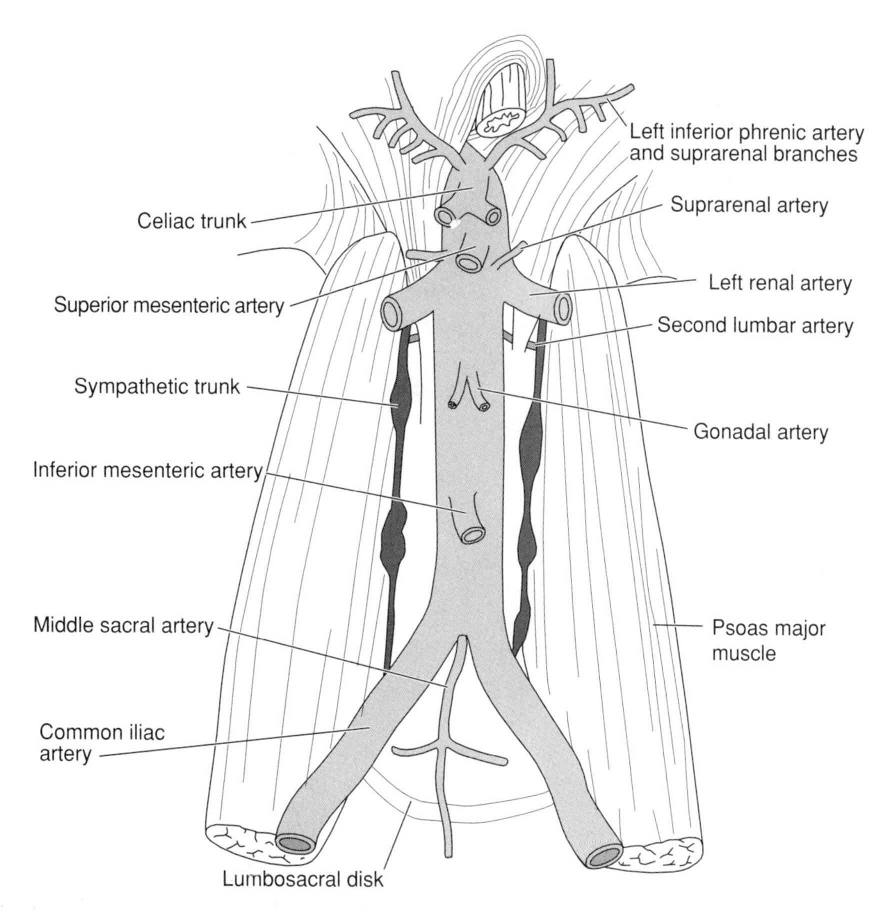


**Figure 6-4.** The abdominal part of the aorta and its branches.

the **right colic** and **ileocolic arteries** pass retroperitoneally to the right toward the cecum and ascending colon. The multiple **intestinal arteries** enter the mesentery and form a continuous system of arterial arcades along the jejunum and ileum.

**Renal arteries**. Arising at about the level of the second lumbar vertebra, these two large arteries pass across the crura of the diaphragm and the psoas major muscles in their transverse courses to the hila of the kidneys. The right artery is longer than the left and passes posterior to the inferior vena cava.

**Gonadal arteries (testicular or ovarian)**. These small arteries arise from the anterior aspect of the aorta 2.5 to 5.0 cm inferior to the renal arteries. The retroperitoneal testicular arteries descend obliquely toward the deep inguinal ring, where they become part of the spermatic cord. The ureteric branches arise as the testicular arteries cross the ureters. The abdominal course of the ovarian arteries is similar to that of the testicular arteries. As they reach the pelvic brim, the ovarian arteries swing medially and pass through the suspensory ligament of the ovary to the ovary.

**Inferior mesenteric**. Arising just below the third part of the duodenum (L3), this unpaired artery is retroperitoneal and passes obliquely downward and to the left. It has left colic, sigmoid, and superior rectal branches. These branches continue the anastomosing network along the large intestine.

**Lumbar arteries**. Usually four pairs, these are the segmental arteries of the abdomen. Each artery passes laterally and disappears into the muscles of the body wall. Each artery separates into a medial branch to the back and a lateral branch to the anterolateral body wall.

**TABLE 6-5. Areas and Structures Supplied by Branches of the Abdominal Aorta**

| Branch(es) of Aorta | Branches of Aortic Branch | Structures and Areas Supplied |
|---|---|---|
| Inferior phrenic | — | Diaphragm |
| Celiac | Left gastric | Stomach, esophagus |
| | Splenic | Pancreas, spleen, stomach |
| | Common hepatic | Liver, stomach, duodenum |
| Superior mesenteric | Inferior pancreaticoduodenal | Duodenum, pancreas |
| | Intestinal arteries | Duodenum, jejunum, ileum |
| | Middle colic | Transverse colon |
| | Right colic | Ascending colon |
| | Ileocolic | Ileum, cecum, appendix, ascending colon |
| Renal | — | Kidney, suprarenal glands, ureter |
| Gonadal (ovarian or testicular) | — | Ovary, uterus, testicle |
| Inferior mesenteric | Left colic | Descending colon |
| | Sigmoid and superior rectal | Sigmoid colon, rectum |
| Lumbar | — | Deep back, contents of vertebral canal, body wall (lateral and anterior) |
| Common iliac | Internal iliac | Pelvis, perineum, gluteal region |
| | External iliac | Lower limb, lower abdominal wall |
| Middle sacral | — | Posterior body wall |

**Common iliac arteries**. The large terminal branches of the aorta, the common iliacs, arise slightly to the left of the midline in front of the body of the fourth lumbar vertebra. The common iliacs divide into the internal and external iliac arteries just lateral to the sacral promontory. The internal iliac arteries enter the pelvis and the external iliacs continue into the thigh, where they become the femoral arteries.

**Middle sacral artery**. The true caudal continuation and termination of the aorta, this vessel arises from the posterior aspect of the aorta just above its bifurcation. It passes inferiorly on the ventral aspect of the lumbar vertebrae, enters the pelvis, and descends on the anterior surface of the sacrum and coccyx. It gives rise to parietal branches in its course.

## Portal Venous System

The portal vein receives blood from the entire gastrointestinal tract and related organs except the liver. This vein is formed posterior to the neck of the pancreas by the junction of the **superior mesenteric** and **splenic veins**. The **inferior mesenteric vein** also supplies portal blood; it joins the splenic vein before the portal is formed. After the portal vein is formed, it ascends through the hepatoduodenal ligament (free edge of the lesser omentum) to enter the liver through the porta hepatis.

## Systemic Venous System

**Inferior vena cava**. The inferior vena cava is formed by the union of the common iliac veins just inferior and slightly to the right of the bifurcation of the aorta. It ascends along the anterolateral aspects of the lumbar vertebrae to the right of the aorta and

**TABLE 6-6. Potential Anastomoses Between the Portal and Systemic Venous Systems**

| Portal Tributaries and Branches | Systemic Tributaries and Branches |
|---|---|
| Esophageal tributaries of left gastric | Esophageal tributaries of azygos system |
| Portal branches in liver | Diaphragmatic veins across bare area of liver<br>Paraumbilical veins in falciform ligament → systemic tributaries in abdominal wall |
| Portal tributaries in mesentery and mesocolon | Retroperitoneal tributaries communicating with renal, lumbar, and phrenic veins |
| Superior rectal tributaries of inferior mesenteric | Inferior rectal tributaries of internal pudendal → internal iliac vein |

passes through the caval opening (T8) in the diaphragm. Its tributaries are the right gonadal, renal, lumbar suprarenal, inferior phrenic and hepatic veins. The hepatic veins enter the inferior vena cava just inferior to the diaphragm and convey the blood delivered to the liver by the portal vein and hepatic arteries.

**Azygos system of veins.** The **azygos vein** begins in the abdomen to the right of the inferior vena cava, within the substance of the psoas major muscle. A similar but smaller and more irregular vessel, the **hemiazygos vein**, begins on the left. These vessels drain posterior body wall structures of the abdomen and then ascend into the thorax via the aortic opening in the diaphragm. In the thorax, the hemiazygos joins the azygos; the azygos then arches over the root of the right lung and joins the superior vena cava just before it reaches the right atrium.

**Venous anastomoses between the portal and systemic systems.** A number of potential alternate venous routes around the liver exist (i.e., anastomoses between the portal and systemic venous systems); these are summarized in Table 6-6.

# NERVE SUPPLY OF THE ABDOMEN

## Body Wall

The skin of the abdominal wall is innervated by the anterior and lateral cutaneous branches of intercostal nerves seven through eleven (T7–T11), the subcostal nerve (T12), and the iliohypogastric branch of the first lumbar nerve. The muscles of the abdominal wall and the parietal layer of peritoneum are also innervated segmentally by the same nerves.

## Viscera

Autonomic innervation of the viscera of the abdominal cavity is provided by sympathetic fibers from spinal cord segments T5 through L1 or L2 and parasympathetic fibers from the vagus nerve and spinal cord segments S2 through S4. Both the sympathetic and parasympathetic fibers are distributed in periarterial plexuses, which are associated with the abdominal aorta and its branches. The **aortic (preaortic) plexus** is a continuous plexus found on the ventral aspect of the aorta and associated with the three large, unpaired arteries (Fig. 6-5). The aortic plexus is divided into the **celiac**,

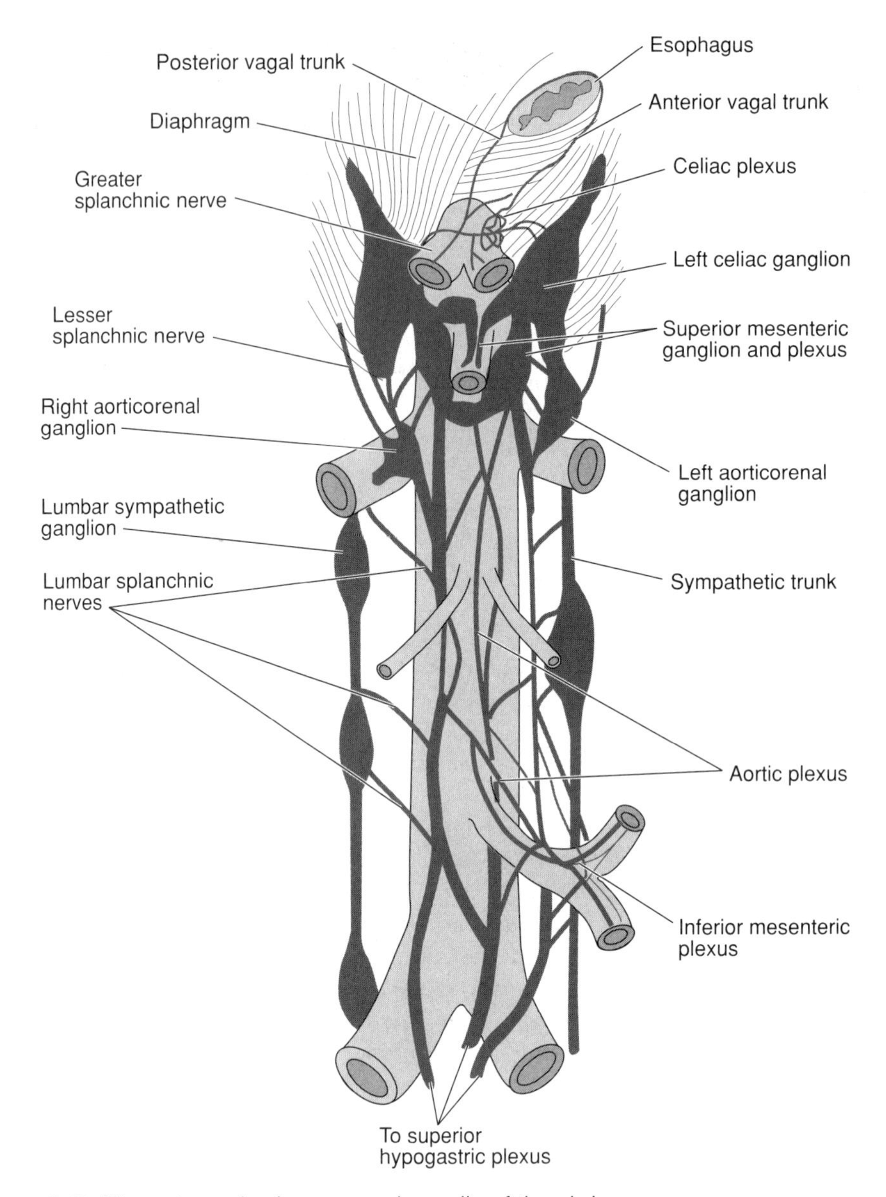


**Figure 6-5.** The autonomic plexuses and ganglia of the abdomen.

**superior mesenteric**, and **inferior mesenteric** plexuses, with the **intermesenteric plexus** connecting the superior and inferior mesenterics and the inferior mesenteric plexus continuing inferiorly as the **superior hypogastric plexus**. Continuations of these plexuses follow various other branches of the aorta to their respective destinations and bear the names of the arteries (e.g., the renal plexus).

The sympathetic input into the celiac and superior mesenteric plexuses is from the greater (T5–T10), lesser (T10–T12), and least (T12) **splanchnic nerves**. These nerves arise in the thorax and enter the abdomen by passing through the crura of the diaphragm.

The intermesenteric, inferior mesenteric, and superior hypogastric plexuses receive additional sympathetic fibers from the lumbar splanchnic nerves, which are branches of the lumbar portion of the sympathetic trunk. The fibers in the thoracic and lumbar splanchnic nerves are preganglionic. They synapse with postganglionic fibers in ganglia associated with the aortic plexuses. The largest and most demonstrable of these is the **celiac ganglion**.

**TABLE 6-7. Lymphatics of the Abdomen**

| Afferents From | Lymph Nodes | Efferents To |
|---|---|---|
| Common iliac nodes (pelvis, lower limb, perineum, gluteal region, lower abdominal wall), kidneys, suprarenals, gonads | Lumbar (aortic, caval) | Lumbar trunks → cisterna chyli → thoracic duct |
| Gastrointestinal tract and associated organs | Celiac and superior mesenteric | Intestinal trunk → cisterna chyli → thoracic duct |

**Vagal (parasympathetic) fibers** supply the organs of the abdomen and the gastrointestinal tract as far distally as the splenic flexure of the large intestine. The vagal fibers enter the abdomen through the esophageal hiatus as **anterior** and **posterior vagal trunks**. The trunks distribute on the anterior and posterior aspects of the stomach and enter the celiac plexus. The **pelvic splanchnics** (parasympathetic fibers from spinal cord segments S2–S4) provide direct retroperitoneal branches to the descending and sigmoid portions of the large intestine. The fibers in both the vagus and pelvic splanchnic nerves are preganglionic. They synapse with postganglionic fibers either in or very near the organ innervated.

Afferent fibers from the abdominal viscera reach the central nervous system through both the vagus and splanchnic nerves. Those sensory fibers in the vagus are concerned with muscular and secretory reflexes. Most pain fibers are thought to be carried in the splanchnic nerves.

## LYMPHATIC SYSTEM OF THE ABDOMEN

The **thoracic duct** conveys most of the lymph from the entire body below the diaphragm. This duct begins as the dilated cisterna chyli, which is formed by the union of the intestinal and right and left lumbar trunks. This cistern is found between the aorta and right crus of the diaphragm, posterior to the left renal vein on the ventral aspect of vertebra L2. The tributaries to the thoracic duct are summarized in Table 6-7.

## CROSS SECTIONS OF THE ABDOMEN

Figure 6-6 represents a section through the **upper portion of the first lumbar vertebra**. The liver occupies most of the right half of the abdomen, with the left lobe extending to the left and anterior to the stomach. The left and quadrate lobes are separated by the fissure containing the ligamentum teres, and the right lobe is most of the large portion on the right. The caudate lobe is posterior (and superior) to the porta hepatis. In the porta hepatis, the portal vein is posterior to both the common hepatic duct and the proper hepatic artery, with the duct to the right of the artery. The superior pole of the right kidney is related anteriorly to the right lobe of the liver and the right suprarenal gland. The right suprarenal gland is bounded anteriorly by the inferior vena cava, posteriorly by the right kidney, laterally by the right

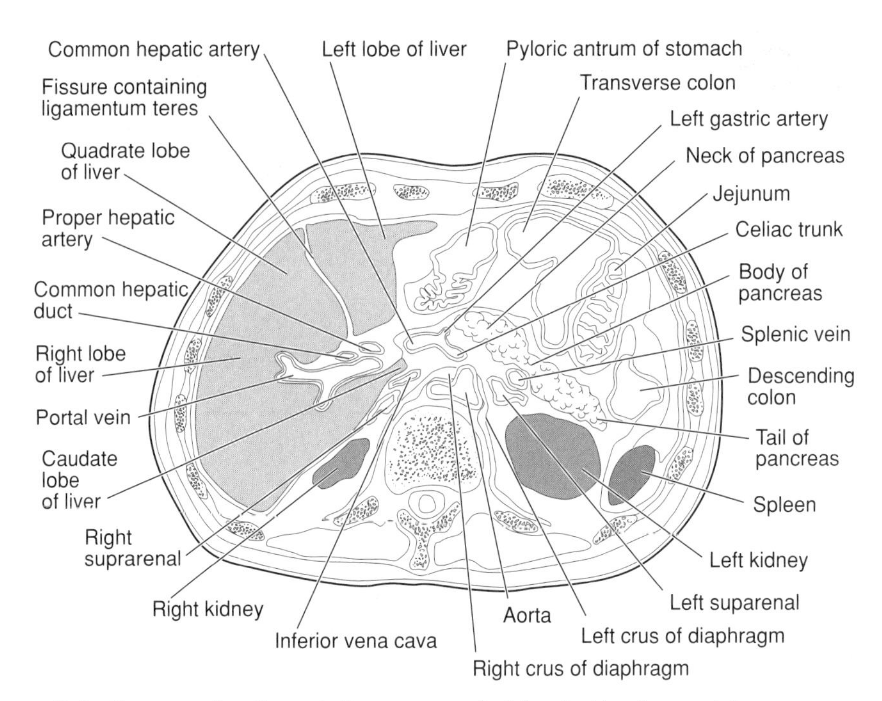


**Figure 6-6.** Cross section through the upper part of the first lumbar vertebra.

lobe of the liver, and medially by the right crus of the diaphragm. The celiac trunk passes anteriorly from the aorta and branches into the left gastric artery, which passes anteriorly, and the common hepatic artery, which passes to the right. The left kidney is related laterally to the spleen and anteriorly to the body and tail of the pancreas, the left suprarenal, and the descending colon. The left suprarenal is posteromedial to the splenic vein and body of the pancreas, anteromedial to the left kidney, and lateral to the left crus of the diaphragm. The splenic vein is positioned between the pancreas anteriorly and the left suprarenal posteriorly. The splenic vein is displaced (splenic vein sign) anteriorly by a mass in the left suprarenal and posteriorly by a pancreatic mass. The pyloric antrum of the stomach is posterior to the left lobe of the liver and anterior to the pancreas.

Figure 6-7 represents a **section that passes through the intervertebral disk between lumbar vertebrae one and two**. Only portions of the right and quadrate lobes of the liver are found at this level, with the gallbladder found in the groove between the two. The inferior vena cava is positioned slightly to the right of the vertebral column; the left renal vein passes to the right anterior to the aorta and posterior to the superior mesenteric artery on its way to the inferior vena cava. The portal vein is seen where it is formed, posterior to the neck of the pancreas and anterior to the uncinate process. The first and second parts of the duodenum are continuous, the second passing posteriorly from the first. The head of the pancreas is related to the first and second parts of the duodenum and has the common bile duct embedded in its posterior aspect. The transverse colon passes from right to left, crossing the duodenum and the pancreas. The descending colon descends along the posterior abdominal wall, passing anterior to the left kidney.

Figure 6-8 represents a section through the **third lumbar vertebra**. The inferior vena cava is anterior and somewhat to the right of the vertebral column; the aorta is anterior and somewhat to the left. The third part of the duodenum passes from right

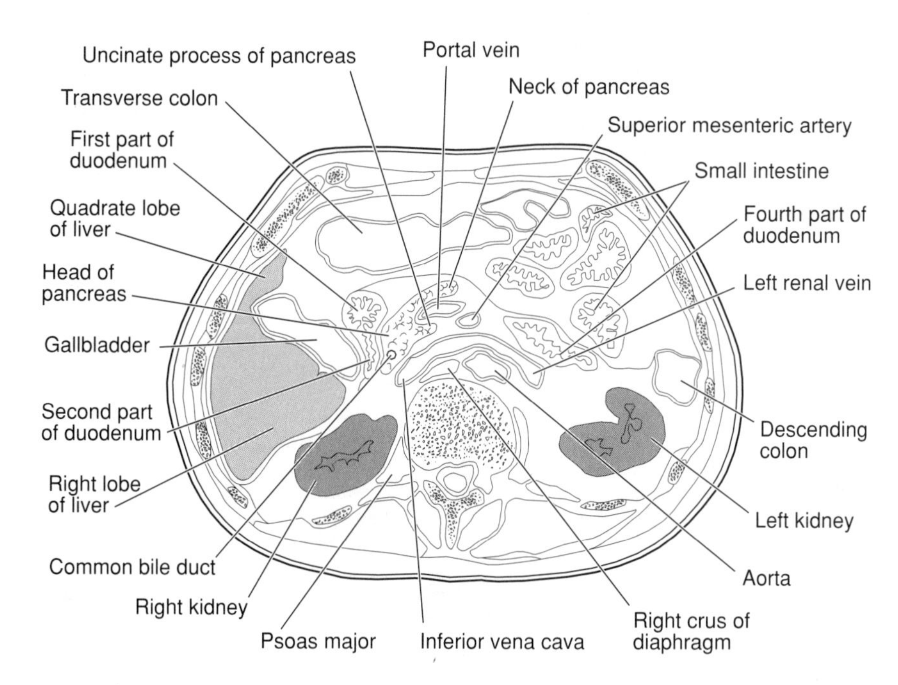


**Figure 6-7.** Cross section through the intervertebral disk between the first and second lumbar vertebrae.

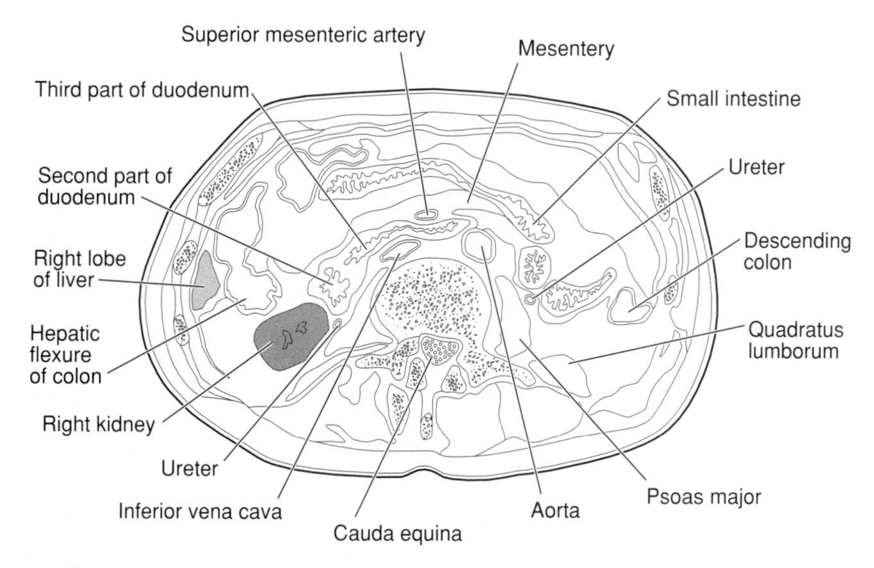


**Figure 6-8.** Cross section through the third lumbar vertebra.

to left across the vertebral column, anterior to the great vessels and posterior to the superior mesenteric artery. The hepatic flexure of the colon is related laterally to the inferior tip of the right lobe of the liver. The right kidney, which is still present at this level, is related anteriorly to the hepatic flexure of the colon and the second part of the duodenum. The ureters descend across the anterolateral aspects of the psoas major muscles. The superior mesenteric artery occupies the root of the mesentery.

Chapter 7

# Pelvis and Perineum

## ORGANIZATION OF THE PELVIS AND PERINEUM

### Pelvis

The bones and articulations that form the pelvis are described in Chapter 2. Discussion here includes a classification of pelvic variability, a definition of the pelvic cavity, and a description of the various planes of the pelvis.

The **gynecoid pelvis** is regarded as the characteristic female pelvis. It is distinguished by an oval inlet with the transverse diameter exceeding the anteroposterior (AP) diameter. This pelvis is shallow with straight walls. The ischial spines are not prominent, and the subpubic arch is wide. It also has the appearance of being flattened from top to bottom, making it shorter and wider than the male pelvis. The **android pelvis** (male) has a heart-shaped inlet, is longer and heavier, and the angle of the subpubic arch is more acute. The **anthropoid pelvis** has an oval inlet with the long axis along the AP diameter. The **platypelloid pelvis** has a flattened oval inlet, which is caused by a marked reduction in the AP diameter.

**Pelvic cavity**. The **pelvic cavity proper (minor or true pelvis)** is that part of the abdominopelvic cavity that is below the pelvic inlet and above the pelvic diaphragm or pelvic floor. The **inlet** is defined by the sacral promontory, arcuate line of the ilium, pecten pubis, and the upper aspect of the symphysis pubis. The area above this plane is the major, or false, pelvis and is the other part of the abdominopelvic cavity. The **floor** of the pelvis is a muscular sling composed of the levator ani and coccygeus muscles. This trough-shaped sling slopes inferiorly from lateral to medial and from posterior to anterior; its lateral attachment extends along a tendinous arch from the symphysis pubis to the ischial spine. The **levator ani muscle** attaches to this arch and, after descending toward the midline, attaches to the levator ani of the opposite side along a median raphe. The **coccygeus** fills the gap between the ischial spine and the sacrum and coccyx. The **lateral wall** of the pelvic cavity consists of the obturator internus and piriformis muscles along with the corresponding portions of the hip bone.

**Pelvic inlet**. The plane of the pelvic inlet has the greatest dimensions. In the female, the transverse diameter exceeds the AP diameter by a considerable amount. The distance between the sacral promontory and the uppermost aspect of the symphysis pubis is the AP diameter of the inlet or the **conjugate vera** (true conjugate). Not strictly part of the pelvic inlet, but important obstetrically, is the **obstetric conjugate**, the distance between the promontory and the most posterior aspect of the symphysis pubis, which represents the true distance between the symphysis and the promontory. The **diagonal conjugate** is the distance between the promontory and the inferior aspect of the symphysis pubis. The **transverse diameter** of the inlet is the greatest distance between the arcuate lines.

**Pelvic outlet**. The **plane of the pelvic outlet** is defined by the inferior aspect of the symphysis pubis, the ischiopubic rami, the ischial tuberosities, the sacrotuberous ligaments, and the tip of the sacrum. The **AP diameter** extends from the inferior aspect of the symphysis pubis to the tip of the sacrum; the transverse diameter is the distance between the inner edges of the ischial tuberosities.

**Midpelvis**. The plane of the midpelvis is the **plane of least dimensions**. The AP diameter of this plane extends from the inferior aspect of the symphysis to the sacrum at the level of the ischial spines. The **transverse diameter** of the midpelvis is the distance between the ischial spines, which is the smallest diameter of the pelvis.

## Perineum

The perineum is the region of the pelvic outlet below the pelvic floor (Fig. 7-1). This diamond-shaped area can be separated into two triangular areas by an imaginary line between the two ischial tuberosities, the anterior urogenital and the posterior anal triangles. The roof of the perineum is the floor of the pelvis, that is, the pelvic sling formed by the levator ani and coccygeus muscles.

**Urogenital triangle**. In the urogenital triangle, the urogenital diaphragm is a horizontal musculofascial shelf that stretches between the ischiopubic rami (Fig. 7-2). The superior and inferior layers of fascia, along with the ischiopubic rami, form the closed **deep perineal space** or **pouch**. The contents of this space are listed in Table 7-1.

The **superficial perineal space** is inferior or superficial to the inferior fascia of the urogenital diaphragm. It is limited externally or superficially by **Colles' fascia**, which attaches to the posterior edge of the urogenital diaphragm and is continuous with the membranous (fibrous) layer of subcutaneous tissue (**Scarpa's fascia**) of the abdomen. This limiting layer also attaches to the ischiopubic ramus and the fascia lata of the thigh so that urine from a ruptured urethra or blood from hemorrhage may extravasate up into the abdominal wall (but not into the thigh) from the superficial perineal space. In the female, the urethral opening is between the clitoris and external vaginal opening, approximately 2.5 cm posterior to the clitoris. The contents of the superficial perineal space are summarized in Table 7-1.

**Anal triangle**. This triangle is posterior to the ischial tuberosities and bounded superiorly by the pelvic sling, inferiorly by the skin, and laterally by the obturator internus muscle and the os coxa. This area or fossa is the **ischiorectal fossa**; in fact, this fossa extends beyond the boundaries of the anal triangle both anteriorly and posteriorly. The fossa is filled with fat and is wedge-shaped, with the apex directed superiorly and the base inferiorly. It has a **posterior recess** that extends posterolaterally under the inferior margin of the gluteus maximus muscle, and an **anterior recess** that extends into the urogenital triangle above the urogenital diaphragm. The **pudendal (Alcock's) canal** is a slit in the obturator internus fascia along the lateral wall of the fossa that conveys the internal pudendal vessels and the pudendal nerve anteriorly as they pass into the urogenital triangle. The inferior rectal arteries and nerves branch from the parent structures in the pudendal canal and pass through the ischiorectal fossa toward the rectum and anal canal.

# VISCERA OF THE PELVIS AND PERINEUM

## Gastrointestinal Tract

**Sigmoid colon**. The sigmoid colon begins at the pelvic rim, descends in the left side of the pelvis, and bends on itself to join the rectum in the midline. It is supported by the sigmoid mesocolon. The sigmoid and its mesocolon are extremely variable in length, exceeding by far the variability in other parts of the gastrointestinal system.

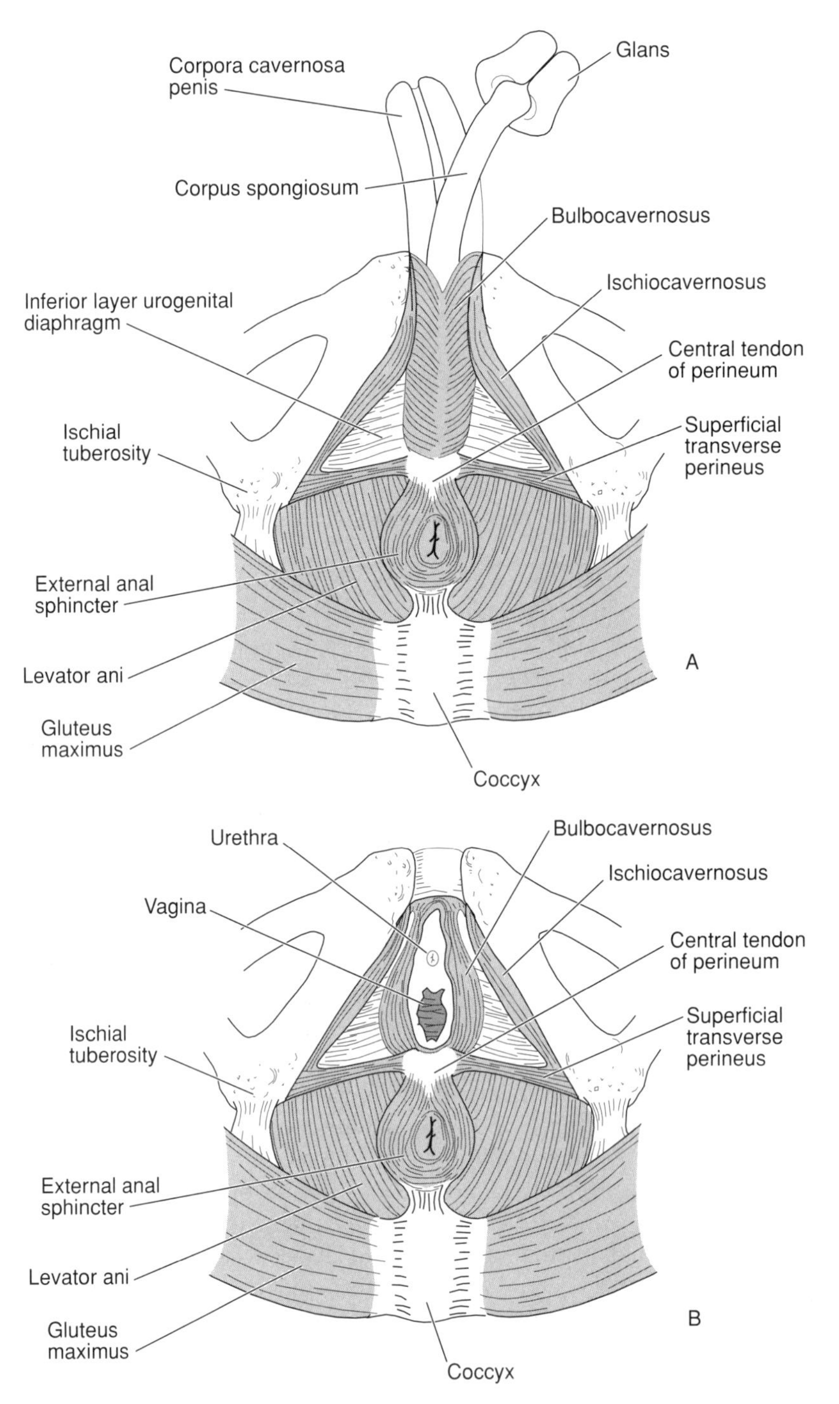


**Figure 7-1.** Inferior view of the perineum in the male **(A)** and female **(B)**.

**Rectum**. The rectum is that portion of the bowel below the midsacral region where the sigmoid mesocolon ceases. The lowest part of the infraperitoneal portion presents a dilated ampulla. Anteriorly, the upper two-thirds of the rectum are in contact with the coils of the ileum. In the male, the lower third is related anteriorly to the trigone of the bladder, the seminal vesicles, the ductus deferens, and the prostate. In the female, the lower third is in contact anteriorly with the vagina and the cervix. Posteriorly, it is related to the sacrum in both sexes. The upper third of the rectum is covered anteriorly and laterally by peritoneum; the middle third, only anteriorly; and the lower third passes below the peritoneum.

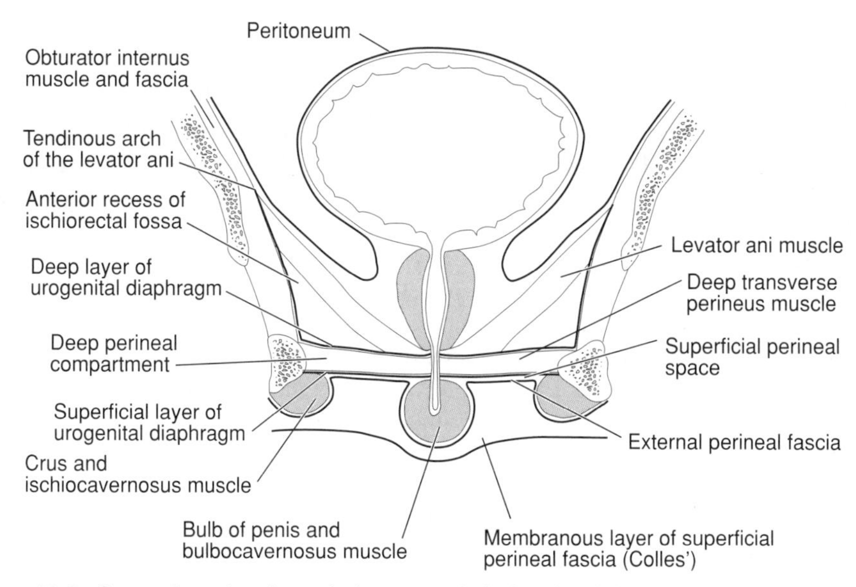


**Figure 7-2.** Coronal section through the urogenital triangle of the male perineum.

**Anal canal**. The anal canal, which is sometimes called the *second portion of the rectum,* is 2.5 to 3.5 cm in length. As it turns dorsally, making a right angle as it passes through the pelvic floor to the anus, it is surrounded by **internal** and **external sphincters**. The following structures can be palpated during a rectal examination in the normal male: the anorectal ring; anteriorly, the prostate; posteriorly, the coccyx and sacrum; and laterally, the ischiorectal fossa and ischial spines. The normal female presents the same structures with the exception of anteriorly, where the perineal body and the cervix of the uterus are palpable.

## TABLE 7-1. Contents of the Superficial and Deep Perineal Spaces

| | Male | Female |
|---|---|---|
| **Superficial space** | Superficial transverse perineal muscle<br>Crura of corpora cavernosa<br>Ischiocavernosus muscles<br>Corpus spongiosum and bulb of penis<br>Bulbospongiosus muscle | Superficial transverse perineal muscle<br>Crura of clitoris<br>Ischiocavernosus muscles<br>Bulbus vestibuli<br>Bulbospongiosus muscle<br>Greater vestibular (Bartholin's) glands |
| **Deep space** | Deep transverse perineal muscle<br>Sphincter of membranous urethra<br>Membranous urethra<br>Bulbourethral (Cowper's) glands | Deep transverse perineal muscle<br>Sphincter of membranous urethra<br>Membranous urethra |

## Urinary System

**Ureter**. The **pelvic portion** of the ureter passes in front of the sacroiliac joint and medial to the internal iliac artery. In the male, it passes posterior and inferior to the ductus deferens. In the female, it passes inferior to the uterine artery and lateral to the cervix of the uterus. The **vesical portion** runs obliquely downward and medially through the bladder about 20 to 25 mm from its counterpart.

**Bladder**. The empty bladder is posterior to the pubic symphysis and totally within the pelvic cavity below the pelvic inlet. The fully distended bladder of the adult projects well into the abdominal cavity. The bladder is extraperitoneal, with only its superior and posterosuperior surfaces covered with peritoneum. Laterally, it is related to the levator ani and the obturator internus muscles. In the male, the bladder is related superiorly to coils of small intestine and the sigmoid colon; posteriorly to the rectovesical pouch, the rectum, the seminal vesicles, and termination of the vas deferens; and inferiorly to the prostate gland. In the female, it is related superiorly to coils of small intestine and the body of the uterus, posteriorly to the vagina and supravaginal portion of the cervix, and inferiorly to the pelvic fascia and the urogenital diaphragm.

**Male urethra**. The male urethra extends from the bladder to the glans penis, traversing the prostate gland, urogenital diaphragm, and the full length of the corpus spongiosum. The **prostatic portion** contains the numerous small openings of the prostatic ducts and the crista urethralis, with the colliculus seminalis containing the openings of the ejaculatory ducts. As the urethra passes through the urogenital diaphragm (**membranous part**), it is somewhat narrowed and is surrounded by the sphincter urethra. In the **cavernous portion**, the urethra is dilated at the openings of the Cowper's glands and also terminally at the fossa navicularis.

**Female urethra**. The female urethra is short, passing only through the urogenital diaphragm, where it is surrounded by the sphincter urethrae. It ends shortly thereafter, opening about 25 mm posterior to the clitoris.

## Male Reproductive System

**Testis and epididymis**. The testis and the epididymis lie in the scrotum and are separated from those of the opposite side by the scrotal septum. The epididymis lies along the posterior border of the testis, is enlarged above to form a head, and tapers to a body and small tail below. It is formed predominantly by the greatly contorted duct of the epididymis, which empties into the beginning of the ductus deferens.

**Ductus deferens**. The ductus (vas) deferens ascends toward the inguinal canal, which it traverses. It leaves the spermatic cord through the internal inguinal ring and passes downward and posteriorly over the lateral surface of the bladder and then medially to the ureter. It then penetrates the prostate gland and opens into the prostatic portion of the urethra through the ejaculatory duct. Just before it enters the prostate, it is joined by the club-shaped seminal vesicles.

**Prostate gland**. The prostate gland surrounds the urethra between the inferior surface of the bladder and the superior surface of the urogenital diaphragm. It is related anteriorly to the symphysis pubis, laterally to the levator ani muscles, inferiorly to the urogenital diaphragm, and posteriorly to the rectum. The posterior lobe of the prostate is readily palpable via rectal examination.

## Female Reproductive System

**Ovary**. The ovaries lie against the lateral pelvic walls just below the pelvic inlet and posteroinferior to the lateral aspects of the uterine tubes. Each ovary is enclosed in a **mesovarium**, a posterior reflection of peritoneum from the broad ligament.

The ovary is suspended from the lateral pelvic wall by the **suspensory ligament** of the ovary, through which the ovarian vessels, nerves, and lymphatics pass to the gland. Each ovary is connected to the uterus just below the uterine tube by the **ovarian ligament**.

**Uterus**. The uterus is a thick-walled muscular organ that consists of a fundus, body, and cervix. The small cavity of the uterus is continuous superolaterally with the narrow lumina of the uterine tubes and inferiorly with the cavity of the vagina. The **fundus** is the superior domed portion that projects above the cavity of the body. The **body** is the major portion of the uterus, and the **cervix** is the inferior portion, part of which projects into the vagina. The uterine cavity is largest within the body. It narrows abruptly at the junction of the body and cervix to form the **internal os**. The lumen of the cervix (cervical canal) is narrow and ends inferiorly as the narrow **external os**, which is readily palpated during a rectal examination. The normally positioned uterus rests on the posterosuperior aspect of the bladder so that the uterovesical pouch is usually empty. The rectouterine pouch (of Douglas) separates the uterus from the rectum posteriorly. Laterally, the uterus is related to the broad ligament and the ureter, the latter passing just lateral to the supravaginal portion of the cervix and inferior to the uterine artery. The normal uterus is both anteflexed and anteverted. **Anteflexion** is a forward bend within the uterus itself at the level of the internal os. **Anteversion** is a forward bend of about 90 degrees at the junction of the uterus and the vagina. This places the uterus in approximately the horizontal plane with its anterior surface resting on the posterosuperior surface of the bladder.

**Uterine tubes**. The uterine (fallopian) tubes extend laterally from the superolateral aspects of the uterus. Each tube consists of a narrow-lumened isthmus, a dilated and long ampulla, and the terminal infundibulum, which is composed of numerous finger-like fimbriae. The tubes curve posteriorly near the lateral pelvic walls, where their fimbriae partially cover the ovaries. The uterine tubes are the most superior structures in the broad ligament.

**Vagina**. The vagina extends from above the inferior aspect of the cervix of the uterus to its external opening in the vestibule. From above, it is inclined downward and anteriorly as it passes through the pelvic floor and the urogenital diaphragm. The upper portion of the vagina, which surrounds the inferior part of the cervix, is divided into anterior, lateral, and posterior **fornices**. The vagina is related anteriorly to the base of the bladder and the urethra; posteriorly, through the very thin wall of the posterior fornix, to the rectouterine pouch, rectum, and anal canal; and laterally, to the levator ani muscle and the ureter, which passes near the lateral fornix.

In a vaginal examination, the urethra, bladder, and symphysis pubis are palpable anteriorly; the rectum and rectouterine pouch posteriorly; the ovary, uterine tube, and lateral pelvic wall laterally; and the cervix in the apex of the vagina.

**Broad ligament**. The broad ligament is a reflection of peritoneum, which passes over the uterus and related structures from front to back and extends laterally to the lateral pelvic walls. As such, it forms a curtain across the pelvis from side-to-side; the broad ligament has an anterior and a posterior layer. Lateral to the uterus, the most superior structure in the broad ligament is the **uterine tube**; below and behind that is the ovarian ligament; and below and anterior is the **round ligament**. The **mesovarium** is the reflection of the posterior layer that suspends the ovarian ligament and ovary. That part of the broad ligament above the mesovarium that suspends the uterine tube is called the **mesosalpinx**. That part of the broad ligament below the mesovarium is the mesometrium. At the base of the broad ligament posteriorly, the posterior layer of **mesometrium** is elevated over the underlying uterosacral ligament as the **rectouterine fold**. The extraperitoneal connective tissue found between the layers of the broad ligament is the parametrium. The **parametrium** in the base of the broad ligament is thickened and forms the **cardinal ligaments**.

## BLOOD SUPPLY OF THE PELVIS AND PERINEUM

The blood supply to the pelvic viscera and to the perineum is provided by branches of the **internal iliac artery**. The **umbilical artery** gives rise to the artery of the ductus deferens and the superior vesical artery, which supplies the bladder. The **inferior vesical artery** also supplies the bladder as well as the prostate and seminal vesicles in the male. The **uterine artery** (homologue to the artery of the ductus deferens) usually arises separately and passes medially to supply the uterus, uterine tube, and upper part of the vagina. The vaginal artery supplies most of the vagina. The middle rectal artery supplies the middle portion of the rectum and anastomoses with the superior and inferior rectal arteries. The **internal pudendal artery** exits from the pelvis through the greater sciatic foramen, passes around the ischial spine, and then enters the perineum through the lesser sciatic foramen. It supplies the somatic and visceral structures of the perineum. Other branches of the internal iliac artery are the **superior** and **inferior gluteal arteries**, which supply the gluteal region; the **iliolumbar** and **lateral sacral arteries**, which supply the posterior body wall; and the obturator, which passes into the medial thigh.

## INNERVATION OF THE PELVIS AND PERINEUM

The nerve supply to the pelvic viscera is provided by the **inferior hypogastric** or **pelvic plexuses**. These plexuses (right and left) are inferior continuations of the hypogastric nerves and spread over the lateral aspects of the pelvic viscera. The sympathetic and parasympathetic inputs are from the lumbar and pelvic splanchnic nerves, respectively.

The nerve supply to the perineum is provided primarily by the **pudendal nerve** (S2–S4). Its branches are the inferior rectal, perineal, and posterior scrotal (labial) nerves. These branches supply all of the muscles and most of the skin of the perineum. The skin of the anterior part of the perineum is supplied by the **anterior scrotal (labial) branches** of the ilioinguinal nerve. Autonomics to the perineum are also provided by the pelvic plexuses.

## LYMPHATICS OF THE PELVIS AND PERINEUM

The lymphatic drainage of the pelvis and perineum is summarized in Table 7-2.

**TABLE 7-2. Lymphatics of the Pelvis and Perineum**

| Afferents From | Lymph Nodes | Efferents To |
|---|---|---|
| Pelvic viscera, perineum, gluteal region | Sacral | Common iliac nodes → lumbar trunks → cisterna chyli → thoracic duct |
| Pelvic viscera, perineum, buttock | Internal iliac | Same as above |
| Pelvic viscera, abdominal wall below umbilicus, lower limb | External iliac | Same as above |
| Sacral, internal iliac and external iliac nodes | Common iliac | Lumbar trunks |

# PART II

# Embryology

# Chapter 8

## Developmental Anatomy

Development begins with the fusion of male and female sex cells called **gametes**. These cells are produced in the gonads and have half the number of chromosomes that regular body cells have. To achieve this haploid number of chromosomes, the cells in the gonad that are to become gametes undergo meiosis, or a reduction division. (The process of meiosis is discussed in Chapter 19.) During ejaculation, the male deposits millions of spermatozoa in the female reproductive tract. The spermatozoa must migrate through the female reproductive tract to find and fertilize the ovum that has been released from the ovary.

Chapter 9

# Early Development

## FERTILIZATION

At **ovulation** a secondary oocyte, zona pellucida, and corona radiata of follicle cells are discharged from the ovary and drawn into the infundibulum of the uterine tube. In the uterine tube, a **spermatozoan** can penetrate the zona pellucida and secondary oocyte. In this process, acrosome enzymes aid in penetration of the corona radiata and zone pellucida, and a cytoplasmic response of the secondary oocyte leads to a zonal reaction that prohibits penetration by other spermatozoa. The union of the spermatozoan and secondary oocyte in the process of fertilization brings about the following major physical consequences: (1) reactivation of the secondary oocyte, (2) completion of the second meiotic division with formation of the second polar body, (3) establishment of a zygote (fertilized ovum) with the diploid number (46) of chromosomes, (4) establishment of the mitotic spindle for the first cleavage division, and (5) the determination of the gender of the new individual.

## CLEAVAGE AND BLASTODERMIC VESICLE FORMATION

During **cleavage**, a series of mitoses occur in the zygote, which result in successive 2-, 4-, 8-, and 16-cell stages. These divisions take place over a period of approximately 3 days as the developing conceptus passes down the uterine tube. At about the time the 16-cell morula reaches the uterine cavity, fluid penetrates between some of the cells and produces a cavity in the solid ball of cells. The conceptus is now called a **blastodermic vesicle (blastocyst)**. It consists of an outer layer of cells called the **trophoblast**, a cavity of the blastodermic vesicle (blastocoele), and an inner cell mass. After 3 days in the uterine cavity, the conceptus "hatches" from the zonal enclosure, and the sticky trophoblastic cells adhere to the endometrium.

## ESTABLISHMENT OF ECTODERM, ENTODERM, AND MESODERM

In the eighth day of development, the inner cell mass cavitates to form an ectodermally lined amniotic cavity. The **ectoderm** of the embryonic disk eventually gives rise to the neural tube, neural crest, and epidermis. An inner **entodermal layer** of cells also differentiates from the inner cell mass. These cells proliferate to form the yolk sac. The dorsal portion of the yolk sac later becomes incorporated into the embryo as the primitive gut. **Embryonic mesoderm** arises from an elongated mass of cells called the **primitive streak**. The mesodermal cells turn inward along the midline and move laterally, insinuating themselves between the ectoderm and entoderm. In its forward, caudal, and lateral movements, the embryonic mesoderm eventually joins the extraembryonic mesoderm, which arises from the **trophoblast**. The notochord arises as a midline forward growth of cells from the primitive **(Hensen's) node**.

In the third week of development, the embryonic mesoderm has differentiated into paraxial (i.e., somite, dorsal), intermediate, and lateral mesoderm. The **paraxial mesoderm** develops further into paired somites whose sclerotome eventually give rise to vertebrae and ribs, whose myotome gives rise to skeletal muscle, and whose dermatome develops into connective tissues. The **intermediate mesoderm** differentiates into much of the urogenital system. The **lateral mesoderm**, like the extraembryonic mesoderm, splits to form a coelom. The lateral mesoderm adjacent to the ectoderm is somatic mesoderm, whereas that next to entoderm is splanchnic mesoderm. Somatic mesoderm later gives rise to body wall tissues. Splanchnic mesoderm further differentiates into the cardiovascular system and into smooth muscle and connective tissues in the walls of most visceral structures, mesenteries, and the spleen.

## DEVELOPMENT OF THE PLACENTA

After the attachment of the blastodermic vesicle to the uterus at about the sixth postfertilization day, the **trophoblast** proliferates rapidly, and the **conceptus** begins to implant into the compacta layer of the endometrium. The conceptus is completely embedded in the uterine stroma by the eleventh postfertilization day. In the rapid proliferation of cytotrophoblastic cells, fusion of the outer cells forms an outer **syncytial trophoblast** over the single inner layer of **cytotrophoblast**. The coalescence of lacunae formed in the syncytial trophoblast leads to the formation of primary stem villi and intervillous spaces.

The **villi** are most extensive in the trophoblast that faces the deeper layers of the endometrium. Most of the nutriments in this region are being supplied to the trophoblast from invaded uterine glands and blood vessels. The primary stem villi consist of a core of cytotrophoblast surrounded by syncytial trophoblast. Later, mesoderm invades these villi, and they become secondary stem villi containing a core of connective tissue. By the end of the third week, blood vessels begin to form in the secondary villi, and they are designated as tertiary villi. With the vascularization of the trophoblast, it is called the **chorion**. That part of the chorion that is the deepest in the uterine wall becomes the chorion frondosum portion of the placenta; the rest loses its villi and is called the **chorion laeve**. An anchoring villus and its free-floating villi constitute a cotyledon. Septa formed from the cytotrophoblastic coating and decidual tissue of the intervillous space project from the decidua and incompletely separate the cotyledons from each other. The functional layer of the endometrium deep to the chorion frondosum is the decidua basalis; that adjacent to the chorion laeve is the decidua capsularis; and the rest is the decidua parietalis. As the embryo enlarges, the uterine cavity is obliterated, and the decidua capsularis and decidua parietalis fuse into a much compressed layer.

After the birth of the newborn, the decidual layers, placenta, chorion laeve, and amnion are discharged as the afterbirth. The **chorion frondosum** consists of a chorionic plate, off which anchoring villi arise and attach to the endometrium. Free villi extend from the anchoring villi into the intervillous spaces.

During the first third of pregnancy, the villi have cores of fetal connective tissue containing fetal capillaries with nucleated red blood cells (RBCs). This core of tissue is covered by an inner cytotrophoblastic and an outer syncytial trophoblastic epithelial layer. Later in pregnancy, the fetal vessels contain nonnucleated RBCs, the cytotrophoblast disappears, and the syncytial trophoblast is thin, except for clumps of syncytial knots where the nuclei are located.

In late pregnancy, an acidophilic fibrinoid material accumulates over the syncytial trophoblast. The "placental barrier" in the late placenta consists of the syncytial trophoblast, the endothelium of the fetal vessels, and the intervening basal laminae of these epithelia, which are fused into one basal lamina in the thinnest portions of the barrier.

In the first trimester of pregnancy, the cytotrophoblast and fetal connective tissue are added layers in this barrier. The cells of the cytotrophoblast produce the syncytial trophoblast and probably gonadotropin-releasing hormone. The syncytial trophoblast cells produce estrogen, progesterone, human chorionic gonadotropin, human placental lactogen or human chorionic somatomammotropin, and human chorionic thyrotropin.

# Chapter 10

# Musculoskeletal System

## DEVELOPMENT OF VERTEBRAE AND RIBS

At the end of the second postfertilization week, the primitive streak gives rise to cells that migrate laterally between the ectoderm and entoderm, forming the intraembryonic mesoderm. At approximately the same time, the **notochord** arises as a cranial midline migration from the primitive node. As development progresses, the intraembryonic mesoderm adjacent to the notochord thickens into longitudinal masses called the **paraxial mesoderm**.

From the twenty-first to thirtieth days, the paraxial mesoderm differentiates into 42 to 44 paired segments called **somites**. This craniocaudal development of somites gives rise to 4 occipital, 8 cervical, 12 thoracic, 5 lumbar, 5 sacral, and 8 to 10 coccygeal somites. Each somite further differentiates so that three distinct cellular regions are apparent. The ventromedial region, **sclerotome**, eventually gives rise to supportive skeletal structures (e.g., vertebrae and ribs); the dorsomedial part, **myotome**, forms the skeletal muscles; and the dorsolateral portion, **dermatome**, gives rise to the dermis of the skin and subcutaneous tissue.

During the fourth week, the sclerotomic mesenchymal mass of each somite begins to migrate toward the midline to become aggregated about the notochord. In this migration, cells of the caudal half of each somite shift caudally to meet the cranially migrating cranial half of the adjacent sclerotome. From each of these joined masses, **mesenchymal processes** grow dorsally around the neural tube to form the **neural arches of the vertebrae** and also give rise to **rib primordia**. Because a vertebra develops from parts of two pairs of adjacent sclerotomes, the original intersegmental arteries will come to pass across the middle of the vertebral bodies. The segmental spinal nerves to the myotomes will come to lie at the level of the intervertebral disks and the myotomes. The notochord degenerates in the region of the vertebral bodies, but persists in the center of the intervertebral disk as the **nucleus pulposus**.

In the cervical region, the migration of sclerotome accounts for the formation of seven cervical vertebrae from eight somites. This is because the cranial half of the first sclerotome becomes part of the occipital bone, whereas the caudal half of the eighth sclerotome becomes part of the first thoracic vertebra. Thus, the first cervical nerve passes between the occipital bone and first cervical vertebra, and the eighth cervical nerve emerges between the seventh cervical and first thoracic vertebrae.

By the seventh week, separate **chondrification centers** develop in the bodies and the lateral half of each neural arch, and these subsequently fuse. Later, ossification centers develop in the vertebral bodies, in each half of the neural arch, and in each rib. These remain as separate centers throughout fetal life. The rib primordia give rise to ribs in the thoracic region, transverse processes in the lumbar region, parts of the transverse processes in the cervical region, and the alae of the sacrum. Excessive growth of the rib primordia can lead to cervical and lumbar ribs.

In **spondylolisthesis**, there is usually a defect in the formation of the pedicles caused by nonunion of ossification centers. In this condition, the spine, laminae, and inferior articular processes of the affected lower lumbar vertebra stay in place while the body migrates anteriorly with respect to the vertebra below it.

In **spina bifida**, the neural arches fail to unite properly in the formation of the spinous process.

## MORPHOGENESIS OF THE SKELETAL MUSCULATURE

In the trunk and extremities, myotomes divide into dorsal epaxial and ventral hypaxial condensations of mesenchyme. Dorsal and ventral rami develop from the segmental spinal nerves and innervate the epaxial and hypaxial portions, respectively. The epaxial masses give rise to the deep muscles of the back. The hypaxial masses develop into anterior and lateral body wall muscles of the cervical and thoracolumbar regions. Muscles of the extremities and those that attach the limbs to the trunk may arise from local somatic lateral mesoderm but are innervated by the ventral rami of spinal nerves. Subsequent migrations of segmental myoblasts, trailing their respective nerves, lead to the formation of complex nerve fiber plexuses from successive spinal cord levels.

In addition to migration, five other basic processes occur in the **establishment of muscles**: (1) **fusions of portions of successive myotomes** (e.g., erector spinae); (2) **change from the original cephalocaudal direction of the fibers** (e.g., transversus abdominis); (3) **longitudinal splitting** of a myotonic mass to form more than one muscle (e.g., rhomboideus major and minor); (4) **tangential splitting** (e.g., intercostals); and (5) **degeneration of parts or all of a myotome** with conversion of the degenerated part to connective tissue (e.g., serratus posterior inferior and superior).

**Muscles of the head** develop from eye (preoptic) and tongue somites and from branchial arch mesenchyme. Some authorities believe that the neural crest is involved in head muscle development.

# Chapter 11

## *Spinal Cord and Spinal Nerves*

The **central nervous system** appears early in the third embryonic week of development as a thickened neural (medullary) plate of ectoderm. This plate is elongated, wider cephalically than caudally, and is located rostral to the primitive node of Hensen. It is continuous laterally with ectoderm that gives rise to the epidermis of the skin. With further development, the lateral edges of the plate elevate to form neural folds that close the neural groove to form the **neural tube**.

The closure of the neural groove begins at the fourth somite and progresses cephalically and caudally, with the anterior and posterior neuropores closing by the twenty-eighth day. Ectoderm arising at the junction of neural ectoderm with general surface ectoderm becomes segmentally arranged as the neural crest material lying dorsolateral to the neural tube.

The cephalic enlargement of the neural tube differentiates into the brain and gives rise to motor components of **cranial nerves**. The caudal portion becomes the spinal cord and also gives rise to the ventral roots of spinal nerves. The neural crest gives rise to sensory neurons comprising the dorsal root ganglia and some sensory cranial nerve ganglia, postganglionic autonomic ganglia of cranial and spinal nerves, Schwann cells, satellite cells, parenchyma of the adrenal medulla, pigment cells, and cartilage and bone cells in the head region.

The early neural tube consists of a neuroepithelium that differentiates into neuroblasts and spongioblasts (glioblasts). **Neuroblasts** differentiate into neurons whose cell bodies are localized into a mantle layer and whose axons contribute to a more peripheral marginal layer. Some of these axons ascend or descend in the marginal layer and become the association fibers of the tracts of the white matter. Others leave the white matter and become motor (efferent) nerve fibers of the ventral roots and spinal nerves. **Spongioblasts** lining the central canal differentiate into ependymal cells, whereas others migrate into the marginal and mantle layers and become astrocytes and oligodendroglia.

# Chapter 12

# Cardiovascular System

## EARLY DEVELOPMENT

The heart and major vessels arise from blood islands of hemangioblastic tissue derived from mesoderm. The **heart** begins to form in the third embryonic week. One embryonic and two extraembryonic (umbilical and vitelline) vascular circuits are completed by the end of the first month of development.

## EARLY DEVELOPMENT OF THE HEART AND VASCULAR CIRCUITS

Two **endocardial tubes** are formed deep to the epimyocardial (myocardial mantle) thickening of splanchnic mesoderm by the coalescence of blood islands. These tubes run longitudinally and are deep to the horseshoe-shaped prospective pericardial cavity. With the lateral folding and forward growth of the embryo, the endocardial tubes are shifted ventrocaudally and fuse in the midline. The adjacent epimyocardium fuses in the midline around the fused endocardial tubes, forming a single hollow heart tube that is suspended in the primitive pericardial cavity by the dorsal mesocardium.

The coalescence of other blood islands in the embryo forms blood vessels that are in continuity with the heart tube. In the embryonic circulation, paired anterior and posterior cardinal veins drain the embryo cranial and caudal to the heart, respectively, and join to form the paired common **cardinal veins (ducts of Cuvier)**. These veins drain into the caudal extent of the endocardial tube at the **sinus venosus**. Blood leaves the cranial extent of the endocardial tube and is distributed into five paired aortic arches that pass dorsally around the foregut in the branchial arches to empty into the paired dorsal aortae. Blood then circulates through branches of the aortae to capillaries that are in continuity with tributaries of the cardinal veins.

Blood vessels developing in the placenta (chorion) are linked to the embryonic circuit to form an **umbilical (i.e., allantoic, placental) circuit**. In this circuit, umbilical arteries arise from the aorta, pass through the body stalk, and go to capillaries of the placenta. Oxygenated and nutritive blood returns by the left umbilical vein to the sinus venosus. The right umbilical vein disappears soon after it is developed.

The **vitelline circuit** involves vascular channels in the yolk sac. Vitelline (omphalomesenteric) arteries arise from the abdominal aorta and pass along the yolk stalk to capillaries in the yolk sac. Blood returns to the sinus venosus by vitelline (omphalomesenteric) veins.

After birth, the umbilical arteries will remain, in part, as a portion of the internal iliac and superior vesical arteries and the lateral umbilical ligaments. The umbilical vein persists as the round ligament of the liver. Portions of the vitelline veins become the portal vein. The vitelline arteries fuse with each other and give rise to the superior mesenteric artery.

# FOLDING AND PARTITIONING OF THE HEART

With fusion of the endocardial tubes, several dilatations become apparent. These are, from cephalic to caudal, the bulbus cordis (truncus arteriosus plus the conus arteriosus), ventricle, atrium, and sinus venosus. Arteries leave the cephalic end of the bulbus cordis from a swelling called the **aortic bulb** (aortic sac). Veins enter at the sinus venosus. With the loss of the dorsal mesocardium, except where the veins and arteries enter and leave, the heart begins to flex into an S-shaped structure.

The first flexure occurs at the junction of the **bulbus cordis** and **ventricle**. The second flexure causes the **sinus venosus** and **atrium** to shift dorsally. The adjacent bulboventricular walls disappear, and this part of the bulbus and primitive ventricle become part of a common **ventricular chamber**. The atrium becomes sandwiched between the pharynx dorsally and the rest of the conus and truncus ventrally, causing the atrium to bulge laterally into right and left swellings. The sinus venosus is shifted to the right and eventually is incorporated into the primitive right atrial swelling. The **pattern of blood flow** is from veins to common atrium, to common ventricle, to conus, to truncus, to aortic bulb, and then to aortic arches.

During the second month of development, the heart is partitioned into **four chambers** (two atria and two ventricles); **atrioventricular (A-V) valves** are formed; and the conus, truncus, and aortic bulb are partitioned into ascending aorta and pulmonary trunk.

In partitioning of the atrium, endocardial tissue from the dorsal and ventral walls fuses into an endocardial cushion that separates the A-V communication into right and left A-V canals. While this is taking place, an endocardial septum primum grows toward the endocardial cushion from the dorsal wall of the atrium. Before fusing with the cushion, an ostium primum exists temporarily between the free margin of the septum primum and the cushion. This ostium will not disappear before an ostium secundum arises from the degeneration of the **septum primum** cephalically. In the seventh week, a **septum secundum** grows dorsocaudally to the right of septum primum and leaves a crescentic free area covered only by septum primum. The communication from the right to the left atrium through the crescentic opening and ostium secundum is the **foramen ovale**. The valve of the foramen ovale is part of the septum primum. The **interatrial septum** thus arises from septum primum and septum secundum.

The **sinus venarum** of the right atrium is formed by the incorporation of the sinus venosus into the right atrium so that the developing great veins enter independently. The smooth-surfaced portion of the left atrium arises after the absorption of the common trunk of the pulmonary veins, thus leaving four **pulmonary veins** entering at the boundaries of this area.

In partitioning of the ventricle, a muscular interventricular septum grows toward the endocardial cushion. Just caudal to the cushion, an interventricular foramen remains for a short time before it is closed by endocardial tissue from the free margin of the interventricular septum, the endocardial cushion, and the conal septum.

In septation of the aortic bulb, truncus arteriosus, and conus arteriosus, a ridge of endocardial tissue develops on opposite walls of each of these structures. These ridges fuse in the middle of the lumen to form a bulbar (i.e., aortic, conal, truncal) septum. This septum spirals about 180 degrees as it descends from the aortic bulb into the conus, thus establishing a pulmonary trunk that intertwines with the ascending aorta. Semilunar valves develop in these vessels as localized swellings of endocardial tissue. The conal septum eventually descends to help close the interventricular septum.

In **development of A-V valves**, subendocardial and endocardial tissues project into the ventricle just below the A-V canals. These bulges of tissue are excavated from the ventricular side and invaded by muscle. Eventually all of the muscle disappears, except that remaining as papillary muscles, and three right cusps of the right A-V valve and two cusps of the left A-V valve remain as fibrous structures.

## DEVELOPMENT OF MAJOR ARTERIAL VESSELS

Five pairs of **aortic arches** develop cephalocaudally in branchial arches. They bridge from the ventral aortic roots to the dorsal aortae. In comparative studies, the five aortic arches represent the first, second, third, fourth, and sixth aortic arches. In humans, the first, second, and distal parts of the right sixth arches (fifth) disappear. The remaining aortic arches, ventral aortic roots, and dorsal aortae give rise to major arteries.

The **internal carotid arteries** develop from the third aortic arches and the dorsal aortae cephalic to the third arches. The **common carotids** arise from the ventral aortic roots and the proximal part of the third arches. The **external carotids** arise in a similar position to the ventral aortic roots lying cephalic to the third arch. The **right subclavian artery** arises from the right fourth arch, the seventh dorsal intersegmental artery, and the intervening portion of the right dorsal aorta. The **left subclavian artery** arises from the left seventh dorsal intersegmental artery. The arch of the aorta develops from the left fourth aortic arch and some septation of the aortic bulb.

The **pulmonary arteries** arise from the proximal portions of the sixth arches along with some new vascular buds. The ductus arteriosus, linking the pulmonary trunk with the aorta, is the distal portion of the left sixth arch. The **brachiocephalic artery** originates from the right ventral aortic root between the fourth and sixth arches. The right dorsal aorta caudal to the right seventh dorsal intersegmental arteries disappears down to the embryonic low thoracic region where the paired dorsal aortae had fused into one midline vessel. The dorsal aortae between the third and fourth arches degenerate.

## DEVELOPMENT OF MAJOR VENOUS CHANNELS

The superior and inferior caval systems and the portal vein arise from early embryonic vessels. The **superior vena cava** forms from the right common cardinal vein and a caudal portion of the right anterior cardinal vein up to the entrance of the left brachiocephalic (innominate) vein. This vessel arises from a thymicothyroid anastomosis of veins. The right brachiocephalic develops from the right anterior cardinal vein between this anastomotic venous attachment and the right seventh intersegmental vein (right subclavian). The left common cardinal vein and part of the left horn of the sinus venosus become the coronary sinus, which drains the heart wall into the right atrium.

The **inferior vena cava**, from heart to common iliacs, arises from the following vessels: (1) a small portion of the right vitelline vein, (2) a new vessel in the mesenteric fold of the degenerating mesonephros, (3) the right subcardinal vein, and (4) a sacrocardinal vein joining the caudal extent of the posterior cardinal veins. The subcardinals and their anastomosis, which developed to drain the mesonephros, also give rise to the renal, gonadal, and suprarenal veins.

The **azygos venous system** arises mostly from the supracardinal veins and their anastomosis. The most cephalic portion of the azygos vein is derived from the right posterior cardinal vein. The portal and hepatic veins arise from the vitelline (omphalomesenteric) veins and their anastomoses.

## FETAL CIRCULATION

The circulation of the blood in the embryo results in the shunting of well-oxygenated blood from the **placenta to the brain and the heart**, while relatively desaturated blood is supplied to the less essential structures.

Blood returns from the placenta by way of the left umbilical vein and is shunted in the ductus venosus through the liver to the inferior vena cava and then to the right atrium. Relatively little mixing of oxygenated and deoxygenated blood occurs in the right atrium because the valve overlying the orifice of the inferior vena cava directs the flow of oxygenated blood from that vessel through the foramen ovale into the left atrium, and the deoxygenated stream from the superior vena cava is directed through the tricuspid valve into the right ventricle. From the left atrium, the **oxygenated blood** and a small amount of **deoxygenated blood** from the lungs pass into the left ventricle and then into the ascending aorta from which it is supplied to the brain and the heart through the vertebral, carotid, and coronary arteries.

Because the lungs of the fetus are inactive, most of the deoxygenated blood from the right ventricle is shunted by way of the **ductus arteriosus** from the pulmonary trunk into the descending aorta. This blood supplies the abdominal viscera and the inferior extremities and is carried to the placenta for oxygenation through the umbilical arteries arising from the aorta.

## CIRCULATORY CHANGES AT BIRTH

When respiration begins, the lungs expand, resulting in increased blood flow through the pulmonary arteries and a pressure change in the left atrium. This **pressure change** brings the septum primum and the septum secundum together and causes functional closure of the foramen ovale. Simultaneously, active contraction of the muscular wall of the ductus arteriosus results in its functional closure. Several months later it will become ligamentous as the **ligamentum arteriosum**. The ductus venosus functionally closes and becomes the ligamentum venosum. The fate of the umbilical arteries and veins was described in the section on extraembryonic circuits (Fig. 12-1).

## CONGENITAL ANOMALIES OF THE HEART AND GREAT VESSELS

The complicated sequence of development in the heart and the major arteries accounts for the many congenital abnormalities that, alone or in combination, may affect these structures.

**Septal defects** include **patent foramen ovale** (approximately 10% incidence) and other atrial or ventricular septal defects. An **ostium secundum (foramen ovale) defect** lies in the interatrial wall and is relatively easy to close surgically. An **ostium primum defect** lies directly above the A-V boundary and is often associated with a defect in the membranous part of the interventricular septum and in the A-V valves. A **high interatrial septal defect** may result, which most likely is an improper shifting and incorporation of the sinus venosus into the right atrium.

**Interventricular septal defects** usually involve the membranous part of the interventricular septum and are due mostly to improper formation of the conal septum. Rarely, the septal defect is so large that the ventricles form a single cavity, resulting in a **trilocular heart (cor triloculare biatriatum)**. This defect is caused by improper development of the primitive muscular interventricular septum. Failure of closure of the interventricular foramen also may be caused by defective development of the septum membranaceum contribution from the fused endocardial cushions.

**Congenital pulmonary stenosis** may involve the trunk of the pulmonary artery and its valve or the infundibulum of the right ventricle. If this is combined with an interventricular septal defect, the compensatory **hypertrophy of the right ventricle** develops sufficiently high pressure to shunt blood through the defect into the left side of the heart. This mixing of blood results in the child's being cyanosed at birth.

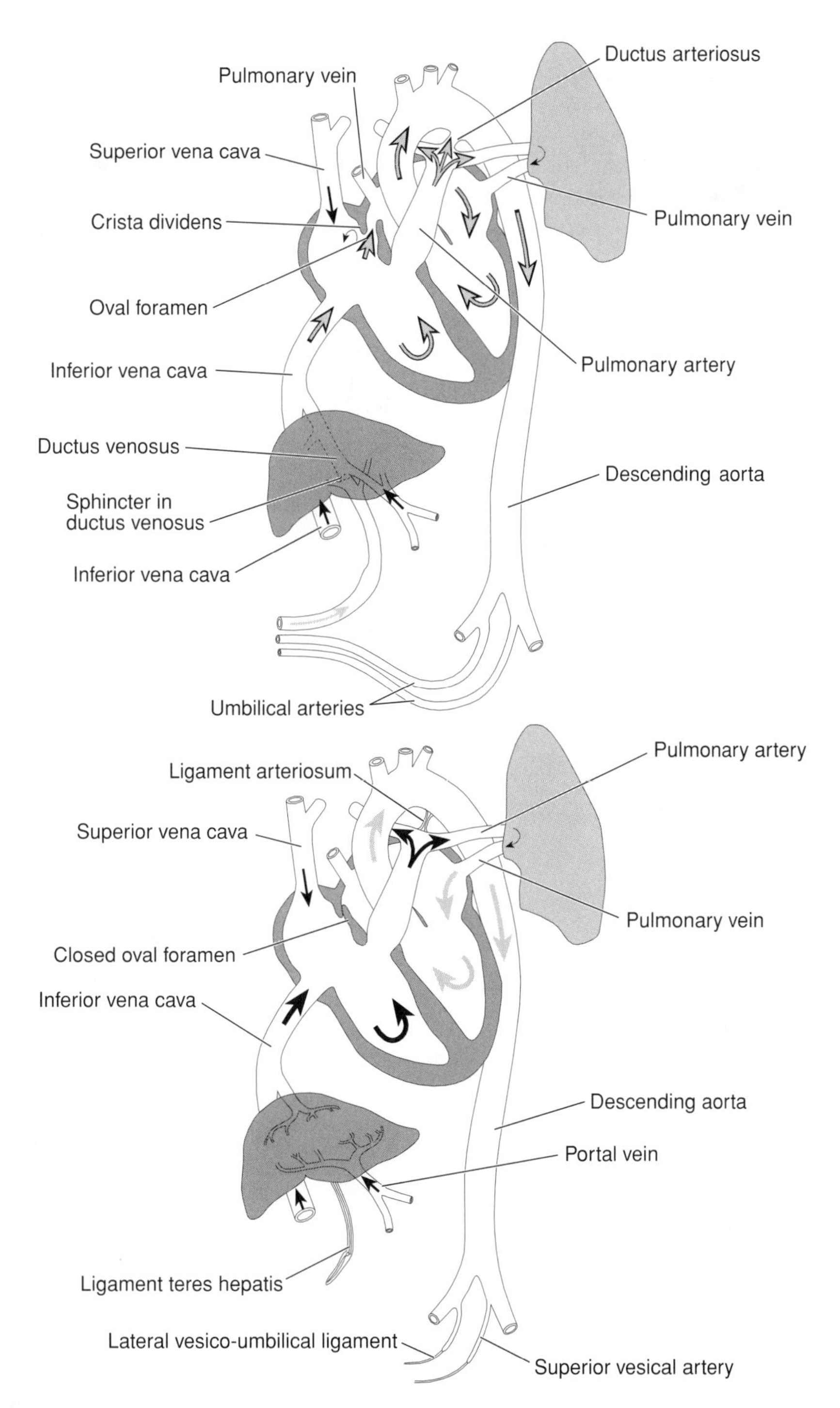


**Figure 12-1.** Two diagrams illustrating the changes in fetal circulation at birth.

**Tetralogy of Fallot** is the **most common congenital abnormality causing cyanosis**. It is comprised of pulmonary stenosis, right ventricular hypertrophy, an interventricular septal defect with an overriding aorta, the orifice of which lies cranial to the septal defect and receives blood from both ventricles.

**Transposition of the great vessels** is caused by improper spiraling of the bulbar septum in the formation of the great vessels. This results in either complete transposition, in which the aorta is from the right ventricle and the pulmonary trunk is from the left ventricle, or in incomplete transposition in which both vessels are reversed but both exit from the right ventricle.

**Aortic stenosis** is caused by either bulbar septum displacement or localized improper growth in supravalvular, valvular, and subvalvular regions of the aorta.

**Patent ductus arteriosus** is a relatively common developmental abnormality. If not corrected, it causes progressive work hypertrophy of the right heart and pulmonary hypertension.

**Aortic coarctation** may be caused by abnormal retention of the fetal isthmus of the aorta or to incorporation of smooth muscle from the ductus into the wall of the aorta. The constriction may occur from the level of the left subclavian artery to the ductus arteriosus (preductal type), which is widely patent and maintains the circulation to the lower part of the body. In other cases, the coarctation may involve a segment below the entrance of the ligamentum arteriosum (postductal type), and the circulation to the lower limb is maintained by collateral arteries around the scapula that anastomose with the intercostal arteries.

**Dextrorotation of the heart** is the most spectacular of the abnormalities. The heart and its emerging vessels lie as a mirror image to the normal anatomy. Dextrorotation may be associated with reversal of all the intra-abdominal organs in situs inversus.

**Abnormal development of the aortic arches** may result in the arch of the aorta lying on the right or actually being double. Rarely, an abnormal right subclavian artery arises from the dorsal aorta and passes behind the esophagus and thus causes difficulty in swallowing (dysphagia lusoria). **Double aorta** is caused by retention of the right dorsal aorta between the seventh dorsal intersegmental artery and the point of fusion of the aortae. If this portion remains and the right fourth aortic arch disappears, then the right subclavian arises from the aorta.

# Chapter 13

# *Respiratory System*

An **entodermal respiratory diverticulum** develops from the floor of the foregut just caudal to the last pharyngeal pouch. The **larynx**, **trachea**, **bronchi**, and **lungs** develop from this diverticulum. The **esophagus** differentiates from the foregut caudal to this outgrowth. In the development of the larynx, the opening is constricted into a narrow T-shaped laryngotracheal orifice by underlying mesodermal arytenoid swellings. A transitory period occurs when the opening is completely obliterated by an overgrowth of epithelium. Persistence of portions of this may lead to webs that obstruct the laryngeal opening.

All cartilage, muscle, and connective tissue of the larynx, trachea, and lungs arise from splanchnic mesoderm. The epithelium and glands develop from branching of the entodermal diverticulum. The **main bronchi** divide dichotomously through 17 generations of subdivisions by the end of the sixth month. An additional seven divisions occur by early childhood.

As the bronchial buds divide, the lung increases in size and bulges laterally into the embryonic coelom. The early **coelom** consists of a more ventral prospective pericardial cavity in continuity with a more dorsal pleural cavity, which is continuous caudally above the septum transversum with the prospective peritoneal cavity. Right and left **pleuropericardial folds** project from the lateral body wall and septum transversum, grow medially between the heart and lungs, and fuse with the primitive mediastinum; thus, these folds become part of the definitive **mediastinum** and separate the pleural from the pericardial cavities. Pleuroperitoneal folds grow from the septum transversum at right angles to the pleuropericardial membranes. These folds invest the esophagus and, along with the septum transversum, contribute to the formation of the **diaphragm**. In addition, the definitive diaphragm receives a major contribution of muscle in its development from the lateral body wall.

**Tracheoesophageal fistulas** may develop from improper separation of the respiratory diverticulum from the foregut, by malformation of the esophagotracheal septum, or by secondary fusion of the esophagus with the trachea. In the most usual fistula, the upper part of the esophagus ends blindly and a lower portion is connected to the trachea by a narrow canal. This may result from dorsal deviation of the esophagotracheal septum in its caudal growth.

**Diaphragmatic hernias** can arise from improper formation of the septum transversum, the pleuroperitoneal folds, or the muscular component from the body wall.

# Chapter 14

# Digestive System

## DEVELOPMENT OF THE GUT

The **entodermal foregut** gives rise to the pharynx, esophagus, stomach, liver, pancreas, and part of the duodenum. The midgut gives rise to the rest of the small intestine, ascending colon, and proximal two-thirds of the transverse colon. The hindgut develops into the rest of the large intestine as far as the upper part of the anal canal. The lower part of the rectum and much of the anal canal is established by the separation of the cloaca into a dorsal anorectal canal and ventral urogenital sinus by the urorectal septum. The rest of the anal canal develops from an ectodermally lined anal pit.

The **stomach** appears in the fourth embryonic week as a dilation of the foregut. As it shifts caudally from its position above the septum transversum, it rotates so that its original left surface faces anteriorly and its dorsal greater curvature extends to the left.

The **intestines** develop from the caudal part of the foregut and the cephalic and caudal limbs of a midgut loop that extend into the belly stalk. The cephalic limb extends from the upper duodenum to the yolk stalk. It gives rise to the rest of the small intestine, except the last portion, which includes 40 to 50 cm of the ileum. The caudal limb extends from the yolk stalk to the hindgut, and it gives rise to the rest of the ileum, the cecum, the ascending colon, and the proximal two-thirds of the transverse colon. As the loop develops, it rotates counterclockwise (in an anteroposterior view) around the omphalomesenteric (vitelline) artery, which becomes the superior mesenteric artery. This rotation places the transverse colon above the jejunum and ileum, anterior to the duodenum, and just below the stomach. By the tenth week, the abdomen enlarges, and the gut loop reenters the abdomen. The cephalic limb enters first, crowding the descending colon to the left. Partial persistence of the yolk stalk may remain as a **Meckel's diverticulum** attached to the ileum, 40 to 50 cm from the ileocolic junction.

## LIVER

The liver arises in the fourth embryonic week as a ventral diverticulum of the foregut. This diverticulum grows through the ventral mesentery and into the caudal face of the septum transversum. The more proximal part of the hepatic diverticulum gives rise to the **common bile duct, cystic duct, gallbladder, and hepatic ducts**. The more distal portions differentiate into the **hepatic plates and the smaller bile ducts**. Because the liver grows in the septum transversum and bulges from its caudal face, it is covered by peritoneum lining the septum transversum, except at the bare area of the liver where it abuts directly against the septum (diaphragm).

**TABLE 14-1. Summary of the Derivatives of the Three Parts of the Primitive Gut**

| Region | Artery | Ventral Mesentery Derivatives | Dorsal Mesentery Derivatives |
|---|---|---|---|
| Foregut | Celiac artery | Falciform ligament, lesser omentum, capsule of the liver | Greater omentum |
| Midgut | Superior mesenteric artery | — | Mesentery part of transverse mesocolon |
| Hindgut | Inferior mesenteric artery | — | Part of transverse mesocolon, sigmoid colon |

## PANCREAS

The pancreas forms from dorsal and ventral entodermal buds located at the level of the duodenum. The dorsal bud grows into the dorsal mesentery. The proximal portion of the ventral bud joins with the common bile duct, whereas the distal portion grows into the dorsal mesentery and fuses with the dorsal bud as the anterior portion of the duodenum grows more extensively than the dorsal part does. The **pancreatic duct (of Wirsung)** develops from the ventral bud and the distal part of the dorsal primordium. The proximal portion of the dorsal bud may give rise to the **accessory duct (of Santorini)**.

## ABDOMINAL MESENTERIES AND SPLEEN

The **abdominal mesenteries** develop primarily from the embryonic dorsal mesentery. The ventral mesentery may give rise to the **lesser omentum** and the falciform ligament, although the **falciform ligament** probably forms from a "shearing" of peritoneum covering the anterior body wall. The peritoneum covering the liver reflects at the bare area of the liver to form the coronary and triangular ligaments of the liver. Much of the dorsal mesogastrium suspending the stomach fuses with the dorsal body wall to form the dorsal lining of the lesser sac. The rest of the dorsal mesogastrium fuses with the embryonic transverse mesocolon to form the definitive mesentery of the transverse colon and then drapes over the small intestine to become the **greater omentum**.

The **spleen** develops from mesoderm of the dorsal mesogastrium. Most of the dorsal mesentery of the duodenum fuses with the dorsal body wall, making it and the pancreas secondarily retroperitoneal. The dorsal mesentery of the jejunum and ileum becomes the mesentery proper. Most of the dorsal mesentery of the ascending and descending colon fuses with the dorsal body wall. The dorsal mesentery of the rest of the hind gut becomes the transverse mesocolon and the sigmoid mesocolon (Table 14-1).

# Chapter 15

# Development of the Head and Neck

## DEVELOPMENT AND FATE OF THE BRANCHIAL ARCHES

During the third and fourth weeks of development, the embryo develops head and tail folds that result in the incorporation of the dorsal portions of the primitive yolk sac entoderm as foregut, midgut, and hindgut. The **rostral portion of the foregut** (primitive pharynx) develops five lateral pairs of pharyngeal pouches. These and the floor of the pharynx give rise to the tongue, pharynx, trachea, lungs, eustachian tube, middle ear cavity, thyroid gland, parathyroid glands, and thymus. The more **caudal portions of the foregut** develop into the esophagus, stomach, part of the duodenum, and the liver and pancreas.

While the pharyngeal pouches are forming internally, five pairs of **branchial (pharyngeal) arches** appear externally. These are numbered 1, 2, 3, 4, and 6. They are separated by branchial (pharyngeal) grooves that are aligned with the pharyngeal pouches to form branchial (pharyngeal) membranes consisting of outer ectodermal and inner entodermal layers (Table 15-1).

Each groove is numbered according to the arch that lies rostral to it. Each branchial arch is comprised of an outer ectodermal layer and an inner entodermal lining, with a vertical bar of mesoderm and a cranial nerve interposed between two layers. Some of the mesoderm will condense to form a cartilage and some will contribute to the formation of the aortic arches. The remainder of the mesoderm will contribute to the formation of muscles.

The **first arch** is divisible into **mandibular** and **maxillary arches**. The surface ectoderm of these arches will become the epidermis of the upper and lower jaws, the epithelium of most of the oral cavity, the parenchyma of the major salivary glands, and the enamel of the teeth. The mandibular and maxillary divisions of the trigeminal nerve course in these arches and supply the skin of the face and the mucosa of the oral and nasal cavities with sensory nerves and the muscles of mastication with motor nerves.

The muscles developing from the mandibular arch are the temporalis, masseter, medial and lateral pterygoids, mylohyoid, anterior belly of the digastric, tensor veli palatini, and tensor tympani. **Mesenchyme** and **neural crest of the mandibular arch** develop into a transitory Meckel's cartilage before forming a mandible, malleus, and incus; mesenchyme of the maxillary arch forms the maxilla.

The first branchial groove gives rise to the external acoustic meatus. The first branchial membrane develops into the tympanic membrane, and the first pharyngeal pouch presages part of the auditory tube (eustachian) and middle ear cavity.

The **second arch**, also known as the **hyoid arch**, grows back over arches 3 to 6 and fuses with them, obliterating branchial grooves 3 to 6 and resulting in a transitory cervical sinus. This sinus normally obliterates, but if it persists, it can develop into a cer-

**TABLE 15-1. Summary of the Derivatives of the Branchial Arches**

| | Arch I | Arch II | Arch III | Arches IV–VI |
|---|---|---|---|---|
| Cartilage | Meckel's | Reichert's | | |
| Skeletal | Malleus, incus | Stapes, styloid | Greater horn and body of hyoid bone | Intrinsic and extrinsic laryngeal cartilages |
| | Zygoma, temporal, maxilla, mandible | Lesser horn and body of hyoid bone | | |
| Muscles | Muscles of mastication, anterior belly of digastric, mylohyoid, tensor tympani, and tensor veli palatini | Muscles of facial expression, posterior belly of digastric, stylohyoid | Stylopharyngeus | Pharyngeal constrictors, internal and external laryngeal muscles |
| Nerve | Trigeminal | Facial | Glossopharyngeal | Vagus |
| Vessel | Maxillary artery | Hyoid and stapedial arteries | Common carotid and proximal part of internal carotid | Subclavian from the fourth right aortic arch, aortic arch from the fourth left aortic arch, ductus arteriosus from the sixth left aortic arch |
| Pouch | Lining of pharyngotympanic tube | Tonsillar fossa | Inferior parathyroids, thymus | Superior parathyroids, parafollicular cells |
| Groove | External auditory meatus | | | |

vical sinus cyst. Cervical (branchial) fistulas may remain if communications are retained externally or internally through branchial membranes.

Because the ectoderm of the second arch gives rise to the epidermis of much of the neck, the openings of external branchial fistulas occur in the neck along the anterior margin of the sternocleidomastoid muscle. **Internal fistulas** most often occur into the second pouch, opening into the tonsillar region because the tonsil develops from the second pharyngeal pouch.

Mesenchyme of the second arch gives rise to the muscles of facial expression (mimetic muscles), stapedius, posterior belly of the digastric, and stylohyoid muscles. Along with neural crest, mesenchyme gives rise to the stapes, styloid process, stylohyoid ligament, and lesser cornua of the hyoid bone. The facial nerve runs in the second arch and supplies innervation to muscles developing from this arch.

The **third pharyngeal arch** gives rise to the **stylopharyngeus muscle**, supplied by the glossopharyngeal nerve, and to the body and greater horn of the hyoid bone. The endoderm of the dorsal part of the third pharyngeal pouch gives rise to the inferior parathyroid glands. The ventral part of the third pharyngeal pouch becomes the **thymus gland**.

The **fourth and sixth arches** give rise to the laryngeal cartilages and the pharyngeal, palatal, and laryngeal muscles. All of these muscles, except the tensor veli palatini and the stylopharyngeus, are innervated by the vagus nerve, with the superior laryngeal nerve supplying the fourth arch and the recurrent laryngeal nerve supplying the sixth. The dorsal part of the fourth pharyngeal pouch gives rise to the superior parathyroid gland. The fate of the ventral part is not certain. The fifth (sixth) pharyngeal pouch gives rise to the ultimobranchial body, which, with neural crest involvement, gives rise to the calcitonin-producing cells of the parathyroid gland.

**Aortic arches** arise from the aortic bulb and enter each branchial arch, where they run chiefly caudal to the cranial nerve in each arch. The first aortic arch contributes to the development of the maxillary artery; the second aortic arch largely disappears. The third aortic arch becomes part of the internal carotid and common carotid arteries. The right fourth aortic arch becomes part of the subclavian artery, and the left fourth arch becomes the arch of the aorta.

The proximal portions of both sixth aortic arches become part of the pulmonary arteries; the left distal part becomes the **ductus arteriosus** (see Table 15-1). Because the recurrent laryngeal nerve is caudal to the sixth aortic arch, retention of the ductus arteriosus as the ligamentum arteriosum accounts for the left recurrent laryngeal nerve looping around the arch of the aorta, caudal to the ligament; the right loops higher around the subclavian artery.

The **tongue** develops from elevations in the floor of the primitive pharynx and by forward migration of developing muscle from occipital somites. The body of the tongue arises from the two lateral swellings and a median tuberculum impar in the floor of the mandibular arch. The root of the tongue develops from a copula of mesenchyme of the second, third, and fourth arches. The epiglottic swelling also comes from mesenchyme of the fourth arch. The **muscles of the tongue** develop from occipital somites and are innervated by the hypoglossal nerve. Because the oral membrane demarcating the ectodermally lined stomodeum from the entodermal primitive pharynx existed just in front of the fauces, most of the epithelium of the body of the tongue arose from oral ectoderm, while the root of the tongue and the foramen caecum area are developed from entoderm. Thus, general sensation of the anterior two-thirds of the tongue is supplied by branches of the **trigeminal nerve**, whereas the posterior third of the tongue is innervated by the **glossopharyngeal nerve**.

The **thyroid gland** develops as an evagination from the floor of the pharynx at the level of the first pharyngeal pouch. It migrates caudally to the region of the larynx, leaving the foramen caecum as the site of original evagination and often leaving thyroglossal duct cysts along its migratory path. **Thyroglossal duct cysts** are found in the root of the tongue, along the neck, and in or behind the hyoid bone.

## DEVELOPMENT OF THE FACE AND NASAL AND ORAL CAVITIES

By the sixth week of embryonic development, a frontal prominence overhangs the cephalic end of the stomodeum. It is bounded laterally by nasal pits surrounded by horseshoe-shaped elevations. The medial portion of the horseshoe-shaped elevation is the **nasomedial process**; the lateral portion is the nasolateral process. The **nasolateral process** is delimited from the maxillary process by the nasooptic (nasolacrimal) furrow. The developing oral cavity is bounded inferiorly by distal fusion of the **mandibular processes** of the first branchial arch.

As the **maxillary processes** become more prominent, they fuse with the nasomedial processes and push them toward the midline. This displaces the frontal prominence upward and leads to fusion of the nasomedial processes in the midline. The fused **nasomedial processes** (i.e., intermaxillary segment, primary palate) give rise to the medial part of the upper lip, distal nose, incisor teeth and associated upper jaw, and the primary palate (median palatine process). The **maxillary process** gives rise to the rest of the upper lip, teeth, and jaw and the palatine shelves, forming the secondary palate. The lower lips, jaw, and teeth are formed from the **mandibular processes**. The nasolacrimal duct is formed at the point of obliteration of the nasooptic furrow by fusion of the nasolateral and maxillary processes. Inability of these processes to fuse leads to an **oblique facial cleft**.

The **nasolateral processes** give rise to the alae of the nose. The nasal pits become deeper and break through the bucconasal membrane into the primitive oral cavity. Toward the end of the second embryonic month, the palatine shelves (lateral palatine processes) of the maxillary processes grow medially and fuse with the primary palate rostrally, and with each other and with the inferiorly growing nasal septum caudally. The lateral **palatine processes** thus give rise to a **secondary palate**, which separates the nasal cavity above from the definitive oral cavity below. The caudal free borders of the palatine shelves project as the soft palate into the pharynx, dividing it into an upper nasopharynx and a lower oropharynx.

## CONGENITAL MALFORMATIONS OF THE FACE AND NASAL AND ORAL CAVITIES

Incomplete degeneration of the bucconasal membrane can lead to **choanal atresia**. Failure of the palatine shelves to fuse in the midline or to fuse with the primary palate produces **cleft palate**. Such clefts can be divided into three groups: (1) those occurring between the palatine shelves and the primary palate (i.e., anterior, primary palate types); (2) those occurring posterior to the incisive foramen at the point of fusion of the palatine shelves with each other (i.e., posterior, secondary palate types); and (3) those involving defects in both the anterior and posterior palate (i.e., complete unilateral or bilateral types). The anterior and complete types may be associated with **cleft lip** if the nasomedial and maxillary processes fail to merge and fuse. **Median cleft of the upper lip and jaw** is caused by the lack of fusion of the nasomedial processes with each other.

# Chapter 16

# Urogenital System

The urinary and reproductive systems originate from the urogenital sinus and the intermediate mesoderm.

## DEVELOPMENT OF THE KIDNEY AND URETER

Three pairs of embryonic kidneys develop in humans: the pronephros, mesonephros, and metanephros. The pronephros and mesonephros will degenerate, but their development is essential for the establishment of the metanephros, which becomes the definitive kidney.

### Pronephros

The pronephros arises at the C3 to T1 vertebral levels by the dorsal proliferation of cords of cells from the intermediate mesoderm. These cords become **pronephric tubules**. They grow caudally and link up with the other pronephric tubules, forming a common pronephric duct that extends caudally toward the cloaca. The pronephric kidney does not function, but the **pronephric duct** is important for the normal formation of the mesonephric kidney.

### Mesonephros

The mesonephros develops by the formation of mesonephric tubules from the intermediate mesoderm of the C6 to L3 vertebral levels. Unlike the pronephric tubules, these tubules do not communicate with the coelom but receive a capillary glomerulus from the aorta, which is encapsulated by the proximal blind end of the tubule. The distal end of the mesonephric tubules tap into the pronephric duct and contribute to its caudal growth. This enlarged pronephric duct taps into the cloaca, and it then becomes the **mesonephric (wolffian) duct**. The extensive growth of the mesonephros produces a large urogenital ridge projecting from the dorsal body wall.

### Metanephros

The metanephros arises from two sources: the ureteric bud (metanephric diverticulum) and the metanephrogenic intermediate mesoderm of the L4 to S1 vertebral levels. The **ureteric bud** arises as a tubular outgrowth from the mesonephric duct near its entrance into the cloaca. It grows toward the intermediate mesoderm, where its blind end becomes capped by metanephrogenic tissue. The ureteric bud elongates as the **ureter**, and its blind end enlarges as the **renal pelvis** and undergoes a series of branchings. These branchings give rise to the major and minor calyces and the collecting tubules. The metanephrogenic condensations capping the blind ends of the collecting tubules develop into **nephrons**. One end of the blind nephron forms a **Bowman's capsule** around a glomerulus of capillaries. The other end taps into the **collecting tubule**.

## DEVELOPMENT OF THE URINARY BLADDER AND URETHRA

The urinary bladder and urethra develop from the entoderm of the urogenital sinus and allantois. In early development, the **allantois** is a diverticulum of the cloaca. A urorectal septum of mesoderm arises between the allantois and hindgut, grows caudally, and divides the cloaca into a dorsal rectum and ventral urogenital sinus. With this division, the mesonephric duct empties into the urogenital sinus. As the urogenital sinus and a small adjacent portion of the allantois enlarge to form the **urinary bladder**, portions of the mesonephric and metanephric (ureter) ducts are incorporated into the wall of the urogenital sinus. This results in the ureters entering the bladder and the mesonephric ducts entering more caudally into the less dilated portion of the urogenital sinus. This distal portion of the urogenital sinus will become the **urethra** of the male and the **urethra**, **vestibule**, and part of the **vagina** of the female.

## DEVELOPMENT OF THE REPRODUCTIVE SYSTEM

Even though the sex of the embryo is determined at fertilization, the gonads, ducts, and external genitalia pass through an indifferent stage of development in which male and female components have the same appearance. This stage lasts until about the sixth week of development.

In the indifferent stage, **gonads** form on the medial wall of the urogenital ridges. Starting in the third week, **primordial sex cells** migrate from the entoderm of the yolk sac to the urogenital ridge. By the sixth week, the coelomic epithelium has proliferated, invaginated, and surrounded the primordial sex cells to form primitive **gonadal (primary sex) cords** in the underlying mesoderm of the gonad.

In the indifferent stage of genital duct formation, both mesonephric and müllerian (paramesonephric) ducts are present. The **müllerian ducts** arise as longitudinal invaginations of the coelomic epithelium on the lateral wall of the urogenital ridge. Cranially this duct remains open to the coelom. Caudally it opens through the dorsal wall of the urogenital sinus. In its craniocaudal course, it lies at first lateral to the **mesonephric duct**, then passes anterior to it, and finally fuses with the opposite müllerian duct medial to the mesonephric ducts. During this fusion, the urogenital ridges of the two sides are brought together to form a **genital cord** (septum) between the developing bladder anteriorly and the rectum posteriorly (Table 16-1).

In the indifferent stage of the development of the **external genitalia**, mesoderm invades the lateral walls of the external opening of the urogenital sinus producing elevations called **urogenital (urethral) folds**. These folds unite anterior (cephalic) to the urogenital opening at the genital tubercle. Labioscrotal swellings develop lateral to the urogenital folds (Table 16-2).

In the development of the **male reproductive system**, the gonadal cords become testis cords, which differentiate into **seminiferous tubules** and **rete testis**. The primordial sex cells become spermatogonia; ingrowing coelomic epithelial cells give rise to supportive (Sertoli) cells. Leydig cells develop from mesenchyme. Efferent ductules develop from adjacent mesonephric tubules. The developing testis produces **müllerian-inhibiting hormone** and **androgens**, which lead to the degeneration of the müllerian

**TABLE 16-1. Summary of Development of Male and Female Genital Ducts**

| Embryonic Structure | Male Derivative | Female Derivative |
|---|---|---|
| Yolk sac primordial germ cells | Spermatogonia | Oogonia |
| Mesonephric cortex | Tunica albuginea | Cortical sex cords > follicle cells of ovary |
| Mesonephric medulla | Medullary sex cords > seminiferous tubules of testis | — |
| Mesonephric tubules | Efferent ductules and rete testis | Epoöphoron |
| Mesonephric duct | Epididymis, ductus deferens | Paroöphoron |
| Paramesonephric duct | Appendix of testis, prostatic utricle | Oviduct, uterus, upper part of vagina |

**TABLE 16-2. Summary of the Origins of Male and Female External Genitalia**

| Embryonic Structure | Male Derivative | Female Derivative |
|---|---|---|
| Genital tubercle | Glans and body of penis | Glans and body of clitoris |
| Urogenital sinus | Penile urethra | Vestibule of vagina |
| Urethral folds | Corpus spongiosum surrounding urethra | Labia minora |
| Labioscrotal swelling | Scrotum | Labia majora |

duct and to the differentiation of the mesonephric duct into the **ductus epididymis**, **vas (ductus) deferens**, and **ejaculatory duct**. The seminal vesicle arises as an outgrowth of the mesonephric duct. The urogenital sinus gives rise to the **urethra** and the **prostate** and **bulbourethral** and **urethral glands**. The genital tubercle enlarges and carries with it inferiorly a urethral plate of entoderm. This plate is transformed into the penile (cavernous) urethra after the lateral urogenital folds fuse ventrally. The urethral plate, urogenital folds, and genital tubercle (phallus) give rise to the definitive **penis**. The **scrotum** is formed by the ventral fusion of the labioscrotal swellings. The **testes** descend late in gestation from their retroperitoneal abdominal location. They are "anchored" in the scrotum by the gubernaculum testis, which is derived from mesoderm of the urogenital ridge caudal to the testis. The path of descent is indicated by the inguinal canal. This follows the embryonic pathway of the processus vaginalis evaginating from the peritoneum.

In the development of the **female reproductive system**, the initial gonadal cords degenerate and a second series of ovarian cords develop from primordial sex cells and coelomic epithelium. This second set of cords splits into groups of follicles near the surface of the developing ovary. Each primitive ovarian follicle consists of a developing sex cell, surrounded by a flattened layer of follicular cells. The sex cells complete the prophase of the first meiotic division and are in the arrested dictyotene stage by the time of birth. From the sixth month until parturition, there is a tremendous rate of degeneration of primitive follicles, the number decreasing from about 6 million to about 400,000 or less.

The unfused portions of the müllerian ducts develop into the **uterine tubes**. The fused portions give rise to the **uterus** and part of the **vagina**. The genital cord remains as the **broad ligament of the uterus**. The proper ligament of the **ovary** and the **round ligament of the uterus** probably arise from the mesoderm of the urogenital ridge caudal to the ovary. The mesonephric duct and tubules degenerate, but some remain as the paroöphoron (tubules), epoöphoron (tubules and duct), and Gartner's duct (duct). Where the fused müllerian ducts empty into the urogenital sinus, some entodermal tissue forms a vaginal plate, which eventually hollows out as the lower two-thirds of the vagina; the upper one-third is thought to arise from the müllerian ducts. The urogenital sinus caudal to the vaginal opening becomes enlarged as the vestibule.

In the female, the genital tubercle remains relatively small as the **clitoris**. The urethral folds become the **labia minora**, and the labioscrotal swellings develop into the **labia majora**. The **hymen** probably forms from the entoderm of the vaginal plate.

## CONGENITAL MALFORMATIONS OF THE UROGENITAL SYSTEM

**Horseshoe kidney** is usually due to fusion of the caudal ends of the two kidneys across the midline. This most likely occurs as they are approximated in their cranial migration out of the pelvis over the umbilical arteries. **Bifid ureter** is usually the result of a premature division of the ureteric bud. When this occurs, it may result in **double pelvis** or **double kidney**. **Exstrophy of the bladder** is caused by failure of mesoderm to invade the area anterior to the developing bladder. This results in improper development of the anterior abdominal wall and bladder with exposure of the posterior mucosal wall to the outside. **Congenital hydrocele** is a collection of fluid in a remnant of the processes vaginalis. In **hypospadias**, the external meatus of the urethra is on the ventral surface of the penis or scrotum. This may be caused by improper closure of the urogenital folds or labioscrotal swellings and by failure of the outer ectodermal cells to grow into the glans and join the penile urethra.

**Improper fusion of the müllerian ducts** leads to many different abnormalities of the uterus. These range from **uterus didelphys**, with its two bodies (bicornis) and two cervices bicollis, to **uterus arcuatus**, in which there is a minor degree of imperfect fusion in the fundus. **Double vagina** is caused by a defect in the canalization of the paired sinovaginal bulbs, which give rise to the vaginal plate.

**Pseudohermaphroditism** is a condition in which the individual has either testes (male pseudohermaphrodite) or ovaries (female pseudohermaphrodite) but possesses external genitalia of the opposite sex. In **hermaphroditism**, both ovarian and testicular tissue are present. In **testicular feminization**, the male duct system and external genitalia are not induced to develop. An immature female duct system and female external genitalia remain. In **adrenogenital syndrome**, a genetic abnormality results in absence of an enzyme necessary for the production of hydrocortisone. This leads to an **excess of adrenocorticotropic hormone**, which causes overproduction of adrenal androgens. In females, the excessive androgen causes hypertrophy of the clitoris and fusion of the labia majora, thus producing **female hermaphroditism**. In males, the overproduction may cause **precocious secondary sexual characteristics**.

# Chapter 17

# *Endocrine System*

## PITUITARY

The pituitary gland, or **hypophysis**, develops from two primordia. The **anterior lobe**, also known as the **adenohypophysis**, is glandular in appearance and develops as a dorsal outgrowth from the primitive oral cavity. This diverticulum, although initially hollow, becomes solid and forms the anterior lobe, pars intermedia, and pars tuberalis of the pituitary gland. The **posterior lobe** is known as the **pars nervosa** and develops as a ventral diverticulum of the diencephalon of the brain. It remains connected to the hypothalamus by way of the infundibular stalk. The cells that produce posterior lobe hormones are located in the hypothalamus. From here, they are transported into the posterior lobe of the pituitary, where they are stored and secreted.

## THYROID GLAND

The thyroid gland begins development as a ventral diverticulum of the primitive pharynx. Initially, it is connected to the floor of the pharynx by way of the thyroglossal duct. This connection is eventually lost as the thyroid migrates caudally in the anterior neck to assume its final position over the upper tracheal rings. The origin of the **thyroglossal duct** is evident in the adult as the **foramen caecum**, which is found on the dorsal surface of the tongue between the anterior two-thirds and posterior one-third of the tongue. The **thyroid** becomes **bilobed**, with the lobes connected across the midline by an isthmus.

In addition to the thyroid follicular cells, which produce **thyroxin**, the thyroid gland contains parafollicular cells, which produce a calcium-regulating hormone called **calcitonin**. These cells develop from neural crest and migrate into the thyroid gland as it is migrating in the neck.

## PARATHYROID GLANDS

The endoderm of the third and fourth pharyngeal pouches is responsible for the development of the inferior and superior parathyroid glands, respectively. The **inferior parathyroids** arise from the dorsal endoderm of the third pharyngeal pouch in conjunction with the thymus. As the thymus migrates caudally to the mediastinum, the parathyroid glands migrate to an inferior position behind the lobes of the thyroid. The **superior parathyroids** arise from the dorsal endoderm of the fourth pouch, but maintain a more superior position behind the lobes of the thyroid.

## SUPRARENAL GLANDS

The suprarenal, or **adrenal**, glands develop from two sources. The outer cortex is derived from condensation of the posterior body wall mesenchyme. It differentiates into the three zones characteristic of the suprarenal cortex. The suprarenal medulla, which consists of norepinephrine- and epinephrine-secreting cells, is derived from neural crest.

# Chapter 18

# Nervous System

## NEURAL TUBE FORMATION

The nervous system begins its development as a thickened midline area of ectoderm called the **neural plate**. A longitudinal neural groove develops in the neural plate. The sides of the groove are flanked by raised neural folds. By the end of the third week of development, the **neural folds** begin to fuse in the midline, beginning in the center of the neural plate and proceeding simultaneously in a rostral and caudal direction. This results in an open area of the neural fold rostrally, known as the **rostral neuropore**, and a similar area caudally, known as the **caudal neuropore**. The **rostral neuropore** closes at about day 26, and the **caudal neuropore** closes at about day 28, thus completing the formation of the **neural tube**.

## NEURAL CREST

As the ectoderm of the neural folds fuses to form the neural tube, some of the lateral ectoderm buds off and migrates throughout the embryo as the neural crest. Neural crest has been identified as the origin of many structures throughout the body including sensory and autonomic ganglia, neurolemmocytes of peripheral nerves, skeletal elements of the head, components of some of the endocrine glands, and melanocytes.

## DIFFERENTIATION OF THE NEURAL TUBE

Three concentric zones develop in the wall of the neural tube. The innermost layer, which lines the space within the neural tube is the ependymal zone. Cells in this layer differentiate into two types of cells: **spongioblasts**, which develop into the glial cells, and **neuroblasts**, which develop into neurons. Both of these cell types migrate into the mantle zone where they complete their development. The area of **gray matter** that is developing adjacent to the ependymal zone is the mantle layer. The outermost zone of developing **white matter** is the marginal zone.

Laterally on each side of the developing neural tube, paired masses of cells develop. These are separated from each other by a longitudinal groove called the **sulcus limitans**. The mass dorsal to the sulcus limitans is the **alar plate**, which will give rise to sensory components of the central nervous system. The mass ventral to the sulcus limitans is the basal plate and will give rise to motor components of the nervous system. The caudal end of the neural tube will develop into the **spinal cord**, and the rostral end of the neural tube will undergo extensive bending and folding to develop into the **brain**.

## DEVELOPMENT OF THE SPINAL CORD

The spinal cord develops from the caudal end of the neural tube. For the first 12 weeks of development, it is equal in length with the vertebral column. Spinal nerves emerge from the intervertebral foramina, with which they are associated. During postnatal growth and development, the vertebral column progressively lengthens, causing an upward shift in the caudal end of the spinal cord, the **conus medullaris**. This results in a lengthening of the caudal nerve roots in order to reach their respective intervertebral foramina. These elongated nerve roots are termed the **cauda equina**. At birth, the **conus medullaris** is located opposite the L3 vertebral level; in the adult, it is opposite the intervertebral disc between L1 and L2.

## DEVELOPMENT OF THE BRAIN

The rostral end of the neural tube develops into the brain. Initially, it is in a straight line and demonstrates the presence of three localized swellings, the primary brain vesicles: the **prosencephalon**, or forebrain; the **mesencephalon**, or midbrain; and the **rhombencephalon**, or hindbrain.

During the fourth week of development, a cervical flexure, which is concave anteriorly, develops between the rhombencephalon and the developing spinal cord. Subsequently, a midbrain flexure develops. This flexure is concave ventrally. A ventrally convex flexure, the **pontine flexure**, also develops in the region of the hindbrain. These flexure allow the developing brain to fit compactly within the cranial cavity.

During the fifth week of development, the three primary brain vesicles differentiate into five secondary brain vesicles. Lateral outgrowths of the prosencephalon constitute the telencephalic vesicles. These subsequently develop into the **cerebral hemispheres**. The caudal part of the prosencephalon becomes the **diencephalon**. The diencephalon will give rise to the **thalamus**, **hypothalamus**, and **pineal body**. The **optic vesicles** also arise as outgrowths of the diencephalon. The mesencephalon remains as the mesencephalon and becomes the adult midbrain. The rhombencephalon differentiates into the metencephalon, which will develop into the **pons**, **cerebellum**, and the **myelencephalon**, which will give rise to the **medulla**.

# PART III

# Histology

# Chapter 19

## Cell Structure

The human body is organized into cells, tissues, organs, and organ systems. **Cells** are the smallest units of structure of the body that have the ability to carry on all of the vital functions of the body. **Tissues** are groups of cells and intercellular material that are specialized for the performance of specific functions. Each **organ** consists of certain arrangements of tissues that join in performing a specific bodily function or functions. An **organ system** is a group of organs that perform related functions. The major organ systems of the body are the integumentary, skeletal, muscular, nervous, circulatory (cardiovascular and lymphatic), respiratory, digestive, endocrine, urinary, and reproductive systems.

The cells of the body are composed of **protoplasm**, which is organized into two major components: the **nucleus** and the **cytoplasm**. The cytoplasm is separated from its surrounding environment by a plasma membrane and from the nucleus by a nuclear membrane (envelope). Table 19-1 presents the common components of the cell and their functions.

## NUCLEUS

Most cells possess one nucleus, which, in the interphase (nondividing) state, consists of chromatin, one or more nucleoli, nucleoplasm, and an investing nuclear membrane. Some cells (e.g., red blood cells) have no nuclei, whereas many cells have more than one nucleus (e.g., skeletal muscle cells, parietal cells of stomach, hepatocytes, osteoclasts).

### Chromatin

Chromatin particles in the nucleus are the threads of **DNA** and proteins of the chromosomes. The DNA double helix (2-nm diameter) represents the genes of the chromosomes and is coiled or folded back on itself to form each chromatin fibrillar thread (10-nm diameter). Chromatin material occurs in two forms: **heterochromatin**, consisting of tightly coiled and condensed threads of DNA and protein; and **euchromatin**, comprised of less coiled, lighter staining, and more dispersed threads. The heterochromatin form may represent the more inactive metabolic state; euchromatin is more active in the synthesis of **RNA**. Although the ratio of euchromatin to heterochromatin is different in various cells of the body, the chromatin content and chromosomal number are constant for somatic cells and for sex cells. Somatic cells contain **46 chromosomes** consisting of 22 pairs of autosomes and 1 pair of sex chromosomes (XX in females and XY in males). Sex cells contain 23 chromosomes made up of half of each pair of autosomal chromosomes and one of the pair of sex chromosomes (i.e., an X or a Y).

**TABLE 19-1. Common Components of the Cell and Their Functions**

| Organelle | Function |
|---|---|
| Nucleus | Contains chromatin, the general material of the cell |
| Nucleolus | Composed of RNA |
| Plasma membrane | Phospholipid bilayer that separates cytoplasm from environment |
| Ribosomes | Contain RNA plus protein; necessary for the production of structural and secretory proteins |
| Rough endoplasmic reticulum | Membranes network with ribosomes in which secretory proteins are synthesized |
| Smooth endoplasmic reticulum | Does not have ribosomes; synthesis of steroid compounds |
| Golgi complex | Packages secretory products into vesicles |
| Mitochondria | Filamentous structures that generate adenosine triphosphate for use as an energy source for the cell |
| Lysosomes | Contain digestive enzymes for removal of worn cell parts and ingested substances |
| Centrosomes | Consists of centrioles and functions in cell division and movement of cilia and flagella |
| Microtubules | Cylinders that contribute to structure of cilia, flagella, and cytoskeleton |
| Filaments and fibrils | Slender threads of protein that contribute to cytoskeleton |
| Vesicles | Spherical organelles that serve as storage reservoirs for secretory product in the cell |
| Inclusions | Generic category for non–membrane-bound materials stored in cells (e.g., glycogen) |

## Nucleolus

One nucleolus, or several nucleoli, is present in each cell. Most of each nucleolus is protein, 5% to 10% is RNA, and a small amount is DNA. The protein and RNA often appear as a meandering thick thread of ribonucleoprotein called a **nucleolonema**. The **nucleolonema** appears as fibrillar (pars fibrosa) and granular (pars granulosa) portions of the nucleolus. A chromosomal portion of intranucleolar chromatin (nucleolar organizer) contains the genes from which **ribosomal RNA (rRNA)** is transcribed and synthesized. The newly formed rRNA is packaged with proteins into ribosomal subunits. These units pass from the pars fibrosa and pars granulosa before being transferred to the cytoplasm, where they are assembled into ribosomes.

## Nucleoplasm

Nucleoplasm is an amorphous substance consisting of proteins, ions, and metabolites. Chromatin and nucleoli appear to be suspended in this substance.

## Nuclear Membrane

The nuclear membrane (envelope) consists of two parallel unit membranes that are separated by a perinuclear space. The unit membrane also is a component of the plasma membrane, Golgi apparatus, lysosomes, mitochondria, coated vesicles, and secretion granules. With the electron microscope, the basic unit membrane is seen as an 8-nm thick complex of an electron-light area sandwiched between two electron-dense areas. The dense areas represent the **membrane proteins** and the polar ends of the phospholipid bilayers that make up the membrane. The light area represents the fatty acids of a bilayer of **phospholipids**. The outer unit membrane of the nuclear envelope is continuous with the endoplasmic reticulum (ER). The inner unit membrane of the nuclear envelope is attached to the chromosomes by a fibrous network of three polypeptides called the **nuclear lamina**. Circular openings (nuclear pores) covered by thin membranes occur at intervals throughout the nuclear envelope. The nuclear pores are important passageways for the transfer of substances between the cytoplasm and the nucleus.

# CYTOPLASM

The cytoplasm contains dynamic "living" structural components of the cell (organelles) and "nonliving" metabolites or products of the cell (inclusions). The **organelles** include the plasma membrane, ribosomes, ER, Golgi apparatus, lysosomes, mitochondria, centrioles (centrosome), fibrils, filaments, microtubules, peroxisomes, and coated vesicles. **Inclusions** include stored glycogen, lipid, and protein and secretion granules.

## Plasma Membrane

The plasma membrane is a unit membrane (8 nm thick) consisting of a fluid bilayer that contains membrane proteins. The lipid component consists of phospholipids, cholesterol, and glycolipids. The **phospholipid molecules** have their hydrophilic glycerol-phosphate heads oriented toward either the outer or the inner surface of the plasma membrane. The **fatty acid tails** of this double layer of phospholipid molecules meet in the center of the plasma membrane to form an intermediate hydrophobic zone. Cholesterol molecules stabilize the membrane, whereas glycolipids are located in the outer portion of the wall and seem to serve in cellular communication.

**Membrane proteins** are responsible for most of the specialized functions of the cell and the types of proteins vary from cell to cell, depending upon its function. Membrane proteins attach the cytoskeletal filaments of the cell membrane. They attach the cell to the extracellular matrix, transport molecules into and out of the cell, act as receptors for chemical signals, and possess specific enzyme activity. Some membrane proteins extend through the plasma membrane (transmembrane proteins) and help to transport specific molecules into and out of the cell. Other membrane proteins extend only from the internal or external surface (e.g., glycoproteins) of the membrane. These proteins may serve as enzymes, attachment sites of the cytoskeleton, or receptor sites. The membrane proteins can move around in the plasma membrane.

A **glycocalyx** coats the outer surface of the cell. It consists of the carbohydrate components of glycolipids and glycoproteins of the membrane and also of glycoproteins and proteoglycans that have been absorbed on the cell surface.

## Ribosomes

Ribosomes are **rRNA plus protein**. They occur in the cytoplasm as free ribosomes or in clusters of polyribosomes or as the granular component of the **rough ER (RER)**. Generally the free ribosomes and polyribosomes are involved in the synthesis of structural proteins and enzymes that stay in the cell. The RER is necessary for the synthesis of proteins that are secreted from the cell and for producing lysosomal enzymes used in cellular digestion. In routine histologic stains, ribosomes are generally stained by the "basic" component of the stain and hence contribute to the "basophilia" observed in some cells.

# ENDOPLASMIC RETICULUM

The **ER** is a membranous network of tubules and flattened sacs that are often continuous with the outer layer of the nuclear membrane. ER that does not have attached ribosomes is called **smooth ER (SER)**; that with ribosomes is **RER**. SER is involved in steroid production in the adrenals and gonads, excitation-contraction mechanisms of muscle, absorption of fats in the intestine, and in cholesterol and lipid metabolism and drug detoxification in the liver. RER is the organelle in which proteins for secretion by the cell are produced. Hence, it is a prominent feature of cells that are active in protein synthesis.

## Golgi Apparatus

One or more Golgi apparatus **(Golgi complex)** usually is found near the nucleus. Each consists of stacks of flattened smooth-surfaced saccules composed of unit membrane. The Golgi apparatus has an immature forming (*cis*) face where new stacks are added, and a maturing (*trans*) face where secretory vesicles may bud off. Protein and sugar molecules manufactured in the RER pass to the forming face through transfer vesicles. Proteins are sorted and condensed in the Golgi apparatus. Sugars may be added to the protein, and the product is packaged into secretory vesicles. The Golgi apparatus also is involved in cell membrane and lysosome production.

## Lysosomes

Lysosomes are unit membrane-bound vesicles of digestive (hydrolyzing) enzymes. In the digestion of worn-out parts of the cell, primary lysosomes fuse with autophagic vacuoles containing the dead portion. In digesting endocytosed (i.e., phagocytized, pinocytosed) material, primary lysosomes fuse with the internalized membrane-bound substance to form a secondary lysosome. If the material is not completely digested, a membrane-bound residual body remains as a cell "inclusion."

## Mitochondria

Mitochondria are filamentous or granular structures comprised of a double unit membrane. The inner membrane, separated from the outer membrane by a space 6 to 10 nm wide is thrown into many transverse folds (**cristae**) or tubules, which project into the fluid matrix in the interior of the mitochondrion. The inner surface of the cristae is studded with many "elementary particles."

Mitochondria are the sites where phosphate bond energy in the form of **adenosine triphosphate (ATP)** is produced. In this process, pyruvate passes through the mitochondrial membranes and into the matrix, where it is converted to **acetyl coenzyme A (acyl-CoA)**. Under the influence of the enzymes of the citric acid cycle in the matrix, the acyl-CoAs are transformed into carbon dioxide ($CO_2$) and electrons (e.g., NADH and $FADH_2$). These electrons are then passed along an electron transport system of enzymes on the inner mitochondrial membrane and its cristae. In this process, the energy that is available is used in forming ATP by adding inorganic phosphate to **adenosine diphosphate**. A coupling factor (enzyme) needed in this final step of **oxidative phosphorylation** is located in the elementary particles of the inner membrane.

Mitochondria contain their own specific mitochondrial DNA, which is localized in the matrix space. Mitochondria are the only organelles capable of self-replication. Mitochondrial morphology varies according to the specific function of the cell in which they are contained. The mitochondria of most cells tend to have shelf-like cristae; those of steroid-producing cells have tubular cristae.

## Centrosome

The centrosome (**cell center**) is located near the nucleus and consists of two centrioles surrounded by homogeneous cytoplasm. The **centrioles** are cylindrical and lie perpendicular to each other. The wall of each centriole consists of nine parallel units, each of which is made up of three fused microtubules. **Basal bodies**, located at the base of cilia or flagella, are similar in structure to centrioles. Centrioles are nucleation centers for microtubule formation, which is apparent in the formation of the spindle during cell division.

## Microtubules

Microtubules are straight or wavy cylinders about 25 nm in diameter with walls made of rows of alpha and beta tubulin. Microtubules function in maintaining cell shape, in intracellular transport, and in cell movement. **Cilia** and **flagella** are motile, microtubular processes that extend from the free surfaces of many different cells. They consist of a core of microtubules, the axoneme, which is arranged as two central microtubules surrounded by nine peripheral doublets, each of which shares a common wall of two or three protofilaments. Dynein arms possessing ATPase activity extend from one doublet toward an adjacent doublet. Motion of cilia and flagella are most likely the result of doublets sliding within the axoneme as the dynein arms of one doublet walk along the adjacent doublet.

## Filaments and Fibrils

**Filaments** are slender threads of protein molecules. Filaments less than 5 nm in diameter are composed of actin and are part of the cytoskeleton. They are involved in cell contraction. Filaments between 8 and 12 nm in diameter are intermediate filaments (tonofilaments), which also lend skeletal support. Bundles of filaments comprise **fibrils**. Thus, **myofibrils** of skeletal muscle are comprised of small myofilaments (actin and myosin). **Actin filaments** also make up the core of microvilli, which project from the free surface of absorptive cells. The actin filaments of the microvilli intermingle with the filaments of the terminal web at the base of the microvilli.

## Peroxisomes

Peroxisomes are small membrane-bound vesicles containing several oxidative enzymes involved in the production of hydrogen peroxide. They are usually approximately 0.5 to 1.0 μm in diameter and contain a dense core when viewed by electron microscopy.

## Vesicles

Vesicles are small spherical **organelles**. They generally develop by a budding off membranes from other parts of the cell. In general, they store and transport material within the cell. The membranes of coated vesicles are coated on their cytoplasmic surface by several proteins (e.g., clathrin). Coated vesicles are involved in intracellular transport or packaging of secretory material, or they are specialized for the receptor-mediated endocytosis of macromolecules from the extracellular fluid. In this process, the macromolecules are internalized when the coated plasma membrane to which it binds is pinched off as a coated vesicle. Cholesterol is internalized by this method. Coated vesicles join to form endosomes, before they fuse with lysosomes.

## Inclusions

Inclusions are a general category of structures found within the cytoplasm of cells; some are non–membrane bound. They include a wide variety of **intercellular substances**. Glycogen, lipid droplets, and pigments (e.g., lipofuscin, melanin, carotene) are stored in many of the cells of the body. Secretion granules are membrane-bound vesicles containing a protein or protein-carbohydrate secretory product.

## Intercellular Junctions

Cells demonstrate variable degrees of cohesion and attachment. These junctions are particularly pronounced in epithelial and muscle tissues. **Tight junctions (zonulae occludens)** form bands around the apical wall of epithelial cells. Because the outer leaflets of adjacent cell membranes fuse in a series of ridges in this type of junction, a relatively tight seal prevents passage of materials from the lumen between the cells. The **zonula adherens** is a band that forms just deep to the tight junction. Microfilaments of the terminal web insert into a dense plaque on the cytoplasmic surface of the zonula membrane. This junction retains a 20-nm intercellular space. **Desmosomes** (maculae adherens) are small spot junctions that are similar in structure to the zonula adherens, except that tonofilaments insert into the attachment plaque and there may be dense material in the intermediate space. The **gap junction** (nexus) is characterized as an apposition of membranes of adjacent cells with a 2-nm intercellular space. This space is bridged by connexons that permit cells to communicate with each other through their 2-nm lumina.

# Chapter 20

# *Protein Synthesis*

The mechanisms for protein synthesis are important because structural proteins comprise most of the important structures of the body and because most of the chemical reactions in the body are catalyzed by enzymes that are proteins. The general sequence of events in protein synthesis is as follows: (1) activation of genes, (2) transcription of DNA to form messenger RNA (mRNA), (3) recognition of amino acids by their specific transfer RNA (tRNA) molecules, and (4) translation of the mRNA by the tRNA at the ribosomes (rRNA).

## ACTIVATION OF GENES

**Structural genes (cistrons)** are specific segments of the DNA molecule that provide coded messages essential for the assembly of certain amino acids in the production of specific structural proteins or enzymes. For different genes to exist, the double helix of DNA that comprises the chromosomes must possess structures whose variation in sequencing permits different proteins to be formed. These components of DNA are sequences of nucleotides. Each **nucleotide** is assembled so that one deoxyribose sugar unit and one unit of phosphoric acid provide part of the DNA strand, and a nitrogenous base forms a side chain that pairs with the base of a nucleotide in the adjacent DNA strand of the double helix.

The only four nitrogenous bases that exist in DNA are two **purines**, adenine (A) and guanine (G), and two **pyrimidines**, cytosine (C) and thymine (T). When the bases pair, adenine bonds only with thymine, and cytosine bonds only with guanine. The sequence of the bases (i.e., the sequence of genetic code letters A, T, C, and G) along a single strand of DNA determines which **amino acids** will be combined to form a particular protein. A sequence of three nitrogenous bases (codon) codes for each amino acid. Several codons in sequence form each cistron; thus, the genetic code for protein synthesis is found in the nitrogenous base sequence of DNA.

Although each somatic cell contains all of the genes for that individual, relatively few genes are activated at any one time. A gene is activated only when a specific mRNA molecule is to be transcribed from the DNA. In this activation process, it appears that a specific gene regulatory protein acts as the activator by binding to the DNA and facilitating the binding of RNA polymerase to a promoter segment of the DNA. The binding of RNA polymerase promotes transcription of a specific mRNA when the specific RNA polymerase recognizes a starting point and moves down the gene to its termination signal. The affinity of the specific gene regulatory protein for the DNA may be increased or decreased by inducer and inhibitory ligands, respectively.

## TRANSCRIPTION OF MESSENGER RNA

mRNA has a structure similar to a single strand of DNA, except that ribose sugar takes the place of deoxyribose, and thymine is replaced by uracil. When genes are activated, the DNA strands of the helix separate in the region of the involved cistron, and the exposed nitrogenous bases serve as templates for the synthesis of an mRNA molecule. In this process, adenine of the DNA pairs with uracil of the developing mRNA molecule, the thymine of DNA pairs with adenine of RNA, the guanine of DNA pairs with cytosine of RNA, and the cytosine of DNA pairs with guanine of RNA. The newly formed mRNA separates from the DNA strand, passes into the cytoplasm, and becomes associated with **tRNA**.

## RECOGNITION AND TRANSPORT OF AMINO ACIDS BY TRANSFER RNA

tRNA molecules are multipolar structures produced in the nucleus. Each tRNA molecule passes into the cytoplasm, where one of its poles recognizes and binds a specific amino acid. Another pole of each tRNA molecule contains a triplet of nitrogenous bases that is able to "read" the complementary codon of an mRNA molecule. A third pole recognizes ribosomes.

## TRANSLATION OF THE MESSENGER RNA MESSAGE

rRNA and tRNA play roles in translating the mRNA message. The small subunit of the ribosome attaches to a start codon on the mRNA molecule and then travels down the molecule from codon to codon. At each codon, the appropriate tRNA that "reads" the codon attaches to the mRNA and deposits its amino acid. As each amino acid is assembled into a new protein molecule, the tRNAs are released and another ribosome may read the mRNA. When each ribosome reaches a **stop codon**, the translation is terminated.

If the protein that is produced is a structural protein, the ribosomes used are **free ribosomes**. If the protein is to be a **secretory product**, the protein molecule passes through the ribosome, into the lumen of the rough (granular) endoplasmic reticulum (ER), and through transfer vesicles to the region of the Golgi apparatus, where it is concentrated and packaged into secretory granules (storage vacuoles). When these membrane-bound granules fuse with the plasma membrane, the secretory product is elaborated by exocytosis. In some cells, the concentration step is eliminated and the product is transferred directly to the plasma membrane from the ER and Golgi region.

# Chapter 21

# *Cell Division*

Somatic cells divide by the **mitotic process** of cell division; sex cells divide by **meiosis**. Before either type of cell division, the stem cell undergoes an **interphase stage** when the DNA is replicated during the synthesis (S) phase. In this replication process, the DNA strands separate (replication fork) when the bonds between nitrogenous bases are broken, and free nucleotides attach to their complementary bases; thus, two new helixes are formed, of which one DNA strand in each helix has served as the template for the new chromatid. If mitosis occurs, the cell will divide into two daughter cells, each with **46 chromosomes**. In meiosis, two divisions (meiosis I and II) take place with no further replication of the DNA, thus producing four cells, each with the **haploid (23) number of chromosomes**.

## MITOSIS

After interphase, the cell enters the **prophase stage**, when the chromosomal strands become coiled as they shorten and thicken. The two chromatids of each chromosome are joined by a centromere, the chromosomes are suspended in a spindle of microtubules that bridges the two centrioles, and the nucleolus and nuclear envelope disappear. During **metaphase**, the chromosomes align on a metaphase (equatorial) plate. In **anaphase**, the chromosomes separate in the region of the centromeres, and each chromatid is pulled toward opposite poles of the cell. During **telophase**, a cleavage furrow in the plasma membrane continues to separate the cytoplasm into two daughter cells, and the nucleoli and nuclear envelope reappear.

## MEIOSIS

In the **first meiotic division (M1)**, cells undergo a lengthy prophase, and the homologous chromosomes come together in a synapsis. This configuration is called a **tetrad of four chromatids**. In metaphase, these synapsed pairs line up on the equator, but they separate during anaphase and telophase so that the two new daughter cells will have only 23 chromosomes, each of which is represented by two chromatids. Thus, M1 is a **reduction division**.

In the **second meiotic division (M2)**, the centromeres separate during metaphase so that each of the new daughter cells receives 23 chromatids, each of which becomes a new chromosome. The result of meiosis is the production of four sex cells from one stem cell, with each of the four cells having the haploid (23) number of chromosomes. In the female, three of the four daughter cells (polar bodies) die and do not become oocytes. In the male, all four daughter cells become spermatids, which transform into spermatozoa.

# Chapter 22

# *Tissues*

All cells of the body are organized into one of four tissue types. These four tissue types in turn comprise the structural elements of all organs. The four basic tissue types are epithelia, connective tissue, muscle, and nerve.

## EPITHELIA AND GLANDS

Epithelium is a tissue that is comprised of sheets of cells and a minimal amount of extracellular matrix. Epithelia line the body cavities and hollow organs and cover the viscera and body surfaces. Sheets of epithelia are also organized into special secretory organs called **glands**.

In general, **epithelia** are classified on the basis of the number of layers of cells in the epithelium (Table 22-1). Those with one layer of cells are **simple** epithelia, those with more than one layer are termed **stratified**. In addition, epithelia are classified by the shape of their surface cells. The most common shapes of epithelial cells are flat or squamous, cuboidal, and columnar.

**Pseudostratified columnar epithelium** is so named because the cells appear to be in layers, although they really form only a single layer of columnar cells. This type of epithelium also possesses surface specializations known as **cilia** and is primarily limited to the lining of the upper respiratory tract.

**Transitional epithelium** is a specialized form of epithelium found in the urinary bladder. It is characterized by several layers of cells, the surface cells of which are rounded. The typical epithelial cell displays an apical surface that faces the lumen of the organ of which the epithelium is a part. The apical surface may possess specializations related to the specific function of the epithelium. The apical surface of columnar absorptive cells is characterized by the presence of **microvilli**, which increase the surface area of the cell.

Cells that are part of an epithelial sheet responsible for the movement of fluids across the surface may possess **cilia**. The lateral surfaces of epithelial cells display specializations that contribute to the adherence of adjacent cells to each other. These include occluding junctions (**zonulae occludens**), anchoring junctions (**zonulae adherens**), and communicating junctions (gap junctions). The occluding junctions prevent the diffusion of molecules between epithelial cells. Anchoring junctions attach cells to each other and stabilize them. Communicating junctions allow for the selective exchange of molecules between cells. The basal surface of epithelial cells rests on a specialized extracellular matrix component called the basement membrane. The **basement membrane** is composed of collagen plus other extracellular proteins and is secreted by both the epithelial cells and the underlying connective tissue cells.

**TABLE 22-1. Types of Epithelia, Characteristics, and Locations**

| Epithelium | Characteristics | Locations |
|---|---|---|
| Simple squamous | Single layer of flat cells | Lining of heart and blood vessels (endothelium)<br>Lining of body cavities (mesothelium) |
| Simple cuboidal | Single layer of cuboidal cells | Collecting tubules of the kidney, thyroid gland |
| Simple columnar | Single layer of columnar cells | Mucosa of gastrointestinal tract |
| Pseudostratified columnar | Single layer of columnar cells; appear stratified | Lining of the respiratory tract, cilia present |
| Stratified squamous | Several layers of cells; surface cells are flat | Epidermis of the skin |
| Stratified cuboidal | Multiple layers of cuboidal cells | Rare, parts of the reproductive tract |
| Stratified columnar | Multiple layers of columnar cells | Rare, parts of the reproductive tract |
| Transitional | Several layers of cells; surface cells are rounded | Urinary tract |

**Glands** are comprised of individual epithelial cells or sheets of epithelial cells that produce and secrete substances used elsewhere in the body. In general, glands are classified as **endocrine** if they secrete into blood vessels, or **exocrine** if they secrete into a duct system or onto an epithelial surface. Exocrine glands are further characterized by the presence and nature of their duct system. **Unicellular glands** are comprised of single cells within epithelial sheets and discharge their secretions onto the surface of the epithelium. **Goblet cells** in the lining epithelium of the gastrointestinal tract are an example of this kind of gland. **Multicellular glands** include secretory epithelial sheets, such as the lining of the stomach, and all glands that possess ducts. These glands are further subdivided into simple and compound gland types. Simple glands have unbranched duct systems and are classified according to the shape of their secretory end-pieces. These may be **tubular** or **alveolar (acinar)**. Compound glands have branched duct systems and their secretory end pieces may be tubular, alveolar, or tubuloalveolar (Fig. 22-1).

# CONNECTIVE TISSUE

Connective tissue functions to provide **protection** and **support**. It is characterized by the presence of cells embedded in an abundant matrix. The adult connective and supportive tissues are connective tissue proper, cartilage, bone, and blood. Connective tissue proper is further classified into loose irregular (areolar) connective tissue and dense regular and irregular connective tissues (Table 22-2). These tissues contain cells and a preponderance of intercellular fibers and ground substance.

## Loose Irregular Connective Tissue

Loose connective tissue is found in the superficial and deep fascia and as the stroma of most organs. It is generally considered the **packing material of the body**. Loose connective tissue contains most of the cell types and all of the fiber types found in the other connective tissues.

The most common cell types are the fibroblast, macrophage, adipose cell, mast cell, plasma cell, and wandering cells from the blood.

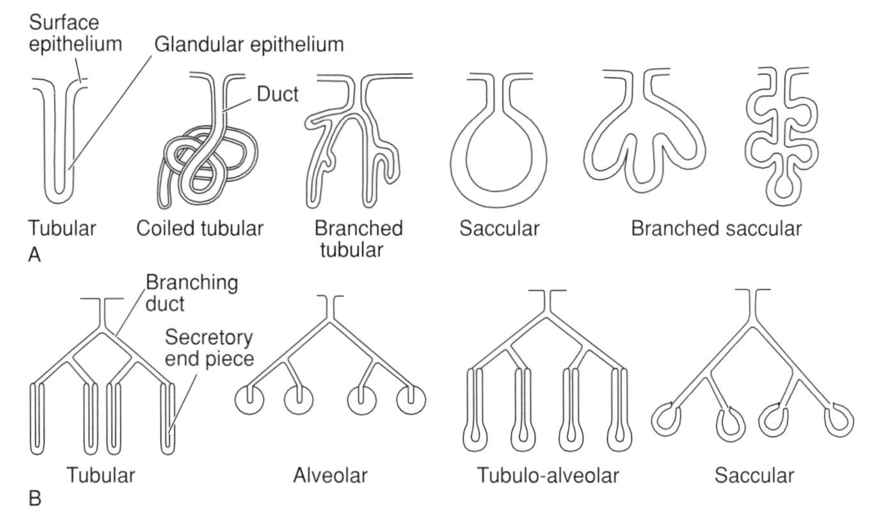


**Figure 22-1.** Diagram illustrating the principal types of glands. Structural types of simple **(A)** and compound **(B)** multicellular glands.

**Fibroblasts.** These cells contain the organelles that permit them to produce all of the fiber types and the intercellular material. In their production of these proteinaceous substances, messenger RNA (mRNA), ribosomal RNA (rRNA), and transfer RNA (tRNA) are produced in the nucleus and pass to the cytoplasm. Amino acids that have been taken into the fibroblast attach to specific tRNA and are translated on the mRNA in the region of the ribosomes of the rough endoplasmic reticulum (RER). The polypeptides produced pass through the cisternae of the RER to the region of the **Golgi complex**, where they are packaged into membrane-bound procollagen macromolecules that

**TABLE 22-2. Principal Types of Connective Tissue and Its Cellular and Fiber Composition**

| Connective Tissue Type | Principal Cells | Matrix |
|---|---|---|
| Loose connective tissue | Fibroblasts, macrophages, mast cells, plasma cells, adipocytes | Collagen, elastic, and reticular fibers plus ground substance of glycoproteins and glycosaminoglycans |
| Dense irregular connective tissue | Same as above | Same as above, but collagenous fibers predominate |
| Dense regular connective tissue | Fibroblasts | Parallel arrangement of predominately collagen fibers; ligament flava are predominately elastic |
| Hyaline cartilage | Chondrocytes | Glycosaminoglycans, chondroitin sulfate, collagenous fibers |
| Elastic cartilage | Same as above | Same as above with addition of elastic fibers |
| Fibrocartilage | Same as above | Same as hyaline cartilage with large bundles of collagen fibers |
| Bone | Osteocytes, osteoblasts, and osteoclasts | Dense collagen bundles and osteomucoid ground substance; calcium phosphate |

attach to the cell surface before discharge from the fibroblast. Outside the cell, the procollagen molecules are cleaved of their registration peptides and form tropocollagen, which is assembled into collagen. The Golgi complex also is responsible for adding the carbohydrate components to the **glycosaminoglycans (GAG)** of the ground substance.

**Macrophages**. These cells are part of the reticuloendothelial system (mononuclear phagocyte system). They possess large lysosomes that contain digestive enzymes, which are necessary for the digestion of phagocytized materials.

**Mast cells**. These cells occur mostly along blood vessels, and their granules contain heparin and histamine, produced by these cells.

**Plasma cells**. These cells are part of the immune system in that they produce circulating antibodies. They are extremely basophilic because of their extensive RER.

**Adipose cells (adipocytes)**. These cells are found in varying quantities. When they predominate, the tissue is called **adipose tissue**. Two kinds of adipocytes exist in the body. The majority contribute to **yellow fat** distributed throughout the body. These fat cells are unilocular because each cell contains a single lipid droplet. Multilocular fat cells are a feature of **brown fat**, which is distributed along blood vessels. These cells contain many droplets of lipid and have a "foamy" appearance in routine histologic preparations.

Collagenous, reticular, and elastic fibers are irregularly distributed in loose connective tissue.

**Collagenous fibers**. These fibers are usually found in bundles of fibers and provide strength to the tissue. Each fiber is made up of fibrils, which are composed of staggered monomers of tropocollagen. Many different types of collagen are identified on the basis of their molecular structure. Of the most common types, collagen type I is the most abundant. It is found in dermis, bone, dentin, tendons, organ capsules, fascia, and sclera. Type II is located in hyaline and elastic cartilage. Type III probably is the collagenous component of reticular fibers. Type IV is found in basal laminas. Type V is a component of placental basement membranes.

**Reticular fibers**. These fibers are smaller, more delicate fibers that form the basic framework of reticular connective tissue.

**Elastic fibers**. These fibers branch and provide elasticity and suppleness to connective tissue.

**Ground substance**. This is the gelatinous material that fills most of the space between the cells and fibers. It is composed of GAG and structural glycoproteins; its properties are important in determining the permeability and consistency of the connective tissue.

## Dense Connective Tissue

Dense irregular connective tissue is found in the dermis, periosteum, perichondrium, and capsules of some organs. All fiber types are present, but collagenous fibers predominate. Dense regular connective tissue occurs as aponeuroses, ligaments, and tendons. In most **ligaments** and **tendons**, collagenous fibers are prevalent and are oriented parallel to each other. **Fibroblasts** are the only cell type present. The ligamenta flava in the vertebral column are considered elastic ligaments because elastic fibers dominate.

## Cartilage

Cartilage is composed of **chondrocytes** embedded in an **intercellular matrix**, consisting of fibers and an amorphous firm ground substance. Three types of cartilage (hyaline, elastic, and fibrous) are distinguished on the basis of the amount of ground substance and the relative abundance of collagenous and elastic fibers.

**Hyaline cartilage**. This type of cartilage is found as costal cartilages, articular cartilages, and cartilages of the nose, larynx, trachea, and bronchi. The intercellular matrix consists primarily of collagenous fibers and a ground substance rich in chondromucoprotein, a copolymer of a protein and chondroitin sulfates. The gel-like firmness of cartilage depends on the electrostatic bonds between collagen and the GAG and the binding of water to the GAG. The GAG are composed of chondroitin and keratan sulfates covalently linked to core proteins, which are bound to hyaluronic acid molecules. Chondrocytes occupy lacunae. During the growth period of the cartilage, these cells exist as chondroblasts and produce the intercellular matrix.

All types of cartilage grow **interstitially** by the mitoses of cells in the center of the cartilage mass; most types also grow **appositionally** by the formation of chondroblasts from undifferentiated cells in the cellular layer of the perichondrium. Unlike the fibrous layer of the perichondrium, the cellular layer and the cartilage are avascular, so they receive nutrients and oxygen through diffusion from blood vessels in the fibrous layer of the perichondrium. Articular cartilages receive nutrients by diffusion from blood vessels in the marrow and from the synovial fluid. As a person ages, there is a decrease of GAG and an increase in noncollagenous proteins; calcification may occur because of degenerative changes in the cartilage cells.

**Elastic cartilage**. This type of cartilage is found in the auricle of the ear, auditory tube, and epiglottic, corniculate, and cuneiform cartilages of the larynx. Elastic fibers predominate and thus provide greater flexibility to this cartilage. Calcification of elastic cartilage is rare.

**Fibrous cartilage**. This type of cartilage occurs in the anchorage of tendons and ligaments, in intervertebral disks, in the symphysis pubis, and in some interarticular disks and ligaments. Chondrocytes occur singly or in rows between large bundles of collagenous fibers. Compared with hyaline cartilage, only small amounts of hyaline matrix surround the chondrocytes of fibrous cartilage.

## Bone

Bone tissue consists of **osteocytes** and an **intercellular matrix** that contains organic and inorganic components. The organic matrix consists of dense collagenous fibers and an osseomucoid substance that contains chondroitin sulfate. The inorganic component is responsible for the rigidity of bone and is composed chiefly of **calcium phosphate** and **calcium carbonate** with small amounts of magnesium, fluoride, hydroxide, and sulfate. Electron microscopic studies show that these minerals are deposited in an orderly fashion on the surface of the collagenous fibrils in their interband areas.

In the basic organization of bone tissue, osteocytes lie in lacunae and extend protoplasmic processes into small canaliculi in the intercellular matrix. The protoplasmic processes of adjacent osteocytes are in contact with each other, and **gap junctions** are present. The matrix is organized into concentric layers or **lamellae**. The number and arrangement of lamellae differ between compact and cancellous bone.

**Compact bone**. This type of bone is found in the shaft of long bones and the cortical bone that makes up the outer surface of most bones. It contains haversian systems (osteons), interstitial lamellae, and circumferential lamellae. **Haversian systems** consist of extensively branching haversian canals oriented chiefly longitudinally in long bones. Each canal contains blood vessels and osteogenic cells and is surrounded by 8 to 15 concentric **lamellae** and **osteocytes**. The collagenous fibers in adjacent lamellae run essentially at right angles to each other and spiral around the canal.

**Nutriments** from blood vessels in the haversian canals pass through canaliculi and lacunae to reach all osteocytes in the system. Interstitial lamellae occur between haversian systems and represent the remains of parts of degenerating haversian and circumferential lamellae. Outer and inner circumferential lamellae occur under the periosteum and endosteum, respectively. **Volkmann's canals** enter through the outer circumferential lamellae and carry blood vessels and nerves that are continuous with those of the haversian canals and the periosteum. **Sharpey's fibers** are coarse perforating fibers that anchor the periosteum to the outer circumferential lamellae (Fig. 22-2).

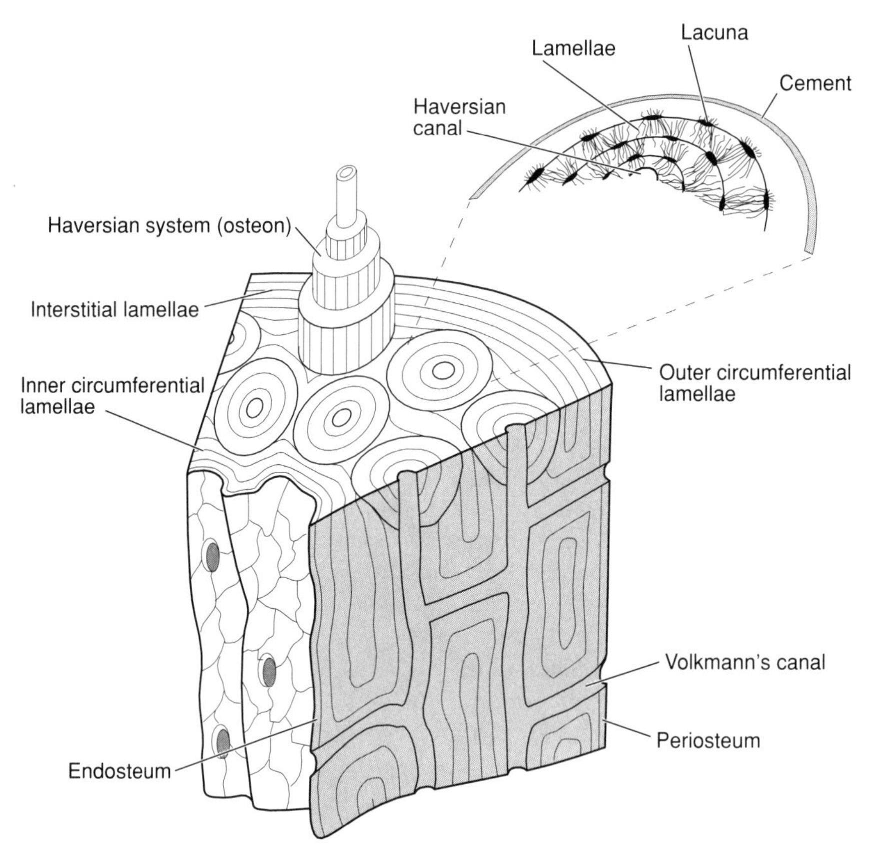


**Figure 22-2.** Diagram illustrating the major features of compact bone. Inset is a detail of a haversian system.

Bones are supplied by a loop of **blood vessels** that enter from the periosteal region, penetrate the cortical bone, and enter the medulla before returning to the periphery of the bone. Long bones are specifically supplied by arteries that pass to the marrow through diaphyseal, metaphyseal, and epiphyseal arteries. In the marrow cavity, some arteries end in sinusoids and others branch and enter the haversian canals, where they supply fenestrated capillaries. The **marrow sinusoids** drain to veins that leave through nutrient canals. The capillaries of the haversian canals drain to veins that pass centrifugally to the periosteum and adjacent muscles.

Bone undergoes extensive **remodeling**, and haversian systems may break down or be resorbed so calcium can be made available to other tissues of the body. **Bone resorption** occurs by osteocytic osteolysis or by osteoclastic activity. In **osteocytic osteolysis**, osteocytes resorb bone that lies immediately around the lacunae. In **osteoclastic activity**, large multinucleated osteoclasts arise from osteoprogenitor cells and abut an osseous surface. Here, their extensive ruffled surfaces and proteolytic enzyme secretions seem to be involved in the resorption of more extensive portions of bone than in osteocytic osteolysis.

**Osteoclasts** are components of the mononuclear phagocyte system. In this way, portions of old haversian systems are resorbed, or longitudinal depressions are formed on the periosteal and endosteal surfaces of the bone. If new haversian systems are to be laid down in the gutters or tubes that remain after the resorptive process is complete, **osteoblasts** differentiate from osteogenic cells of the enlarged haversian canal or periosteum and begin to lay down a lamella at the periphery of the space. Successive new concentric lamellae are laid down inside this initial lamella.

**Cancellous bone.** This type of bone differs from compact bone in that the lamellae are organized into trabeculae or spicules. Few haversian systems are present, and most osteocytes are generally closer to the blood supply than in compact bone.

## Histogenesis of Bone

Two basic patterns of bone formation exist: intramembranous and endochondral. In both types, the process of forming bone tissue and the histologic structure of the bone formed are identical. The major difference between these two types of development is the environment within which bone tissue is laid down.

**Intramembranous bone formation.** This type of bone formation occurs in flat bones of the skull and face and the clavicle. In this type of development, **ectomesenchymal** (mesenchyme derived from neural crest) **cells** differentiate into osteoblasts in a region where mesenchymal cells have produced a fine-fibered vascular membrane. The osteoblasts lay down lamellae of collagenous fibers and ground substance in the form of a meshwork of trabeculae within the membrane. Some osteoblasts become entrapped as osteocytes in this osteoid tissue. When organic osteoid tissue becomes impregnated with inorganic salts, it is called **osseous tissue**. Some intratrabecular spaces become marrow cavities when their mesenchyme differentiates into reticular connective tissue and blood-forming cells. Others become haversian canals as concentric lamellae are formed.

At the periphery of the entire developing bone (e.g., outer and inner surfaces of skull bones), the bone becomes quite compact in its development. This is accomplished by a mesenchymal condensation around the bone that differentiates into a **periosteum**, whose inner cells become osteoblasts and lay down compact bone; thus, the bone takes on an appearance of outer and inner tables of compact bone, between which is the dipole of spongy trabecular bone. Osteoclasts are associated with bone resorption, which takes place chiefly on the inner surfaces of the tables and trabeculae. The membranous junction between two developing flat bones is eventually ossified as a suture.

**Endochondral bone formation.** This type of bone formation is characterized by a cartilage model of the bone preceding bone histogenesis. In the formation of the cartilage model of a long bone, the oldest cartilage is found in the center of the **shaft (diaphyseal) region**. Cells in this region hypertrophy, produce phosphatase, and bring about calcification of the surrounding cartilaginous matrix. The result of this calcification is inhibition of diffusion of nutrient materials to the chondrocytes, and they die or they may become osteoprogenitor cells.

While the cartilage in the center of the shaft is **calcifying**, the chondrogenic layer of the perichondrium is becoming **vascularized**. In this new environment, the undifferentiated mesenchymal cells of the chondrogenic layer start to differentiate into osteoblasts, which lay down a bony collar around the shaft of the cartilage model. The perichondrium is now a periosteum. Osteogenic tissue containing osteoprogenitor cells and blood vessels from this osteogenic layer of the periosteum passes between the trabeculae of the bony collar and penetrates into the degenerating calcified cartilage. This periosteal bud of tissue is instrumental in resorption of some of the smaller calcified cartilage spicules between lacunae and in the laying down of bone on remnants of the calcified cartilage. The **center of the shaft** now consists of bone marrow precursor cells, osteogenic tissue, and bony trabeculae that contain remnant cores of calcified cartilage. This area in the diaphysis is called the **primary ossification center** (Fig. 22-3).

Because the newer cartilage lies toward the epiphyses, the **metaphyses** (epiphyseal plates, physes) demonstrate the following developmental gradient as the diaphysis is approached: (1) a layer of tissue where **cells are not dividing** (zone of resting cartilage), (2) a layer where **chondrocytes are dividing** mitotically and interstitially in an

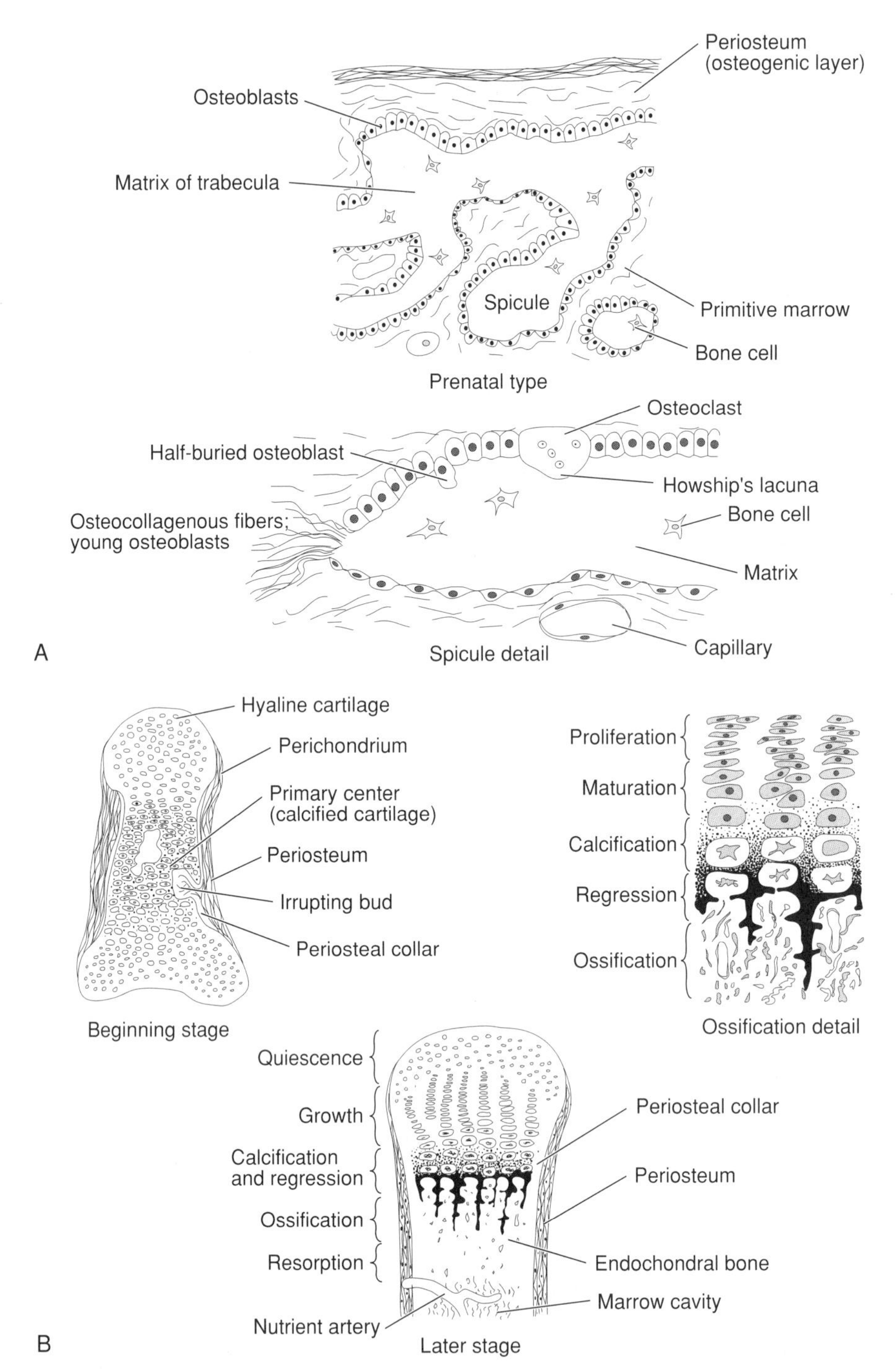


**Figure 22-3.** Diagram illustrating the major features of intramembranous **(A)** and intracartilaginous **(B)** bone development.

axial orientation (zone of multiplication), (3) a layer where **cells are enlarging** (zone of cellular hypertrophy and maturation), and (4) a layer where the **intercellular material is calcifying** and **cells are dying** (zone of calcification).

The shaft grows in length by the multiplication of cartilage cells at the zones of multiplication in each metaphyseal region and by osseous tissue being laid down on the remnants of calcified cartilage in the zone of calcification. This process also brings about an increase in length of the primary marrow cavity. An increase in width of the marrow cavity takes place by resorption of bone on the inner surface of the periosteal bony collar. Because this resorption is not as rapid as the appositional laying down of bone on the outer surface of the bony collar, the compact bone of the shaft increases in width.

**Secondary ossification centers** develop later in fetal life, or after birth, in the epiphyses. These are usually characterized by **hypertrophy of chondrocytes** and **calcification of cartilage** in the centers of the epiphyses, where the older cartilage cells exist. Vascular and osteogenic buds of tissue enter the area from the metaphyseal region. A thin layer of dense bone is laid down on the surfaces of the epiphyses where a periosteum is present. On articular surfaces, no periosteum or perichondrium exists, and hyaline cartilage is retained as a covering to the underlying epiphyseal bone.

# MUSCLE

Three types of muscle tissue exist: smooth, skeletal, and cardiac. All three types are comprised of muscle cells (fibers) that contain myofibrils possessing contractile filaments of actin and myosin. Table 22-3 presents the major types of muscle, characteristics, and locations.

## Smooth Muscle

Smooth muscle cells are spindle-shaped and are organized chiefly into sheets or bands of smooth muscle tissue. This tissue is found in **blood vessels** and other **tubular visceral structures**. Smooth muscle cells contain both actin and **myosin filaments**, but the **actin filaments** predominate. The filaments are not organized into patterns that give cross striations as in cardiac and skeletal muscle. Filaments course obliquely in the cells and attach to the plasma membrane. Electron microscopy shows the plasma membrane as a "typical" trilaminar membrane. In specific regions where smooth muscle cells appose each other, leaving only narrow 2-nm intercellular gaps, specialized zones of contact occur, which are known as nexuses or **gap junctions**. These junctions most likely facilitate the transmission of impulses for contraction. In other intercellular regions, a glycoprotein coat and a small amount of collagenous and reticular fibers is found.

**TABLE 22-3. Major Types of Muscle, Their Characteristics, and Their Locations**

| Muscle Type | Characteristics | Location |
|---|---|---|
| Smooth muscle | Spindle-shaped cells, nonstriated, involuntary, single centrally placed nucleus | Wall of hollow viscera, blood vessels, iris of the eye |
| Skeletal muscle | Cylindrical cells, multinucleated with peripherally placed nuclei, striated, and voluntary | Muscles of the body, pharynx, and upper esophagus |
| Cardiac muscle | Cylindrical cells that branch, single centrally placed nuclei, striated and involuntary | Heart |

## Skeletal Muscle

Skeletal muscle fibers are characterized by their peripherally located nuclei and their **striated myofibrils**. The cross striations are caused by the organization and distribution of actin and myosin filaments. These striations are organized within each muscle fiber into fundamental contractile units called **sarcomeres**, which are joined end to end at the **Z lines**. The striations in a sarcomere consist of an **A band** bordered toward the Z lines by **I bands**. The I bands of a sarcomere are really half I bands that join at the Z lines with the other half I bands of adjacent sarcomeres. The midregion of the A band contains a variable light **H band** that is bisected by an **M line**. The light I band contains actin filaments that insert into the Z line. These filaments interdigitate and are cross-bridged in the A band with myosin filaments, forming a hexagonal pattern of one myosin filament surrounded by six actin filaments. In the contraction of a muscle fiber, a chemical reaction takes place in the region of the cross-bridges, causing the actin filaments of the I band to move deeper into the A band, thus resulting in a shortening of the I bands.

Each skeletal muscle fiber is invested with a **sarcolemma** (plasmalemma) that extends into the fiber as numerous small **transverse T tubules**. These tubules ring the myofibrils at the A-I junction and are bordered on each side by **terminal cisternae** of the sarcoplasmic (endoplasmic) reticulum. This arrangement of one T tubule with two terminal cisternae is called a **triad**. In **excitation-contraction coupling**, acetylcholine released from the motor end-plate causes depolarization of the muscle membrane, which is propagated to the T tubule–sarcoplasmic reticulum junction. This brings about **release of calcium** from the terminal cisternae of the sarcoplasmic reticulum, catalyzing the chemical reaction between the actin and myosin filaments in the region of the cross-bridges. In this process, calcium attaches to the **troponin C** subunit of troponin, resulting in movement of tropomyosin and uncovering of the active sites for the attachment of actin to the cross-bridging heads of myosin. Because of this attachment, adenosine triphosphate (ATP) in the myosin head hydrolyzes, producing energy, inorganic phosphate, and adenosine diphosphate, which results in a bending of the myosin head and a pulling of the actin filament into the A band. The **actin-myosin bridges** detach when myosin binds a new ATP molecule and when calcium returns to the terminal cisternae at the conclusion of neural stimulation.

## Cardiac Muscle

Cardiac muscle contains **striations** and **myofibrils** similar to those of skeletal muscle. It differs from skeletal muscle in several major ways. Cardiac muscle fibers branch and contain **centrally located nuclei** and large numbers of **mitochondria**. Individual cardiac muscle cells are attached to each other at their ends by **intercalated disks**. These disks contain several types of membrane junctional complexes, the most important of which is the **gap junction**. This junction electrically couples one cell to its neighbor so that electrical depolarization is propagated through the heart by **cell–T cell** contacts rather than by nerve innervation to each cell. The sarcoplasmic reticulum–T tubule system is arranged differently in cardiac muscle than in skeletal muscle. In cardiac muscle, each T tubule enters at the Z line and forms a diad with only one terminal cisterna of sarcoplasmic reticulum.

## Histogenesis of Skeletal Muscle

Skeletal muscle cells develop from mesenchyme that arises from the myotomes of somites or from the mesoderm of branchial arches. Stellate mesenchymal cells differentiate into **myoblasts** that fuse and elongate into multinucleate myotubes containing peripherally located **myofibrils** and centrally located nuclei. Later in development, myofibrils increase in size and number, and the nuclei migrate peripherally. In the limited regeneration of muscle, new fibers may be formed from satellite cells that lie between the skeletal muscle cell and its basement membrane.

# NERVE TISSUE

## Neuron

The basic functional cell type of nerve tissue is the neuron. Each neuron consists of a **nerve cell body (perikaryon)** and one or more **nerve processes (fibers)**. The cell body of a typical neuron contains a nucleus, Nissl material of RER, free ribosomes, Golgi apparatus, mitochondria, neurotubules, neurofilaments, and pigment inclusions. The cell processes of neurons occur as axons and dendrites. Dendrites contain most of the components of the cell body except the nucleus and Golgi apparatus, whereas axons contain the major structures found in dendrites except for the Nissl material. At the synaptic ends of axons, the presynaptic process contains vesicles from which excitatory or inhibitory substances are elaborated. The functional dendrites of some neurons, such as the sensory pseudounipolar neurons of spinal nerves, are structurally the same as axons.

Table 22-4 lists the cell types of the central nervous system (CNS) and the peripheral nervous system (PNS).

## Neurolemmocytes

All nerve fibers in both the CNS and the PNS are ensheathed by specialized cells known as **neurolemmocytes (Schwann cells)** in the PNS and **oligodendrocytes** in the CNS. Unmyelinated fibers in peripheral nerves lie in grooves on the surface of **neurolemma** (Schwann) cells and are incompletely invested by the plasmalemma of these cells. Myelinated peripheral neurons are invested by numerous "jellyroll-type" layers of Schwann cell plasma membrane that constitute a myelin sheath. The Schwann cell cytoplasm and nucleus lie peripheral to the myelin sheath. There are many Schwann cells along each myelinated fiber. In the junctional areas between adjacent Schwann cells, there is a lack of myelin. These junctional areas along the myelinated process constitute the nodes of Ranvier. In the CNS, a single oligodendrocyte may myelinate several axons; in the PNS, a neurolemmocyte may myelinate only one axon. In the PNS, the neurolemmocytes provide a basal lamina on the outside of the myelin sheath. This basal lamina is not present in the CNS.

## Spinal Nerves

Spinal nerves have an outer **epineurial connective tissue** investment and an inner, more **cellular perineurial covering** that extends internally to surround nerve bundles. The cells of the perineurium form an epithelioid sheath wherein the cells are joined by occluding junctions and the layers of cells are separated by basal lamina material. This perineurial layer seems to be an effective barrier against material entering or leaving the nerve. A loose endoneurial connective tissue separates nerve processes and lies next to the basement membranes of the Schwann cells.

Spinal nerves contain the processes of neurons whose cell bodies are located in sensory dorsal root ganglia (**pseudounipolar neurons**), sympathetic ganglia (**multipolar neurons**), and in the gray matter of the spinal cord (**multipolar neurons**). Each spinal nerve contains myelinated and unmyelinated fibers that are invested by Schwann cells. In the ganglia, each cell body is surrounded by supportive satellite cells.

**TABLE 22-4. Cell Types of the Central Nervous System (CNS) and Peripheral Nervous System (PNS)**

| Cell Type | Characteristics | Function | Location |
|---|---|---|---|
| Multipolar neuron | Possesses cell body, many dendrites, single axon, and synapses; may be myelinated | Transmit and process information in the nervous system | Majority of neurons in the CNS, neurons in autonomic ganglia in the PNS |
| Pseudounipolar neuron | Possesses cell body and single process that divides into central and peripheral processes | Transmits sensory information from the periphery to the CNS | Sensory ganglia of the PNS |
| Bipolar neuron | Possesses cell body and a single central and a single peripheral process | Functions in the special sensations of olfaction, vision, hearing, and equilibrium | Retina of the eye, olfactory mucosa, ganglia of the vestibulocochlear nerve |
| Astrocytes | Stellate shape with multiple cytoplasmic processes | Isolate neurons from the extracellular space and capillaries; may have nutritional role | Fibrous astrocytes in white matter of CNS, protoplasmic astrocytes in gray matter of CNS |
| Microglia | Minute cells, resemble macrophages | Macrophages of CNS, remove cellular debris | CNS |
| Oligodendroglia | Small cells with few cytoplasmic processes | Produce myelin on axons in the CNS | CNS |
| Ependymal cells | Resemble simple columnar epithelium | Line the ventricles in the CNS | CNS |
| Schwann cells (neurolemmocytes) | Small rounded cells with prominent nuclei | Produce myelin in the PNS | Peripheral nerves |

## Neuroglia

**Glial cells** are the supportive cells of the CNS. They consist of ependymal cells, astrocytes, oligodendrocytes, and microglia. Protoplasmic astrocytes are found mostly in the gray matter, and fibrous astrocytes are located mostly in the white matter. **Astrocytes** provide structural support for nerve tissue and may help to isolate groups of nerve endings from each other. **Oligodendroglia** invest nerve fibers to form the myelin of the CNS. One oligodendrocyte may envelope and myelinate several axons. Microglia, unlike the other glia, arise from mesodermally derived monocytes and are components of the mononuclear phagocyte system.

# Chapter 23

# Integument

## STRUCTURE OF THE INTEGUMENT

The **skin**, or integument, consists of an outer epidermis of stratified squamous keratinized epithelium and an underlying dermis of dense irregularly arranged fibroelastic connective tissue. Beneath the dermis is a subcutaneous layer of loose connective tissue. The epidermis varies in structure and thickness, depending on the region of the body. For example, on the palmar surface of the hands and on the soles of the feet, the epidermis consists of four layers. These layers are, from the dermis to the surface, the stratum germinativum, stratum granulosum, stratum lucidum, and stratum corneum (Fig. 23-1).

### Epidermis

**Stratum germinativum**. This is the deepest layer of the epidermis, which is further divisible into a **stratum basale** and **stratum spinosum**. Epidermal cells in both of these layers are capable of mitoses. Cells of the stratum spinosum (prickle cells) are joined at numerous desmosomes (macula adherens) where many tonofilaments converge toward the plasma membrane. **Desmosomes** are dense bodies where plasma membranes of adjacent cells appear thickened because of a dense layer on their cytoplasmic surfaces. A thin intermediate lamina occurs in the intercellular space at a desmosome. Tonofilaments give rise to the fibrous protein **keratin**. Melanin granules are found in the deeper cells of the stratum germinativum. **Melanin** is produced in melanocytes, which arise from neural crest. The melanin granules are distributed to the epidermal cells by a process of cytocrine secretion. The color of the skin is dependent on the amount of melanin, carotene, and the vascularity of the skin.

**Stratum granulosum**. The stratum granulosum represents cells that are older and further differentiated in the keratinization process than those of the deeper layers. **Keratohyalin granules** predominate and give rise to the interfilamentous amorphous matrix that is prevalent in cells of the stratum corneum. Membrane-coating granules in the granulosal cells provide the intercellular "sealing" cement between cells of the stratum corneum.

**Stratum lucidum**. The stratum lucidum contains flattened translucent cells that have lost their nuclei. The cytoplasm contains an orderly array of tonofilaments, and it is often not visible in regions of the body covered by "thin skin."

**Stratum corneum**. Cells of the stratum corneum also have lost their organelles and contain soft keratin, which consists of tightly packed filaments embedded in an amorphous matrix.

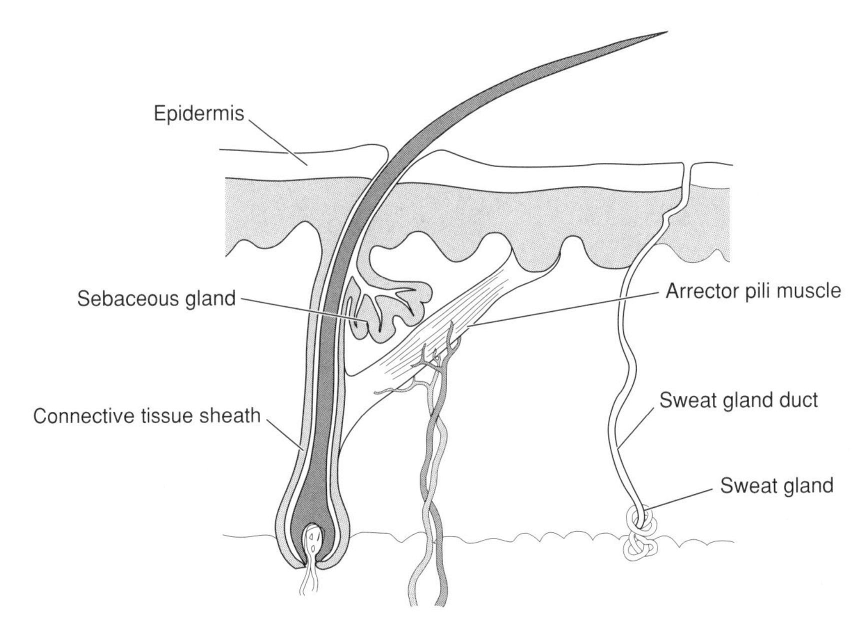


**Figure 23-1.** Diagram of the basic structure of hairy skin.

## Dermis

The dermis is divisible into papillary and reticular layers.

**Papillary layer**. The papillary layer is immediately below the epidermis and contains dense fine collagenous fibers, blood vessels, and free and encapsulated (e.g., Meissner's tactile corpuscles) nerve endings.

**Reticular layer**. The reticular layer contains coarser bundles of collagenous fibers. Smooth muscle can be found in this layer in the nipple and scrotum. Hair follicles, arrector pili smooth muscles, sebaceous glands, and the ducts of sweat glands are located in the reticular layer. The secretory portions of simple tubular sweat glands, the roots of hairs, and pacinian corpuscles (deep pressure receptors) are constituents of the subcutaneous layer. Extensive accumulations of fat in this layer define it as the panniculus adiposus.

# INTEGUMENTARY GLANDS

## Sweat Glands

Secretory portions of sweat glands consist of a tall cuboidal epithelium invested by contractile myoepithelial cells. Because no part of the cell is lost in the secretory process, it is classified as a **merocrine type of secretion**. The ducts are stratified cuboidal epithelium, except for the intraepidermal portion, which is lined by the epidermal epithelium.

## Sebaceous Glands

Sebaceous glands have short ducts that empty into hair follicles, except for those in the labia minora and glans penis, where they open directly to the surface. The secretory portion of the sebaceous gland is stratified cuboidal epithelium; the lumen is generally filled with rounded cells containing numerous fat droplets. The secretion of sebum involves the holocrine discharge of sebaceous cells. New cells proliferate and differentiate from more basal regions of the gland (see Fig. 23-1).

# APPENDAGES OF THE SKIN

## Hair

Hair is most prevalent in the scalp and is lacking in such areas as the palms of the hands and soles of the feet. Each hair consists of a root and a **shaft**. The **root** is enclosed by a tubular hair follicle that consists of inner and outer epithelial root sheaths, both derived from epidermis, and an outer connective tissue root sheath that corresponds to the dermis. At its lower end, the follicle and root expand into a hair bulb, which is indented by a connective tissue hair papilla.

The hair, consisting of an inner medulla, a cortex, and an outer cuticle, grows upward from the differentiation of matrix cells in the hair bulb. Pigment in the cortical layer and air in the cortex and medulla determine the color of the hair. Arrectores pilorum of smooth muscle run in the obtuse angle between the follicle and epidermis, and they attach to the follicle deep to the **sebaceous glands**. Contraction causes erection of the hair and dimpling of the skin (goose flesh) at the smooth muscle attachment in the dermis. In the scalp and face, skeletal muscles are found in the superficial fascia.

## Nails

Nails consist of horny plates that overlie the nail bed. They are bordered laterally by the nail wall. The posterior aspect of the nail plate is the nail root, from which nail growth occurs. The eponychium or cuticle is located above the nail plate. The hyponychium is found below the free end of the nail plate.

# Chapter 24

# Blood and Hematopoiesis

## CIRCULATING BLOOD

**Blood cells** (formed elements) constitute 45% of the total volume of circulating blood; **plasma** comprises the remaining 55%. Of the 45% cell volume, erythrocytes (red blood corpuscles, RBCs) make up 44%, and the remainder is composed of leukocytes (white blood cells, WBCs). The plasma, minus its blood-clotting factors, is called **serum**.

Table 24-1 is a list of the formed elements of peripheral blood and their characteristics and frequencies.

### Plasma

Plasma acts as a medium for metabolic substances and circulating cells. Like tissue fluid, its primary components are water, inorganic salts, and a number of proteins. **Albumin**, the most abundant plasma protein, maintains the colloid blood pressure. **Gamma globulins** are also important because they include the circulating antibodies. **Beta globulins** transport lipids, hormones, and metal ions. **Prothrombin** and **fibrinogen** are products of the liver that are essential components of the clotting process. **Chylomicrons** are microscopic particles of fat that are especially prominent in the plasma after a fatty meal.

### Erythrocytes

Erythrocytes, when mature, are anucleate biconcave disks approximately 8 nm in diameter and 2 nm thick. There are about 4.8 and 5.5 million erythrocytes per cubic millimeter of blood in the normal female and normal male, respectively. Erythrocytes lack the usual complement of organelles, and they do not have the capacity for protein synthesis. Each RBC exists for approximately 120 days. It lacks the mechanism to reproduce itself. About 17% of the erythrocytes possess some residual ribosomal material. Because of their stained appearance, they are called **reticulocytes**; they are considered immature erythrocytes. The reticulocyte count provides a rough index of the rate of erythrocyte development.

### Leukocytes

Leukocytes are divisible into granular leukocytes and nongranular leukocytes. The granular leukocytes are further classified as eosinophils, basophils, and neutrophils on the basis of the affinity of their granules for different stains. The nongranular leukocytes are the lymphocytes and monocytes.

**Neutrophils**. These cells are about twice the size of erythrocytes and make up 60% to 70% of the WBC. Their nuclei consist of three to five lobes that are interconnected

**TABLE 24-1. Formed Elements of the Peripheral Blood, Their Characteristics, and Their Frequencies**

| Cell Type | Characteristics | Function | Frequency in Blood |
|---|---|---|---|
| Erythrocyte | Biconcave disk, most numerous formed element of the blood, no organelles, red in color | Filled with hemoglobin and carries oxygen to all body tissues | 4.8–5.5 million/$mm^3$ of blood |
| Neutrophil | Granulocyte, multilobed nucleus, neutral granules | First line of defense against pathogens, undergoes phagocytosis, contains lysozymes | 60–70% of white cells |
| Eosinophil | Granulocyte, bilobed nucleus, red-staining granules | Contains lysozymes, responds to infestations | 1–3% of white cells |
| Basophil | Granulocyte, S-shaped nucleus, dark staining blue granules | Contains heparin and histamine | 0.5% of white cells |
| Lymphocyte | Agranulocyte, large round nucleus, bluish cytoplasm, classified as T and B cells | Functions in immunity | 20–35% of white cells |
| Monocyte | Agranulocyte, oval to kidney-shaped nucleus, grayish cytoplasm | Becomes macrophages in the connective tissues | 3–8% of white cells |
| Thrombocytes | Basophilic cytoplasmic fragments of bone marrow cells called megakaryocytes | Aid in the formation of blood clots and constriction of vessels | 250,000–300,000/$mm^3$ of blood |

by fine filaments of nuclear material. Two types of granules are present in the cytoplasm: the **specific granules**, which stain with neutral dyes, and **nonspecific granules**, which are azurophilic. Both types of granules contain hydrolytic enzymes that are used by the cell in the digestion of phagocytized materials.

**Eosinophils**. These cells are about the size of neutrophils, but constitute only 1% to 3% of the total leukocyte population. The nucleus is usually bilobed, and the chromatin is dense. The eosinophilic granules are membrane-bound vesicles containing lysosomal enzymes.

**Basophils**. These cells are about the same size as the other granular leukocytes. They constitute only 0.5% of the WBC. The nucleus is usually S-shaped, and its chromatin is less dense than that of the other granular leukocytes. The basophilic membrane-bound cytoplasmic granules contain histamine and heparin.

**Lymphocytes**. These cells constitute 20% to 35% of the WBC population. Most of the mature lymphocytes are the size of erythrocytes, but larger cells traditionally called **large** and **medium lymphocytes** are occasionally seen in circulating blood. The small lymphocyte has a relatively large round or slightly indented nucleus of dense chromatin. The nucleus is surrounded by a thin rim of cytoplasm containing a few ribosomes and some nonspecific azurophilic granules. The small lymphocytes are further designated as T and B lymphocytes. **T lymphocytes** are cytotoxic cells of the cell-mediated response that "kill" foreign cells and sensitizing agents that enter the body. They also assist the **B lymphocytes** (and their subsequent plasma cells) in their humoral antibody response to such invasive organisms as bacteria and viruses.

**Monocytes**. These cells range from 9 to 20 nm in diameter and make up 3% to 8% of the circulating leukocytes. The nucleus is oval, kidney-shaped, or horseshoe-shaped. The cytoplasm is more abundant than that of lymphocytes, and contains a few azurophilic granules, a Golgi complex, polyribosomes, and some glycogen. Monocytes give rise to macrophages when they pass into connective tissues.

**Platelets (thrombocytes)**. These are small, irregular disk-shaped structures that are 1 to 2 nm in diameter. They are basophilic fragments of megakaryocytes containing a variety of granules. There are 250,000 to 300,000 platelets per cubic millimeter of blood. They have a natural tendency to cling to each other and to all "wettable" surfaces they contact when blood is shed. Platelets contain **serotonin**, which helps to constrict small blood vessels during vascular injury. They also contain **thromboplastin**, a substance released by platelets and injured endothelial cells. **Thromboplastin** helps convert **prothrombin** of the plasma to thrombin. The **thrombin** then converts plasma **fibrinogen** to **fibrin**, which forms a network trapping blood cells and platelets; thus, a blood clot, or **thrombus**, is formed (see Table 24-1).

# HEMATOPOIESIS

The main hemopoietic tissue in the body is **bone marrow**. Because all of the erythrocytes, platelets, and granular leukocytes are produced in bone marrow, these blood components are called the **myeloid elements**. The specific development of these elements is referred to as **myelopoiesis**. Although the nongranular elements are produced in both lymphatic tissues and bone marrow, they are referred to as **lymphoid elements**; their development is termed **lymphopoiesis**.

The first blood cells develop in the third embryonic week from yolk sac and body stalk mesoderm. During the second month of development, hemopoietic sites arise in the liver, spleen, and mesonephric kidneys. In later months of fetal development, bones are established, and the bone marrow becomes the dominant hemopoietic tissue. **Red bone marrow** consists of a reticular fiber meshwork, which contains and supports reticular cells, myelopoietic (blood-forming) cells, adipose cells, and thin-walled sinusoids.

The **myelopoietic cells** occur in many stages. Some are relatively undifferentiated stem cells from which all myeloid elements develop. Some are mature erythrocytes, granular leukocytes, and nongranular leukocytes, which are about ready to leave the bone marrow through the sinusoids and veins. The majority of cells, however, are in the numerous stages of differentiation that stem cells go through during erythropoiesis, granulopoiesis, and thrombopoiesis.

## Erythropoiesis

Erythropoiesis is the formation of **erythrocytes** from stem cells. In this process, pluripotent stem cells, which have the potential to give rise to any blood cell type, differentiate into proerythroblasts. Proerythroblasts divide into **basophilic proerythroblasts**, which contain free polyribosomes, a condensed nucleus, and **no nucleoli**. Without nucleoli, the cells cannot produce ribosomes; thus, when they divide into smaller polychromatophilic erythroblasts, their basophilia disappears and their acidophilia increases—caused by accumulating hemoglobin. When these cells have acquired their full amount of hemoglobin and their nuclei become very small and concentrated, they are called **normoblasts** (orthochromatic erythroblasts). When the nuclei are extruded, they become erythrocytes. Erythropoietin, produced by the kidney, regulates proerythroblast formation. The maturation of erythrocytes is regulated by **extrinsic factor** (vitamin $B_{12}$) and **intrinsic factor** (a mucoprotein produced in the stomach).

## Granulopoiesis

Granulopoiesis is the formation of **basophils**, **eosinophils**, and **neutrophils** from stem cells that differentiate through myeloblast, promyelocyte, myelocyte, and metamyelocyte stages. The myeloblast has a large nucleus with several prominent nucleoli; its cytoplasm is basophilic. When these cells acquire azurophilic granules, they are called **promyelocytes**. As promyelocytes mature into myelocytes, their nuclei become more dense, nonspecific azurophilic granules increase in number, and specific granules make their appearance. If specific granules are neutrophilic, then the cell is a neutrophilic myelocyte; if basophilic or eosinophilic granules are present, the cell is a basophilic myelocyte or eosinophilic myelocyte. All of the cells from myeloblast through myelocyte are capable of mitosis. This ceases when the myelocyte nucleus becomes dense and more deeply indented as a metamyelocyte is formed.

During maturation of the **metamyelocyte** into a mature granulocyte, the nucleus becomes more deeply indented and then becomes lobated or S-shaped. As this takes place, certain juvenile forms of cells are detected. In **neutrophil formation**, the horseshoe-shaped nucleus often designates the cell as a band or stab cell. Because the life span of granular leukocytes is considerably shorter (about 14 hours) than that of erythrocytes (120 days), more developmental forms of granular leukocytes exist in the marrow than of erythrocytes.

## Thrombopoiesis

Thrombopoiesis is the formation of **blood platelets** from megakaryocytes. In this process, plasma membranes of megakaryocytes partition off cytoplasmic fragments, which are released from the cell and pass into the bloodstream as **platelets**. The megakaryocyte may then die and is replaced by a stem cell in the marrow. Megakaryocytes are very large cells with multilobed nuclei. Blood platelets live for only about 8 to 11 days.

## Lymphopoiesis

Lymphopoiesis is the formation of **lymphocytes** from a stem cell. In this process, a stem cell differentiates into a large lymphocyte (**lymphoblast**), which further divides and matures into medium lymphocytes and then into small lymphocytes. The sites of these lymphopoietic changes are in the bone marrow and in the lymphatic tissues of the spleen, thymus, lymph nodes, tonsils, and mucous membranes of the body. It appears that the small lymphocyte can be the stem cell for large lymphoblasts that produce other small lymphocytes and antibody-producing plasma cells.

In the establishment of the immune system, stem cells are sent to the thymus, the mucous membrane of the gut, and return to the bone marrow. In the thymus, the stem cells become T-lymphocyte precursors, which pass to other tissues and become small cytotoxic "killer" lymphocytes. Stem cells that return to the bone marrow become lymphocytes that develop into antibody-producing B lymphocytes.

## Monopoiesis

Monopoiesis is the formation of **monocytes** from stem cells. Monocytes appear to develop in the marrow from pluripotent stem cells. Within a few days after developing in the marrow, monocytes pass into the circulation for 1 or 2 days before entering the connective tissue and becoming macrophages.

# Chapter 25

# Cardiovascular System

## HEART

The heart wall consists of three layers: an inner endocardial layer, a middle myocardial layer, and an outer epicardial layer.

### Endocardium

The endocardium consists of simple squamous epithelium or endothelium and a subendothelial connective tissue layer of fine collagenous and elastic fibers and some smooth muscle fibers. The endocardium of the atria is thicker than that of the ventricles. A subendocardial layer of loose connective tissue and blood vessels binds the endocardium to the myocardium. This layer in the ventricles contains the specialized muscle fibers of the conduction system.

### Myocardium

The myocardium consists of spiraling bundles of cardiac muscle that originate from the **annuli fibrosi**, fibrous rings that serve as part of the "skeleton" of the heart. In the atria, the myocardium is a thin layer of fibers with a simple arrangement. The muscle of the ventricles is more complex and consists of several layers. The **ventricular bands of muscle** originate from the **fibrous annulus** and course in a helical manner from right to left and toward the apex. Some of the more superficial fibers can be traced in a mantle covering both ventricles. Some intermediate fibers weave from ventricle to ventricle by way of the septum. The more numerous deeper fibers pass into either of the ventricular walls and end by piercing deeply and becoming the papillary muscles. The microscopic structure of cardiac muscle was reviewed in Chapter 4 (see Muscle) as one of the four basic tissue types. Research now indicates that the atrial myocytes contain granules that have been identified as one or several polypeptides called **atriopeptins**, which are responsible for a profound natriuresis.

### Epicardium

The epicardium consists of **mesothelium** and an underlying **connective tissue layer**. A subepicardial layer of loose connective tissue containing blood vessels, nerves, and fat binds the epicardium to the myocardium. The epicardium is the visceral pericardium.

### Cardiac Valves

The **atrioventricular (A-V) valves** consist of a core of dense connective tissue, which is continuous with the annuli fibrosi, and an outer layer of **endocardium**. The endocardium on the atrial surface of the valves is thicker than that on the ventricular side. **Chordae tendineae** are composed of dense, regularly arranged connective tissue. The **aortic** and **pulmonary semilunar valves** resemble the A-V valves, but they are much thinner.

## Cardiac Conduction System

The cardiac conduction system is composed of specialized cardiac muscle found in the sinoatrial (S-A) node and in the A-V node and bundle. The heartbeat is initiated in the **S-A node (pacemaker of the heart)**, located in the right atrium in the upper part of the crista terminalis, just to the left of the opening of the superior vena cava. Cells of the S-A node are slender and fusiform. From the S-A node, the cardiac impulse spreads throughout the atrial musculature to reach the **A-V node**, lying in the subendocardium of the atrial septum directly above the opening of the coronary sinus. The A-V node has small, irregularly arranged branching fibers that contain few myofibrils. Thereafter, the impulse is conducted to the ventricles by passing through the specialized tissue of the **A-V bundle (of His)**. This bundle consists of a crus commune and right and left bundle branches. The common bundle travels from the A-V node into the membranous part of the interventricular septum. It divides into right and left bundle branches that pass in the subendocardium along the muscular part of the septum and distribute to the ventricles as Purkinje tissue. **Purkinje cells** are large specialized cardiac muscle cells that usually are binucleate and contain much centrally located sarcoplasm.

# VESSELS

Blood and lymphatic vessels generally consist of three tunics: tunica intima, tunica media, and tunica adventitia. These tunics are most pronounced in the larger vessels and vary with the type and size of vessels as to their constituents. In larger arteries and veins, the three tunics are more pronounced. Table 25-1 lists the major types of blood vessels and their characteristics.

## Tunica Intima

The tunica intima consists of endothelium, subendothelial connective tissue with smooth muscle, and an internal elastic membrane. In larger arteries, the internal elastic membrane is fenestrated, often doubled, and highly developed (see Table 25-1).

## Tunica Media

The tunica media is more pronounced in arteries than in veins. In medium-sized (muscular, distributing) arteries, it contains mostly circularly arranged smooth muscle and some elastic fibers. In large arteries (elastic, conducting), elastic lamellae predominate, but smooth muscle is present. The media of arterioles consist of several layers of circularly arranged smooth muscle (see Table 25-1).

## Tunica Adventitia

The adventitia of arteries is not as pronounced as the tunica media. It is comprised of elastic and collagenous fibers and, in larger vessels, vasa vasorum that supply the outer layers. Veins have a poorly defined tunica media, but the tunica adventitia is well developed. In medium-sized veins, this outer layer contains mostly collagenous fibers. Longitudinally running smooth muscle fibers are found in the adventitia of large veins. Venous valves are local foldings of the intima. Venules have relatively thinner walls than arteries of similar diameter (see Table 25-1).

**TABLE 25-1. Major Types of Blood Vessels and Their Characteristics**

| Vessel Type | Tunica Intima | Tunica Media | Tunica Adventitia |
|---|---|---|---|
| Elastic arteries | Endothelium, thick subendothelial connective tissue, internal elastic lamina not distinct | Up to 70 fenestrated elastic laminae and several layers of smooth muscle | Underdeveloped and containing elastic and collagen fibers |
| Muscular arteries | Prominent endothelium, distinct internal elastic lamina | Up to 40 layers of smooth muscle cells intermingled with elastic laminae, distinct external elastic lamina | Fairly well developed; containing collagen, elastic fibers, and adipocytes |
| Arterioles | Endothelium, thin subendothelial layer, no internal elastic lamina | 1–5 layers of smooth muscle cells, no elastic laminae | Thin with no elastic laminae |
| Venules | Endothelium | Zero to a few layers of smooth muscle cells | Thickest layer, rich in collagenous fibers |
| Small- or medium-sized veins | Endothelium with thin subendothelial layer | Small bundles of smooth muscle cells intermingled with reticular fibers | Well-developed collagenous layer |
| Large veins | Endothelium with a well-developed subendothelial layer | Thin media with few smooth muscle cells | Best developed layer, thick and with longitudinal bundles of smooth muscle |

## Capillaries

Capillary walls consist of an endothelium, basal lamina, and surrounding connective tissue. The endothelium may be continuous and contain no pores, or it may be fenestrated with pores either opened or closed by a thin membrane. Either type of endothelium contains tight intercellular junctions. **Fenestrated** capillaries can be found in the kidney, muscle, gastrointestinal tract, and various endocrine organs. **Continuous** capillaries are a feature of the central nervous system, where the tight junctions of their endothelial cells constitute an important anatomic substrate of the blood-brain barrier.

## Lymphatic Vessels

Large lymphatic vessels are microscopically similar to veins. Lymphatic capillaries appear as endothelium-lined clefts in connective tissue and have very thin walls. The thoracic duct has a thick tunica media consisting of longitudinal and circular smooth muscle bundles. Its tunica intima is prominent, but the adventitia is poorly defined.

# Chapter 26

# *Lymphatic Tissue*

## LYMPH NODES

Microscopically, lymph nodes are bean-shaped, possess a hilum, and are surrounded by a capsule that sends trabeculae into a stroma of reticular tissue containing **lymphocytes**. Nodules of dense lymphatic tissue are located peripherally in the cortical region. If the node is in the "active" stage, the nodules contain germinal centers that consist of medium-sized lymphocytes and larger undifferentiated lymphocytes. These areas produce small lymphocytes that are pushed peripherally in the nodule and then into surrounding lymphatic sinuses.

Nodes receive lymph peripherally through afferent vessels that penetrate the capsule and drain into a subcapsular sinus. This sinus drains along cortical peritrabecular sinuses to medullary sinuses (which lie between trabeculae and medullary cords of lymphatic tissue) before exiting from the lymph node at the hilum through efferent lymphatic vessels. Reticular cells and macrophages lining the sinuses perform a filtering function by phagocytizing dead cells and particulate matter and by offering antigens to the lymphocytes.

Lymph nodes play a major role in the **immune response**. The **B lymphocytes** are found in the subcapsular cortical tissue, in germinal centers, and in medullary cords. In **bacterial infections**, antigens pass to the B lymphocytes and trigger them to form blast cells in the germinal centers that proliferate and differentiate into antibody-producing lymphocytes and plasma cells. These cells pass to the medullary cords, where antibodies and B lymphocytes pass into efferent lymphatic vessels and are transported by the circulatory system to the site of infection.

**T lymphocytes** are located in deep cortical (paracortical) areas known as the **thymus-dependent zone**. This zone contains many postcapillary venules, whose cuboidal endothelium permits a recirculating pool of T lymphocytes and some B lymphocytes to enter the lymph node. Uncommitted lymphocytes from the thymus and bone marrow may enter the lymph nodes through these venules and react with antigens from foreign cells. In this response, T lymphocytes in the deep cortex become blast cells, proliferate, and form small long-lived memory cells and short-lived effector (killer) cells that enter the circulation.

## TONSILS

### Palatine Tonsil

The palatine tonsil is located between the anterior and posterior pillars of the fauces. It bulges into this depression and is covered by a mucosal fold of the anterior pillar;

there is a depressed supratonsillar fossa above the tonsil. The free surface of the tonsil, lined by stratified squamous nonkeratinized epithelium, dips into the underlying lymphatic nodular aggregation as 10 to 20 branching primary and secondary crypts. Lymphocytes from the underlying diffuse and nodular lymphatic tissue often heavily infiltrate the epithelium. The tonsils produce lymphocytes.

The presence of plasma cells indicates that tonsils are involved in antigen-antibody reactions. The presence of many neutrophils is characteristic of **tonsillar inflammation**. Each tonsil is partially invested basally by a connective tissue capsule that sends septa around the aggregations of nodules that invest each crypt. Some mucous glands and the superior pharyngeal constrictor and styloglossus muscles lie peripheral to the capsule.

### Pharyngeal Tonsil

The pharyngeal tonsil is also known as the **adenoids**. It is located in the median dorsal wall of the nasopharynx and as such is covered by pseudostratified ciliated columnar epithelium of the nasopharynx. It contains lymphatic nodules with germinal centers, but has no true crypts.

### Lingual Tonsils

The lingual tonsils constitute aggregates of lymphoid tissue located within the posterior aspect of the tongue.

## THYMUS

The thymus is larger in the infant than it is in the adult. It consists of two lateral lobes invested by a connective tissue capsule that sends septa into the gland and divides it into lobules. The gland lies in the anterior part of the superior mediastinum and extends from the fourth costal cartilage to the lower border of the thyroid gland. It lies anterior to the great vessels and fibrous pericardium.

Microscopically, the thymus is divisible into a **central medulla** and an **outer cortex**. The reticular cell meshwork contains lymphocytes (thymocytes) and differs from the other lymphatic tissues in that the reticular cells are derived from entoderm (of the third pharyngeal pouch). The cortex contains more lymphocytes and is less vascular than the medulla. **Thymic (Hassall's) corpuscles** in the medulla are concentric arrangements of flattened, and often hyalinized, cells. The corpuscles vary in size and occurrence and seem to be indicative of degeneration of reticular cells of the thymus.

Branches of the internal thoracic and thyroid arteries pass through thymic septa and enter the cortex-medulla junctional area as arterioles. Here, the arterioles give off direct branches to the medulla and also feed capillaries that loop into the cortex before draining into postcapillary venules of the medulla and cortex-medulla junction. Macromolecules cannot pass through the walls of the capillary loops, because the endothelial cells have a thick basement membrane bounded by reticular cells; thus, the cortical lymphocytes seem to be protected from circulating antigens by a blood-thymus barrier. Large numbers of lymphocytes pass through the walls of the postcapillary venules and drain by way of thymic veins into the left brachiocephalic, inferior thyroid, and internal thoracic veins.

The thymus is essential for the production of thymus-dependent (T) lymphocytes, which are involved in **cell-mediated immunologic responses** and also assist B lymphocytes in **humoral responses**.

# SPLEEN

The spleen is the largest lymphoid organ in the body, and it is specialized for **filtering blood**. The spleen consists of white pulp (splenic nodules) and red pulp. The **white pulp** is dense lymphatic tissue; the **red pulp** is looser and consists of lymphatic splenic cords of tissue and venous sinusoids.

Arterial blood enters the spleen at the hilum in the splenic artery. Branches of this artery pass in connective tissue trabeculae that radiate from the capsule at the hilar region, which are trabecular (interlobular) arteries. At the ends of the trabeculae, the adventitia of the arteries takes on the character of reticular tissue and becomes infiltrated with lymphocytes forming splenic nodules (white pulp). These arteries are eccentrically located in the nodules and are called **central arteries**. After numerous branchings, these arterioles leave the white pulp and enter the reticular connective tissue of the red pulp that surrounds the splenic nodules. In the red pulp, the pulp arterioles divide into sheathed arterioles, which empty into sinusoids via terminal arterial capillaries or empty first into the pulp reticulum and then filter between the lining cells of the sinusoids.

Many **macrophages** of the reticuloendothelial (mononuclear phagocytic) system lie outside the walls of the sinusoids and phagocytize "worn-out" red blood cells (RBCs). The venous sinuses empty into pulp veins, which pass to trabecular veins before blood is emptied by way of the splenic vein.

In addition to the filtering of blood, the spleen **controls the blood volume** by storing RBCs and by periodically discharging the blood through the contraction of smooth muscle and the action of elastic fibers in the capsule and trabeculae. The spleen also produces lymphocytes, monocytes, plasma cells, and antibodies. Most of the lymphocytes that leave the spleen are from the recirculating pool of lymphocytes; relatively few new lymphocytes are formed in the spleen.

The spleen is involved in both the **cell-mediated** and **humoral responses to antigens**. In the white pulp, T lymphocytes are located in the periarterial sheath, and B lymphocytes are located more peripherally. When B cells are activated, they move to the germinal center and give rise to plasma cells, which elaborate antibodies in the red pulp. Activated B cells also return to the general circulation by way of the red pulp sinuses. In the secondary responses to antigens by memory cells, the spleen is one of the most active organs in antibody secretion.

# Chapter 27

# Digestive System

## ORAL CAVITY AND TEETH

### Walls of the Oral Cavity

The mouth or **oral cavity** consists of a vestibule and the mouth proper. The vestibule of the mouth lies between the lips and cheeks externally and the gums and teeth internally. It receives the parotid duct opposite the second upper molar.

The **mouth proper** is bounded laterally and in front by the alveolar arches and the teeth; posteriorly, it communicates with the pharynx through the fauces. It is roofed by the hard and the soft palates. The floor is composed of the tongue and the reflection of its mucous membrane to the gum lining the inner aspect of the mandible. The midline reflection is elevated into a fold called the **frenulum linguae**. On each side of this fold is the caruncula sublingualis, which contains the openings of the **submandibular (Wharton's) ducts**. Behind these are the openings of the ducts of the sublingual glands.

**Lips**. The lips are muscular folds covered externally by skin and internally by mucosa (mucous membrane). Microscopically, the cutaneous surface consists of a thin skin, which possesses hairs and sweat and sebaceous glands. The vestibular surface consists of stratified squamous nonkeratinized epithelium, lamina propria, and a submucosa rich in mucous and mixed seromucous labial glands. The orbicularis oris muscle lies between the dermis and submucosal layers. The red area of the lip lies in the free margin of the lip at the junction between skin and mucosa. It is covered by stratified squamous epithelium containing variable degrees of keratinization and deeply indenting vascular papillae. No glands are present, and the epithelium is kept moist by licking of the lips.

**Cheeks**. The cheeks are similar in structure to the lips. Thick submucosal fibers tightly bind the mucosa to the buccinator muscle, thus reducing the chance of chewing on mucosal folds. Mixed buccal glands occupy the submucosa.

### Tongue

The mucosa over the anterior two-thirds of the tongue is characterized by filiform and fungiform papillae. **Filiform papillae** are the most numerous and uniformly distributed. They have a slender vascular core of connective tissue and are covered by a partially keratinized epithelium. **Fungiform papillae** are knoblike projections that are larger and more scattered than the filiform papillae. Their epithelium is mostly stratified squamous nonkeratinized and contains taste buds. **Vallate papillae** are the largest and least numerous of the papillae (9 to 12). They are oriented parallel to the sulcus terminalis. Each papilla is surrounded by a trench into which underlying serous glands of von Ebner empty. The sides of the papillae and trench contain many taste buds, which extend intraepithelially from the basement membrane to the surface. Taste buds contain spindle-shaped neuroepithelial cells that receive sensory nerve endings. Anterior lingual glands are located near the tip of the tongue and are mixed seromucous glands.

The mucosa over the root of the tongue is stratified squamous nonkeratinized epithelium overlying connective tissue, lymphatic nodules, and mucous glands. Aggregations of lymphatic nodules around single crypts constitute lingual tonsils. Beneath the epithelium of the tongue are numerous interlacing bundles of skeletal muscle, which constitute the intrinsic and extrinsic muscles of the tongue. Variable amounts of fat are also present.

## Teeth

Teeth, gums, and alveolar bone provide a wall between the vestibule and the mouth proper. Twenty deciduous teeth and 32 permanent teeth are equally distributed between the upper and lower jaws. Each tooth consists of a free crown, a root buried in an alveolus (socket) of the jaw, and a neck between the crown and root at the gum margin. **Dental pulp** of connective tissue, vessels, and nerves occupies a pulp chamber in the crown and root. The root is suspended in the alveolar bone by a **periodontal membrane**, which is similar in structure to the periosteum of bones.

The wall of the tooth consists of **enamel**, **dentin**, and **cementum**. Enamel is the hardest structure in the body, and it covers the crown. It consists of radially arranged rodlike enamel prisms that were elaborated by ameloblasts before the tooth erupted. Ameloblasts develop from the enamel organ, which differentiates from a dental ledge of oral ectoderm. Each of the enamel prisms is invested by a prism sheath, which is rich in organic matter. Adjacent prism sheaths are cemented together by interprismatic substance. Dentin, which is similar in its structure to mature bone, lines the pulp chamber and lies internal to enamel in the crown and internal to cementum in the root. Dentin consists of a meshwork of collagen fibers oriented parallel to the surface of the tooth and a calcified ground substance composed of glycosaminoglycans (GAG) and mineral salts. This dentin matrix is permeated by radially arranged dentinal tubules containing dentinal fibers (of Tomes), which are processes of odontoblasts lining the pulp chamber. These cells are necessary for the production of the dentin matrix. Cementum is like a bony covering of the dentin: It is more acellular and avascular than bone, but collagen lamellation and bone cells (cementocytes) are present. Cementocytes and odontoblasts arise from ectomesenchyme of neural crest origin.

## Gingiva

The gingiva or gums constitute the part of the oral mucosa that is attached to the alveolar processes of the upper and lower jaws. They consist of a stratified squamous epithelium that may be partially keratinized. The underlying lamina propria consists largely of dense collagenous connective tissue binding the gums to the cementum of the tooth and the underlying alveolar bone.

# PHARYNX

The wall of the pharynx consists of a mucosa, muscularis, and fibrosa. A submucosal layer exists only in the superior lateral region and near the junction with the esophagus. The epithelium of the **nasopharynx** is pseudostratified ciliated columnar epithelium with goblet cells; that of the **oropharynx** and **laryngopharynx** is stratified squamous nonkeratinized epithelium. The lamina propria contains many elastic fibers that constitute a dense elastic layer immediately adjacent to the muscularis. **Mucous glands** are found beneath the stratified squamous epithelium; mixed glands occupy the lamina propria under the pseudostratified ciliated columnar epithelium. Aggregations of lymphatic nodules in the posterior nasopharyngeal mucosa constitute the **pharyngeal tonsils (adenoids)**. The superior, middle, and inferior constrictor muscles and the stylopharyngeus and salpingopharyngeus muscles constitute the skeletal muscles of the muscularis layer. The fibrosa layer is a tough fibroelastic layer that attaches the pharynx to surrounding structures.

# PALATE AND TONSILS

The palate forms the roof of the oral cavity and consists of hard and soft portions.

## Hard Palate

The **bony palate** is covered inferiorly by a mucoperiosteum, which is much like that of the gingiva in that it consists of a stratified squamous epithelium that demonstrates keratinization, parakeratinization, or nonkeratinization in different regions. Parakeratinized epithelia are similar to keratinized epithelia, except the surface cells retain their nuclei. An accumulation of fat is found anteriorly in the submucosa. Mucous glands are plentiful in the submucosa of the posterior two-thirds of the hard palate. Transverse corrugations of the mucosa in the anterior region and a median raphe also are characteristic features.

## Soft Palate

The soft palate is a **muscular organ** that extends posteriorly from the hard palate. It is lined on the nasopharyngeal side by pseudostratified ciliated columnar epithelium and on the oral side and free margin by stratified squamous nonkeratinized epithelium. The submucosa contains mucous glands on the oral side and mixed seromucous glands on the nasal side. Most of the skeletal muscles of the soft palate insert into either the palatine aponeurosis, which is continuous with the pharyngobasilar fascia, or are continuous with their opposite partner in the midline. The lowest or most anterior layer of muscle is the palatoglossal muscle. As indicated previously, this muscle and the mucosa covering it constitute the anterior pillar of the fauces (glossopalatine arch). The most posterior layer of muscle is formed by the palatopharyngeus muscle, which, with its mucosa, constitutes the posterior pillar of the fauces (palatopharyngeal arch). The uvular muscles extend from the hard palate to the tip of the uvula.

The **isthmus of the fauces** is the communication between the oral cavity proper and the oral pharynx. It is bounded above by the soft palate, below by the tongue, and laterally by the glossopalatine arch.

# ESOPHAGUS

The esophagus demonstrates the general microscopic plan of the gastrointestinal system. The wall consists of four layers: (1) mucosa, (2) submucosa, (3) muscularis externa, and (4) adventitia or serosa (Fig. 27-1). The **mucosa** consists of stratified squamous nonkeratinized epithelium, a lamina propria with some mucous glands at the upper and lower extents of the esophagus, and a well-developed muscularis mucosae of smooth muscle. The **submucosa** contains some mucous glands and the autonomic submucosal nerve plexus. The **muscularis externa** consists of an outer longitudinal and inner circular layer of muscle with the myenteric plexus sandwiched between the two layers. The submucosal and myenteric plexuses contain postganglionic parasym-

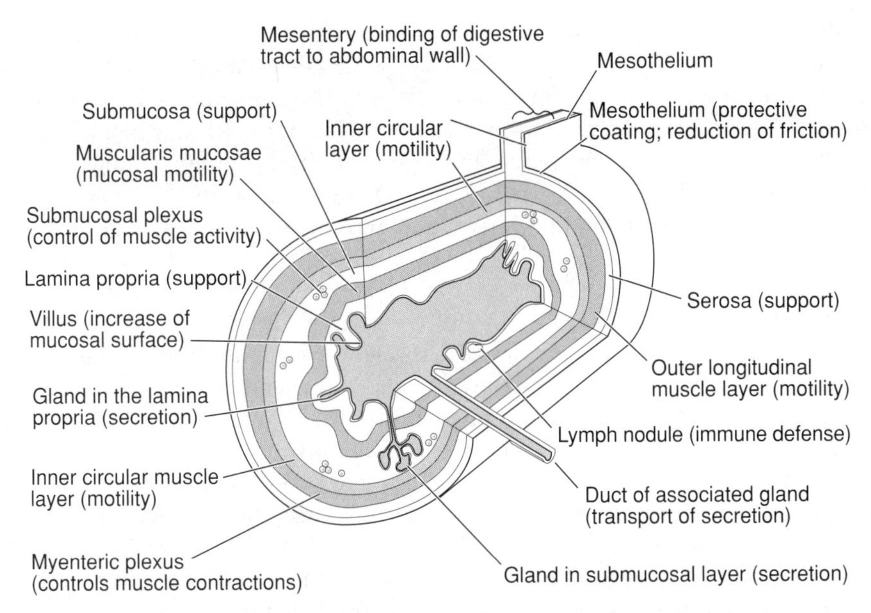


**Figure 27-1.** Diagram illustrating the basic structural plan of the wall of the gastrointestinal tract, noting the relationships between the layers.

pathetic neurons and, in some configurations, cells of the enteric nervous system. In the upper esophagus, the muscle is skeletal; in the lower portion, it is smooth muscle; and in the middle, it is mixed. The **adventitia** is connective tissue that merges imperceptibly with that of the surrounding mediastinum. In the short abdominal portion of the esophagus, the outer layer is peritoneum, and thus it consists of mesothelium and underlying connective tissue and is called a **serosa**.

## STOMACH AND INTESTINE

The general histologic plan of the **gastrointestinal tract** consists of an inner mucosa, a submucosa, a muscularis externa, and an outer serosa (reviewed in the previous section). Different parts of the gastrointestinal tract retain the basic plan but differ as to their internal configuration of the mucosa, epithelial lining, type and extent of mucosal and submucosal glands, and by their thickness and configuration of muscle.

**Stomach**. The stomach is structurally modified for the **production of hydrochloric acid (HCl)** and **pepsin** and for the **mixing of food** with these substances. In the empty stomach, the mucosa is thrown into longitudinal folds called **rugae**. Throughout the stomach, the simple columnar lining epithelium produces mucus and indents into the mucosa as gastric pits. One or more mucosal glands empty into the base of each pit. The length of glands and pits and the glandular cell types differ in the cardiac, body (fundus), and pyloric regions of the stomach.

**Gastric glands** of the body and fundus are the most prevalent and contain the most diverse cell types. In this type of gland, chief cells are located mostly in the base (fundus), parietal cells are located in the neck and isthmus region, mucous neck cells are in the neck, and argentaffin [enteroendocrine, enterochromaffin, amine precursor uptake and decarboxylation **(APUD) cells**] are scattered. Undifferentiated columnar cells in the neck region of the gland differentiate and move upward to replace the lining epithelium, which is replaced every 3 to 7 days. Other undifferentiated cells move into the glands, where they differentiate into the gland cells.

**Chief cells** produce **pepsin** and are typical enzyme-secreting cells in that they contain much rough endoplasmic reticulum (RER), a well-established Golgi apparatus, and membrane-bound zymogen granules. In these cells, amino acids attached to transfer RNA (tRNA) are carried to ribosomal RNA (rRNA) of the RER, where the pepsinogen code is translated from messenger RNA (mRNA). The protein product passes into the cisterna of the RER and is carried to transfer vesicles near the Golgi apparatus. In this region, the **membrane-bound pepsinogen granule** is formed. The granule passes to the apical surface of the cell, where the product is discharged when the membrane of the granule fuses with the plasma membrane and then opens to the lumen (exocytosis, merocrine secretion).

**Parietal cells** produce **HCl** and are also thought to produce **intrinsic factor**, which prevents the development of pernicious anemia. These cells are rounded or pyramidal cells whose apices open to the lumen through an elaborate intercellular canal between adjacent chief cells. An extensive infolding of the surface membrane forms intracellular canaliculi into which microvilli project. Smooth endoplasmic reticulum (SER) and mitochondria are prevalent, and the cells are often binucleate.

In the production of HCl, sodium chloride most likely passes to the intracellular canaliculi where the hydrogen ion, supplied by carbonic acid in the cell, is exchanged for sodium. Thus, HCl passes into the lumen of the gland, and bicarbonate passes from the cell into the blood.

**Mucous neck cells** produce **mucus** of a different nature than that of the surface epithelium. Argentaffin cells are sandwiched between the other gland cells and the basement membrane. Some of these cells produce **serotonin**, and other types produce **cholinesterase**. Both of these secretions are discharged into the bloodstream. Other enteroendocrine (APUD) cells seem to produce a **glucagon-like substance**, and still others produce **gastrin**.

**Pyloric** and **cardiac glands** contain only cells that are similar to the mucous neck cells. In the pyloric region, the gastric pits are deep, and the glands are very tortuous and appear shorter.

The **muscularis externa layer** of the stomach contains three layers: an outer longitudinal, a middle circular, and an inner oblique layer. The two inner layers are thickened in the pylorus as the **pyloric sphincter muscle**. The serosa is continuous at the greater curvature, with the two layers of the greater omentum, and at the lesser curvature, with the two layers of the lesser omentum.

**Small intestine**. The small intestine is structurally modified for the **absorption of nutritive substances**. The absorptive surface is large because of (1) mucosal and submucosal folds called **plicae circulares**, (2) mucosal projections called **villi**, and (3) microvilli forming a striated border on the simple columnar lining epithelium. Mucus is secreted by goblet cells in the lining epithelium. These cells increase in number at progressively lower levels of the gastrointestinal tract where drier wastes are accumulating.

**Crypts of Lieberkühn** are mucosal glands found throughout the intestine and empty at the bases of the villi. Their cells replace the lining cells of the villi every 7 to 8 days, and thus show mitoses and gradations of differentiation. Some of the enteroendocrine gland cells produce **secretin** and **cholecystokinin**. **Argentaffin cells** and **Paneth's cells** are present in the crypts of Lieberkühn.

In the **upper part of the duodenum**, **Brunner's glands** occupy the submucosa and empty into the crypts or the bases of intervillous spaces. Their cells are similar to those in the pyloric glands and produce an alkaline glycoprotein secretion. The lower **ileum** contains aggregations of lymphatic nodules called **Peyer's patches**. These are located opposite the mesentery attachment, chiefly in the mucosa. The epithelium of the ileum over the areas of Peyer's patches becomes dome-shaped rather than the usual villous format of the rest of the intestine. The domed epithelium of this region demonstrates the presence of special cells called **M cells**. These cells have a close association with lymphocytes and are thought to be antigen-presenting cells of the intestinal mucosa.

The **muscularis externa** consists of inner circular and outer longitudinal smooth muscle layers. Myenteric and submucosal nerve plexuses containing autonomic postganglionic cell bodies and cells of the enteric nervous system are present throughout the intestine and stomach.

The **absorption** of **fats**, **carbohydrates**, **proteins**, and **water** in the small intestine takes place through the simple columnar lining cells of the villi because the tight junctions (zonula occludens) of the junctional complexes (zonula occludens, zonula adherens, and macula adherens) do not permit intercellular passage; thus, carbohydrates, fats, and proteins must be broken down in the lumen before absorption can take place. Some of this is accomplished by pancreatic enzymes and liver bile salts. Intestinal juices produced by crypt glands and the lining epithelium are also involved in the terminal hydrolytic digestion of carbohydrates and proteins. The active sites of much of this activity seem to be in the microvillus region near the glycoprotein "fuzz" coat of the plasma membrane. Substances absorbed through the surface epithelium pass to capillaries or the central lacteal of the villi for distribution in the portal vein or thoracic duct, respectively. In fat absorption, bile salts and lipase produce micelles of fatty acids and monoglycerides, which passively enter the cell. In the cytoplasm, the SER resynthesizes triglycerides and the RER produces proteins in the production of chylomicrons. These are discharged into the intercellular space, where they pass mostly to the lacteals.

**Colon (large intestine)**. The large intestine or colon **absorbs much water** and **produces mucus** that lubricates the feces. No villi are present, and the crypts open directly on the surface. The lining epithelium is simple columnar with a striated border, and it contains many goblet cells. The outer longitudinal layer of the muscularis externa has three longitudinally running thickened bands of smooth muscle (taenia coli).

The **appendix** is microscopically similar to the colon, except that it is smaller, does not have taenia, and possesses a prominent ring of lymphatic nodules, which occupy most of the lamina propria and the submucosa.

The **anal canal** functions to **retain** and **eliminate wastes**. The colon-like mucosa of the upper portion forms longitudinal anal columns just above the horizontally oriented anal valves. Epithelium over the anal valves is stratified squamous nonkeratinized epithelium. It becomes keratinized about 2.5 cm below the valves. The lamina propria contains large internal hemorrhoidal veins and circumanal glands. The inner circular layer of the muscularis becomes the internal anal sphincter. The outer longitudinal layer disappears and is replaced in position by skeletal muscle of the external anal sphincter.

# EXTRAMURAL GLANDS OF THE DIGESTIVE SYSTEM

## Salivary Glands

The major salivary glands are located in the head and open into the oral cavity. They include the parotid, sublingual, and submandibular glands. In addition, numerous microscopic glands are embedded in the mucosae of the cheeks, lips, palate, and tongue.

**Parotid glands**. The parenchyma of the parotid glands consists of serous acini and ducts. The acini are grouped into lobules and lobes by connective tissue septa. Pyramid-shaped cells with apical accumulations of zymogen granules and basal concentrations of RER line the small lumina of the serous acini. Myoepithelial cells lie between the acinar cells and the basal lamina. The serous secretion of the acinar cells passes into **intercalated ducts**, which empty into **striated ducts**. Both of these ducts are intralobular ducts. Intercalated ducts have small lumina and are lined by simple cuboidal epithelium. Striated (salivary) ducts are larger and are lined by a simple columnar epithelium whose cells have extensive infoldings of the plasma membrane on their basal surfaces. These basal striations with their interposed mitochondria are like those in cells of the distal convoluted tubules of the kidney and perform similar functions in the reabsorption of sodium and water from the luminal fluid. From the striated ducts, the secretion passes successively to interlobular, interlobar, and the main excretory (Stensen's) ducts. The simple columnar epithelial lining of the ducts gets taller as the main duct is approached. **Stensen's duct** is lined by pseudostratified columnar epithelium that contains some goblet cells. At the opening of the duct into the oral cavity, the epithelium becomes stratified squamous nonkeratinized epithelium.

**Submandibular glands**. These are mixed seromucous glands whose acini are preponderantly serous. Mucous alveoli are frequently capped by serous demilunes or have some serous cells lining their terminal portions. The secretion from serous demilunes passes between mucous cells to reach the lumen. Mucous cells contain basally flattened nuclei, RER, and apical membrane-bound mucigen droplets. The ducts of the submandibular gland are microscopically similar to those of the parotid, but striated ducts are longer and more numerous. The **main duct (Wharton's)** opens into the mouth beneath the tongue.

**Sublingual glands**. These are mixed glands, but the mucous alveoli predominate. Striated (salivary) ducts and intercalated ducts are few in number. The main excretory ducts open into the mouth at the side of the frenulum and, like the main ducts of the parotid and submandibular glands, are lined by pseudostratified columnar epithelium.

## Liver and Gallbladder

**Liver**. The liver is surrounded by a tough connective tissue capsule that penetrates it at the **porta hepatis** to produce many septa. These septa provide support for the parenchyma and divide the liver into lobes and lobules. Branches of the portal vein and hepatic artery further subdivide within the septa and supply the lobules. These vessels and bile ducts constitute portal triads (portal canals, spaces) at the junction of adjacent lobules.

The **hepatic lobules** consist of a central vein from which anastomosing hepatic plates of cells, usually one cell thick, radiate toward the periphery. Between the plates are hepatic sinusoids that connect the portal vein and hepatic artery peripherally with the central vein centrally. Blood drains from the central vein to sublobular veins and leaves the liver by way of the hepatic veins. The sinusoids are lined by discontinuous endothelial cells and phagocytic **Kupffer cells** of the reticuloendothelial system. The lining cells abut against microvilli of the hepatocytes, leaving a **space (of Disse)** between the base of the endothelial cell and the hepatocyte. This space is continuous with the sinusoids, thus providing for efficient interchange between blood in the sinusoid and the hepatocyte.

**Bile canaliculi** lie between adjacent cells in the hepatic plates and are expansions of the intercellular spaces between the cells. They are separated from the rest of the intercellular space by occluding junctions. Microvilli of the hepatic cells extend into the canalicular lumen. These canaliculi receive **bile** produced by the hepatocytes and conduct it peripherally into small ducts that open into the bile ducts of the portal triad.

Bile is transported out of the liver in **hepatic ducts**. It then traverses the cystic duct and is stored and condensed in the gallbladder. Bile is discharged from the gallbladder through the **cystic duct** and **common bile duct** and empties at the sphincters of Oddi and Boyden into the duodenum.

The epithelium lining the bile ducts grades from low cuboidal to high columnar, with the increasing caliber of the ducts. Hepatocytes are polyhedral, with one or more large rounded nuclei. They contain a wide range of organelles that are consistent with the many and diverse functions these cells perform. The SER is involved in the synthesis of glycogen, in the inactivation and detoxification of drugs, in the synthesis of bile acids, and in the production of water-soluble bilirubin glucuronide. The RER synthesizes lipoproteins, prothrombin, albumin, and fibrinogen. Hepatocytes also store lipids, carbohydrates, and vitamins; recirculate bile acids; and can carry out gluconeogenesis. Kupffer cells, like other cells of the mononuclear phagocyte system, assist hepatocytes by producing bilirubin by the breakdown of hemoglobin from phagocytized worn-out erythrocytes.

Two methods of classification of liver lobules are used other than the classic hepatic lobule mentioned above. The **portal lobule** has a portal triad at the center, with the periphery of the lobule being those adjacent portions of hepatic plates that drain into the bile duct of the portal triad. The hepatic (Rappaport) acinus is a diamond-shaped area that drains into an interlobular vein between adjacent hepatic lobules. The periphery of the acinus extends to the central veins of the two adjacent lobules.

**Gallbladder**. The gallbladder is lined by a mucous membrane that is thrown into folds. It possesses a simple columnar epithelium whose sodium pump transports sodium chloride through the epithelium and extensive lateral intercellular spaces to underlying capillaries, leading to the **passive reabsorption of water** and the **concentration of bile**. A circularly arranged smooth muscle layer is present and is surrounded by a prominent connective tissue layer.

## PANCREAS

The pancreas is divided into lobules by connective tissue septa. The lobules are packed with serous acini consisting of **enzyme-secreting pyramidal cells** and **centroacinar cells**. The serous-secreting cells are similar in structure to other protein-secreting cells (e.g., chief cells of the stomach). The cells are arranged in acini with small centroacinar cells lining the lumen. The pyramidal secretory cells are regulated by **cholecystokinin**, which causes secretion of proteases, nucleases, amylase, and lipase. The centroacinar and intercalated duct cells, when stimulated by secretin, produce high concentrations of sodium bicarbonate. The intercalated ducts drain acini to larger intralobular ducts, which empty successively to interlobular ducts and the main pancreatic duct (of Wirsung) or accessory duct. The larger ducts have simple columnar epithelium, goblet cells, and mucous glands; the smaller ducts are lined with simple cuboidal epithelium.

**Islets of Langerhans** are heavily vascularized groups of cells scattered throughout the pancreas and constitute the **endocrine portion of the pancreas**. Several cell types are present in the islets; the most common type is the **insulin-producing beta cell**. Absence or malfunction of these cells leads to **diabetes mellitus**. Another prominent cell type of the islets is the **alpha cell**, which produces **glucagon**, a hyperglycemic-glycogenolytic factor. **Delta cells** in the islets produce **somatostatin** and possibly gastrin.

# Chapter 28

# Respiratory System

The **conducting portion** of the respiratory system includes the nasal cavity, nasopharynx, laryngopharynx, trachea, bronchi, and bronchioles, down to and including the terminal bronchioles. The oropharynx is shared with the digestive system. All but the bronchioles and oropharynx are characterized by (1) a mucosa of pseudostratified ciliated columnar epithelium with goblet cells and an underlying connective tissue containing mixed seromucous glands, and (2) usually a cartilaginous or bony support. The **respiratory portion** consists of respiratory bronchioles, alveolar ducts, alveolar sacs, and alveoli. A respiratory bronchiole and its branches constitute a lobule. All respiratory portions contain alveoli in their walls. Table 28-1 presents a summary of the features of the upper respiratory passes.

## NASAL CAVITY

The nasal cavity extends from the base of the anterior cranial fossa to the roof of the mouth (palate) and is divided into right and left sides by the **nasal septum**. The septum is formed by the perpendicular plate of the ethmoid, the vomer, and the septal cartilage. The nasal cavity is related superiorly to the anterior cranial fossa; laterally to the ethmoid air cells, maxillary sinus, and orbit; inferiorly to the oral cavity; and posterosuperiorly to the sphenoid sinus. It opens anteriorly on the face by way of the vestibule and nares and is continuous posteriorly by the choanae with the nasopharynx. The superior, middle, and inferior conchae divide the cavity into superior, middle, and inferior meatuses. The area posterosuperior to the superior concha is the **sphenoethmoidal recess**.

Most of the nasal cavity is lined by a **mucoperiosteum** consisting of a pseudostratified ciliated columnar epithelium, seromucous glands, and an extensive blood supply. The venous plexus of the conchae and septum can become engorged with blood, thus restricting the nasal passage by swelling of the mucosa. The rostral direction of the arterial blood flow in the mucosa aids in warming the air. The superior portion of the nasal cavity constitutes the olfactory mucosa, whose pseudostratified columnar olfactory epithelium consists of modified bipolar neuroepithelial cells and basal and supporting cells. Nonmotile cilia of the bipolar cells are considered to be the olfactory receptor mechanism of the functional dendrite, and its axon passes to the olfactory bulb. Serous glands of Bowman empty to the olfactory surface, where their secretion washes the cilia and prepares them to respond to new stimuli (see Table 28-1).

## LARYNX AND TRACHEA

The **tubular larynx** is composed of nine cartilages connected by elastic membranes and intrinsic skeletal muscles. It is lined with a mucosa whose folds form the true

**TABLE 28-1. Summary of the Features of the Upper Respiratory Passages**

| Part | Epithelium | Goblet Cells | Glands | Cartilage | Smooth Muscle | Elastic Fibers |
|---|---|---|---|---|---|---|
| Nasal cavity | Pseudostratified ciliated columnar | Many | Many | Hyaline in nasal septum | None | None |
| Nasopharynx | Pseudostratified ciliated columnar | Many | Many | Hyaline in pharyngotympanic tube | None | Some |
| Larynx | Pseudostratified ciliated columnar | Many | Many | Hyaline and elastic cartilage | None | Some |
| Trachea | Pseudostratified ciliated columnar | Many | Many | C-shaped hyaline cartilage rings | On posterior aspect of cartilage rings | Some |
| Bronchi | Pseudostratified ciliated columnar | Some | Some | Hyaline cartilage plates | Spiral bundles | Some |
| Bronchioles | Pseudostratified ciliated columnar | Scattered | None | None | Spiral bundles | Abundant |
| Terminal bronchioles | Ciliated simple columnar | Absent | None | None | Spiral bundles | Abundant |
| Respiratory bronchioles | Ciliated simple cuboidal | Absent | None | None | Spiral bundles | Abundant |

and false vocal folds. The mucosa of each true vocal fold covers the vocal ligament (free margin of the conus elasticus) and vocalis muscle. The lining epithelium of the true vocal fold and of most of the epiglottis is stratified squamous nonkeratinized epithelium. Some taste buds may be found in the epiglottic epithelium. The rest of the lining is pseudostratified ciliated columnar epithelium, which contains goblet cells and is underlain by a lamina propria that contains mixed seromucous glands.

Inhaled particulate matter is entrapped in the sticky mucous secretion, which is transported to the **pharynx** by ciliary movement in the more fluid serous medium. There are no glands in the true vocal folds, but the surface is kept moist by the secretions that arise from numerous glands lining the ventricles. Lymphatic tissue in the ventricles may constitute the laryngeal tonsil. The thyroid, cricoid, and most of the arytenoid cartilages are hyaline cartilage. The epiglottis, cuneiform, corniculate, and tips of the arytenoid cartilages are elastic cartilage.

The **trachea** is a tubular structure whose wall, from the luminal surface outward, consists of a mucosa, submucosa, and adventitia. The mucosa is similar to that of most of the larynx in that it is comprised of a pseudostratified ciliated columnar epithelium with goblet cells, the most prominent basement membrane in the body, and a lamina propria that contains many longitudinally directed elastic fibers. The indistinct submucosa contains seromucous glands that also extend between the C- and Y-shaped hyaline cartilage rings in the adventitial layer. The open interval of the C-shaped cartilages faces the esophagus and is bridged with fibroelastic tissue and a trachealis smooth muscle that runs circularly, attaching at the inner surface of the ends of each cartilage ring (see Table 28-1).

# BRONCHI AND LUNGS

The **main bronchi** and **segmental bronchi** are similar to the trachea. The smaller bronchi have cartilaginous plates instead of rings. In bronchioles, the cartilage disappears, circular smooth muscle becomes more prominent, and the ciliated epithelium becomes simple columnar and simple cuboidal. Glands are no longer present in the terminal bronchioles.

**Respiratory bronchioles** have simple cuboidal epithelium, except at those sites where alveoli are present. Alveolar ducts are completely lined by alveolar sacs and alveoli, and their lumina are marked by spiraling bundles of smooth muscle. Alveoli are separated from each other by interalveolar septa that contain an extensive capillary net, reticular and elastic fibers, blood cells, macrophages, lymph nodes, and nodules. The alveolus is lined by an extremely attenuated simple squamous epithelium. Blood in the capillaries is separated from the air in the alveoli by nonfenestrated endothelial cells and their basal lamina and the simple squamous (alveolar type I) cells and their basal lamina. These four structures constitute the **respiratory membrane**, through which gas exchange takes place. The basal laminae of the alveolar and endothelial epithelia are fused in the thinnest blood-air transport regions. Great alveolar (septal, pneumocyte type II) cells bulge between the squamous cells into the alveolar lumen and produce **surfactant**, which lowers the surface tension of fluid in the alveoli. Alveolar phagocytes (dust cells) migrate into alveolar spaces and engulf debris (see Table 28-1).

**Bronchial arteries** carrying nourishment to the lungs course along the bronchi to the respiratory bronchioles. Venous return of this blood is mainly through pulmonary veins, but some blood returns by way of the bronchial veins to the azygos system. **Pulmonary arteries** branch and follow the air tubes to the capillary plexuses in the alveoli. Oxygenated blood is returned through pulmonary veins that travel in the interlobular connective tissue septa.

The **outer surface of the lung** is covered by a serous mesothelium called the **pleura**. Directly underlying the pleura is a dense layer of collagenous and elastic fibers.

# Chapter 29

# *Urinary System*

## KIDNEY

The kidney is divisible into an outer cortex and an inner medulla. The **medulla** consists of renal pyramids, with the broad base of each pyramid facing the cortex and the apex (renal papilla) opening into a minor calyx. The 10 to 16 minor calyces open into 2 or 3 major calyces, which in turn empty into the funnel-shaped renal pelvis at the hilus of the kidney. The cortex lies peripheral to the medulla and extends between the pyramids as renal columns. The **cortex** consists of medullary rays (pars radiata) and cortical labyrinths (pars convoluta). The medullary rays are parallel accumulations of collecting tubules and thick and thin limbs of the **loop of Henle**, which radiate toward the medulla. Each of these is surrounded by cortical labyrinth tissue of glomeruli and convoluted tubules that empty into the tubules of the medullary ray. A **medullary ray** and its surrounding cortical labyrinth constitutes a lobule. A kidney lobe is a renal pyramid with its overlying cortex and renal columns.

The functional unit of the kidney is the **uriniferous tubule**, which consists of a nephron and collecting tubule (Fig. 29-1). (Note: Some authors define the uriniferous tubule as the nephron.) The **nephron** is comprised of the renal corpuscle (of glomerulus and Bowman's capsule), the proximal convoluted tubule, the thick descending limb (thin portion) and thick ascending limb of the loop of Henle, and the distal convoluted tubule.

The **glomerulus** is a tuft of fenestrated capillaries fed by an afferent arteriole and drained by a smaller efferent arteriole. It is invested by podocytes of the visceral layer of **Bowman's capsule**. The podocytes have pedicles that interdigitate with those from adjacent podocytes and attach to a basal lamina between the podocyte and the capillary endothelium. Filtration slits 25 nm wide exist between pedicles and communicate with the lumen of Bowman's capsule. At the basal lamina surface, these slits are connected by a thin slit membrane. The podocytes, basal lamina, and fenestrated endothelium constitute the filtration barrier. Mesangial phagocytic (stalk) cells in the glomerulus probably remove filtration residues from the basal lamina, which is the main filter for large molecules. The parietal layer of Bowman's capsule consists of simple squamous epithelium.

The **proximal convoluted tubule** is the longest and widest portion of the nephron. It is comprised of a simple pyramidal epithelium possessing a brush border (microvilli) and indistinct lateral borders. The proximal convoluted tubule is located in the cortical labyrinth. It is continuous with, and histologically similar to, the thick descending limb of Henle's loop that courses in the medullary ray. Both the proximal and straight descending tubules resorb 85% or more of the water and sodium chloride of the glomerular filtrate. Glucose and amino acids also are resorbed by the epithelium of these tubules. The thin portion of Henle's loop is comprised of simple

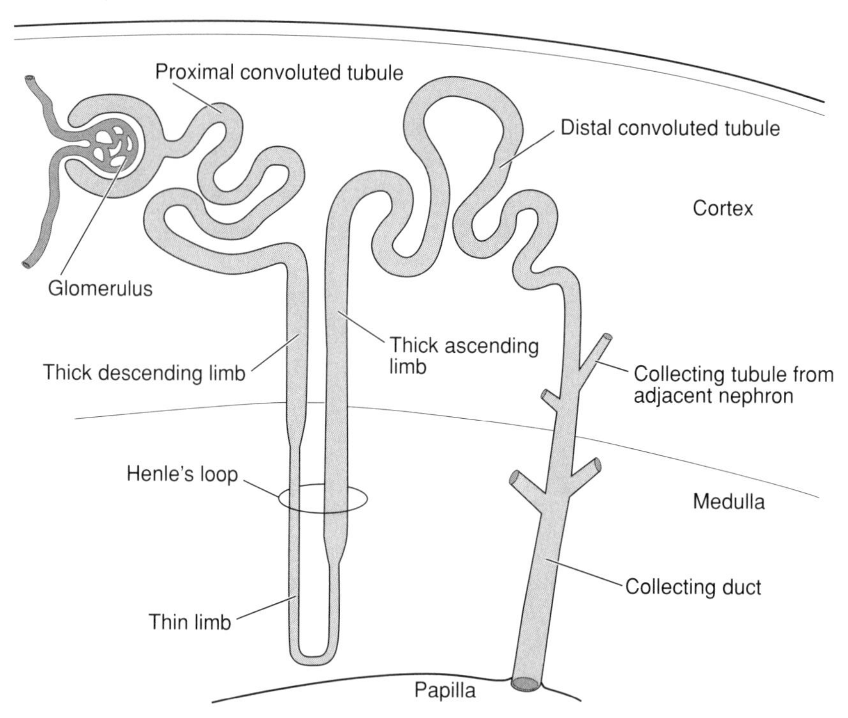


**Figure 29-1.** Diagram of the components of the nephron, indicating the parts of the uriniferous tubule and the collecting ducts.

squamous epithelium. Its descending and ascending portions function similarly to the thick descending and ascending portions, respectively.

The **thick ascending and distal convoluted tubules** have a low cuboidal epithelium with scattered microvilli and an extensive infolding of the basal plasma membrane. In these tubules, sodium is transported out of the cells, and the filtrate becomes hypotonic and acidic. Where the distal tubule abuts the afferent glomerular arteriole, the epithelium is columnar and constitutes the macula densa. The muscle cells of the adjacent afferent glomerular arteriole are replaced by large pale juxtaglomerular cells containing granules. These cells produce renin and, along with cells of the macula densa and some interposed cells, constitute the **juxtaglomerular complex**. Renin acts on its substrate, angiotensinogen, producing **angiotensin I**, which is converted to angiotensin II. **Angiotensin II** causes increased secretion of aldosterone by the adrenal cortex. **Aldosterone** acts on the distal convoluted tubule to bring about reabsorption of sodium and thus reduce sodium loss in the urine.

The **collecting tubules** consist of pale, clear, simple cuboidal epithelium with distinct lateral boundaries. These tubules join to form papillary ducts lined by simple columnar epithelium. Under the influence of antidiuretic hormone from the neurohypophysis, the epithelium of the collecting tubules becomes more permeable to water, which is passively removed from the urine.

Arterial blood is carried to the hilar region of the kidney by the **renal artery**. This artery branches into interlobar branches, which give rise to arcuate arteries passing along the corticomedullary junction. Interlobular arteries arise from the arcuate arteries and pass peripherally in the cortical labyrinths to give rise to afferent glomerular arterioles supplying the glomeruli. From the glomeruli, efferent arterioles supply the capillary plexuses around the tubules. Capillaries of the medulla are supplied by arteriolae rectae from the efferent arterioles. Venous drainage is by venae rectae, and interlobular, arcuate, interlobar, and renal veins that accompany the arteries.

## URETERS

The ureters are lined with transitional epithelium. External to the lamina propria is a muscularis layer consisting of inner longitudinal and outer circular smooth muscle layers. An additional outer longitudinal layer is added near the bladder.

## URINARY BLADDER

The urinary bladder is histologically similar to the lower part of the ureter in that it has a mucosa with transitional epithelium, a three-layered muscularis layer, and an outer connective tissue layer. The superficial (facet) transitional cells have a luminal plasma membrane of thick plates separated by thinner membrane. During contraction of the bladder, the thick areas invaginate and form vesicles of reserve membrane.

# Chapter 30

# Endocrine System

The endocrine system consists of numerous **ductless glands** that are distributed throughout the body. These glands secrete substances known as **hormones** into blood capillaries rather than into ducts. The hormones are then transported by the blood to affect tissues or cells elsewhere in the body. The affected tissues are called the targets or **target tissues**.

## PITUITARY GLAND

The pituitary gland, or **hypophysis**, is located in the sella turcica of the body of the sphenoid bone. It is attached to the hypothalamus by its pituitary stalk, which penetrates the diaphragma sellae, a fold of dura mater that covers the sella turcica. The hypophysis is lateral in relation to the internal carotid artery and the other contents of the cavernous sinus. It is composed of the adenohypophysis and the neurohypophysis. The gland is about 1.5 cm in diameter and about 1 cm in its rostrocaudal extent.

### Adenohypophysis

The adenohypophysis consists of a **pars tuberalis**, which forms an anterolateral cuff to the infundibulum (infundibular stalk); a **pars distalis** (anterior lobe), which produces most of the adenohypophyseal hormones; and a **pars intermedia**, which is interposed between the pars distalis anteriorly and the pars nervosa posteriorly. The parenchyma of the pars distalis comprises cords of cells that are closely opposed to fenestrated sinusoidal capillaries, constituting a secondary capillary plexus. The secondary capillary plexus receives venous trunks originating from capillary loops (primary capillary plexus) that extend into the pars tuberalis and infundibular stalk. This vascular arrangement constitutes the hypophyseal portal system. The primary capillary plexus is supplied by superior hypophyseal arteries, which arise from the internal carotid and posterior communicating arteries.

The cell cords of the pars distalis are composed of chromophils and chromophobes. Chromophobes are considered to be degranulated chromophils. Chromophils are divisible by the staining reaction of their secretory granules into two basic classes: basophils and acidophils. **Acidophils** are further divided into somatotrophs, which produce growth hormone, and mammotrophs, which produce prolactin. **Basophils** include gonadotrophs, which stimulate the activity of the gonads, thyroid, and adrenal cortex. The release of these hormones into adjacent capillaries is controlled by releasing and inhibiting hormones produced in hypothalamic neurons and released into the portal system in the infundibular stalk. Here, the factors are taken up in primary capillaries and traverse the venous trunks and the secondary capillary plexus, where they influence the secretory activity of cells in the pars distalis. The pars tuberalis consists primarily of basophils (Table 30-1).

**TABLE 30-1. Major Cell Types in the Hypophysis, Their Secretions, and Their Targets**

| Hormone | Region Secreted From | Cell Type | Target | Effect |
|---|---|---|---|---|
| Antidiuretic hormone | Pars nervosa | Supraoptic nucleus of hypothalamus | Kidney tubules | Water absorption |
| Oxytocin | Pars nervosa | Paraventricular nucleus of hypothalamus | Mammary gland, uterine smooth muscle | Contraction of smooth muscle and myoepithelial cells |
| Growth hormone | Pars distalis | Acidophil | Epiphyseal plates, muscle, and adipose tissue | Bone growth, hyperglycemia, and elevated free fatty acids |
| Prolactin | Pars distalis | Acidophil | Mammary gland | Milk secretion |
| Adrenocorticotropic hormone | Pars distalis | Basophil | Adrenal cortex: zonae fasciculata and reticularis | Secretion of adrenal cortical hormones |
| Thyroid-stimulating hormone | Pars distalis | Basophil | Thyroid gland | Thyroxine production |
| Follicle-stimulating hormone | Pars distalis | Basophil | Ovary in female, testis in male | Follicle development in female, spermatogenesis in male |
| Luteinizing hormone | Pars distalis | Basophil | Thecal cells in ovary in female, interstitial cells in testis in male | Progesterone secretion in female, testosterone secretion in male |

## Pars Nervosa

The **neurohypophysis** consists of the infundibular stalk and the pars nervosa; by some definitions, it includes the median eminence and secretory neurons of the hypothalamus. The secretory cells of the neurohypophysis are hypothalamic neurons. Cell bodies in the supraoptic and paraventricular nuclei of the hypothalamus give rise to unmyelinated axons that help make up the hypothalamohypophyseal tract of the infundibular stalk before termination in adjacent fenestrated capillaries in the pars nervosa. Glial supportive cells of the pars nervosa are called **pituicytes**. The hormones produced and transported in the neurons, their binding protein (neurophysin), and adenosine triphosphate constitute a neurosecretory material that may accumulate in the axons and their endings as Herring bodies before discharge into the capillaries. Secretory neurons from other hypothalamic nuclei (e.g., preoptic, arcuate, dorsomedial) send axons to the infundibular stalk, where they empty their releasing or inhibitory hormones into the primary capillary plexus of the portal system (see Table 30-1).

# THYROID AND PARATHYROID GLANDS

## Thyroid Gland

The thyroid gland is invested by a thin capsule of connective tissue that projects into its substance and divides it imperfectly into lobes and lobules. The **parenchyma** consists of follicles that are closed epithelial sacs lined by simple cuboidal or simple columnar epithelium, the cells being low when the gland is underactive and taller when the gland is overactive. Follicles vary from 50 to 500 nm in diameter. The size of each follicle depends somewhat on the degree of distention by the stored colloid in the lumen of the follicle. In the production of thyroxine ($T_4$) and triiodothyronine ($T_3$), the follicular cells receive iodide and amino acids from extensively distributed fenestrated capillaries lying adjacent to the basement membrane.

Through the mechanisms of a basally located rough endoplasmic reticulum and apically oriented Golgi apparatus, a glycoproteinaceous thyroglobulin is produced. This substance is discharged at the apical end of the cell into the follicular lumen, where it is stored along with nucleoproteins and proteolytic enzymes as colloid. Under the influence of thyroid-stimulating hormone, portions of thyroglobulin are taken into the apex of the follicular cell by pinocytosis, and the droplets are hydrolyzed by lysosomal activity into $T_4$ and $T_3$. These substances are then secreted from the basal aspect of the cell into the surrounding capillaries.

**Parafollicular cells** located between follicle cells and the basement membrane, and also found between follicles, differ structurally from follicle cells by being larger and lighter staining. They produce **calcitonin**, a hormone that lowers blood-calcium levels by inhibiting the production of bone.

## Parathyroid Gland

A connective tissue capsule separates parathyroid glands from the thyroid gland. Fine connective tissue septa penetrate the parathyroid glands and divide the parenchyma into irregular cords of chief (principal cells) and oxyphil cells. Chief cells produce parathyroid hormone and are found in two functional states as light and dark chief cells. **Dark chief cells** contain membrane-bound argyrophilic secretory granules, a relatively large Golgi complex, and large filamentous mitochondria. **Light chief cells** have a smaller Golgi complex and few secretory granules. **Oxyphilic cells** are very acidophilic, engorged with mitochondria, have a small nucleus, and do not appear until the end of the first decade of life. Their function is unknown.

# SUPRARENAL GLANDS

The suprarenal glands are divisible into two parts: a mesodermally derived cortex and a neural crest–derived medulla. A thick connective tissue capsule sends radially directed trabeculae of reticular fibers into the underlying cortex. The cortex consists of an outer zona glomerulosa, a middle zona fasciculata, and an inner zona reticularis.

## Suprarenal Cortex

**Zona glomerulosa**. Columnar cells of the zona glomerulosa are arranged in arches. They produce mineralocorticoids (e.g., aldosterone).

**Zona fasciculata**. The zona fasciculata consists of cords of cells that radiate inward from the zona glomerulosa. These cords are two cells thick, contain cuboidal cells that are often binucleate, and are separated from adjacent cords by fenestrated capillaries that radiate inward from the capsule. Fasciculata cells often contain much lipid and appear vacuolated. Because of this appearance, they are often referred to as **spongiocytes**.

**Zona reticularis**. The zona reticularis consists of irregularly arranged cords of cells that may contain lipofuscin pigment. Fasciculata and reticularis cells are under the control of **adrenocorticotropic hormone (ACTH)** from the adenohypophysis, and produce **glucocorticoids** (e.g., cortisol, corticosterone) and the sex hormone dehydroepiandrosterone (**DHEA**). The most prominent organelle in adrenal cortical cells is an extensive smooth endoplasmic reticulum, which is indicative of steroid-secreting cells. In addition, these cells are characterized by mitochondria with tubular cristae.

## Suprarenal Medulla

The suprarenal medulla produces **epinephrine** and **norepinephrine** in its polyhedral basophilic cells. These cells receive terminations of preganglionic sympathetic nerve fibers, and each cell is located between a venule and a capillary. The cells exhibit the **chromaffin reaction**. The blood supply of the suprarenal gland is by branches of the suprarenal arteries that go *directly* to the medulla from the capsule and *indirectly* to the medulla through capsular arterioles and their radiating capillary plexuses, which pass between the cortical cords of cells before reaching the medulla.

# Chapter 31

# Female Reproductive System

In both the female and the male reproductive systems, there is a gonad, which produces specialized reproductive cells called gametes, and a system of ducts and glands, which facilitate the movement of the gametes to the outside of the body. The gonads also serve as endocrine organs that produce the necessary hormones to maintain the secondary sex characteristics associated with males and females.

Table 31-1 summarizes of the major components of the female reproductive system.

## OVARY

The ovary is an **exocrine** organ secreting secondary **oocytes**, and it is an **endocrine** organ producing **estrogens** and **progesterone**. It is divided into an outer cortex and inner medulla. The cortex basically consists of a rather cellular connective tissue stroma, ovarian follicles, and a simple cuboidal surface epithelium (germinal epithelium). The medulla consists of loose connective tissue, blood vessels, lymphatics, nerves, some smooth muscle, and a few vestigial tubular structures called **rete ovarii**.

An **ovarian follicle** consists of a developing ovum and an investment of follicle cells and connective tissue. Follicles originate in the embryo and undergo extensive changes during the childbearing years when they are under the influence of adenohypophyseal hormones. In the embryo, primordial sex cells develop into oogonia, which differentiate into primary oocytes. Primary oocytes go through the prophase of the first meiotic division before they enter an arrested dictyotene stage. Each cell is surrounded by a single layer of follicle cells from the germinal epithelium. Many of these embryonic primordial follicles die, but 70,000 to 400,000 survive in the cortex of the ovary until the time of puberty.

At puberty, **hypothalamic nerve** cells produce **gonadotropin-releasing hormone (GnRH)**. This hormone stimulates **follicle-stimulating hormone (FSH)** release by basophils of the anterior pituitary gland. Under the influence of FSH and other factors, ovarian follicles develop periodically. In this process, some oocytes emerge from the arrested dictyotene stage and start to complete the first meiotic division. Coincident with this, follicle cells enlarge and proliferate to form a stratified cuboidal epithelial layer around the oocyte, thus forming a primary follicle. As the follicle enlarges, spaces between follicle cells coalesce to form an antrum filled with liquor folliculi.

The growing follicle is now called a **secondary (vesicular) follicle**. As this follicle differentiates, the follicular cells around the antrum form a stratified epithelial membrana granulosa that sits on a prominent basal lamina. This in turn is surrounded by an inner, richly vascular theca interna and an outer, more fibrous theca externa. The oocyte is surrounded by a protein-polysaccharide layer called the **zona pellucida**. Cytoplasmic processes of the oocyte and of the immediate surrounding follicle cells of the

**TABLE 31-1. Summary of the Major Components of the Female Reproductive System**

| Component | Associated Epithelium | Other Features |
|---|---|---|
| Ovary | Covered with germinal epithelium, no serosa | Follicles in differing stages of development: corpus luteum or corpora albicans |
| Oviduct (fallopian tube) | Lined with simple columnar epithelium consisting of ciliated cells and secretory cells | Divided into infundibulum with finger-like projections called fimbriae, ampulla, isthmus, and intramural parts |
| Uterus | Simple columnar epithelium with ciliated and secretory cells and many glands | Epithelial lining is endometrium, muscle layer is myometrium, and outer covering is serosa |
| Vagina | Stratified squamous epithelium, nonkeratinized | Longitudinal smooth muscle |
| Mammary gland | Stratified cuboidal or columnar epithelium with myoepithelial cells | Lactiferous ducts give rise to secretory cells under stimulation of hormones |

corona radiata are closely aligned in the substance of the zona pellucida. The combined oocyte, zona pellucida, and corona radiata project into the antrum of the follicle as the cumulus oophorus.

The **mature vesicular follicle** occupies much of the thickness of the cortex, and it causes a bulge (stigma) on the surface of the ovary. Its primary oocyte completes the first meiotic division and becomes a secondary oocyte. The secondary oocyte is relatively metabolically inactive, but it possesses more extensive protein-producing mechanisms than are found in the primary oocyte. It starts its second meiotic division and reaches the metaphase stage at about the time it is ovulated from the ovary along with the zona pellucida and corona radiata of granulosa cells. Just **before ovulation**, large amounts of **estrogen** and small amounts of **progesterone** are produced by the follicle, and more **luteinizing hormone** (LH) is produced by the adenohypophysis.

**After ovulation**, a small amount of blood accumulates in the collapsed follicular remains, and a clot is formed in the antrum region. Under the influence of LH, the granulosa and theca interna cells enlarge, accumulate lipid, and become lutein cells of a corpus luteum. The **granulosa lutein cells** constitute the bulk of the **corpus luteum** and produce **progesterone** and **estrogen**. The **theca lutein cells** are smaller, fewer, more deeply staining, found at the periphery, and may produce **estrogens**. Lutein cells possess major characteristics of steroid-secreting cells in that they have relatively more smooth endoplasmic reticulum (SER) and have mitochondria containing "tubular" lamellae rather than cristae. After clot formation, the thecal connective tissue penetrates into the developing corpus luteum and replaces the blood clot in the central core.

If the ovulated secondary oocyte is fertilized and implantation takes place, the corpus luteum will survive for about 6 months under the influence of **human chorionic gonadotropin (hCG)** from the placenta before it starts to regress. If the secondary oocyte is not fertilized, the corpus luteum will survive for approximately 14 days. When a corpus luteum degenerates, the lutein cells become swollen and then pyknotic, and a hyalinized scar of connective tissue replaces the dead lutein cells. This white scar is called the **corpus albicans**.

Usually only one follicle reaches maturity and is involved in **ovulation** during each cycle. The other maturing follicles are no longer supported by the waning levels of

FSH after ovulation, and they degenerate. In small follicles, the oocyte degenerates and the stroma invades the follicle, leaving no trace of the follicle. In larger follicles, cells of the theca interna enlarge, and the basement membrane becomes a distinct glassy membrane before coarser stromal fibers penetrate the degenerating follicle, giving the atretic follicle the appearance of a small corpus albicans.

## UTERINE TUBE

The uterine tube (**oviduct**, **fallopian tube**) has four regions: **infundibulum**, **ampulla**, **isthmus**, and **interstitial** (intramural) **portion**. The wall of each of these parts consists of a mucosa, muscularis, and serosa. The mucosa of the trumpet-shaped infundibulum and, to a lesser extent, that of the ampulla are characterized by many elongate fimbriae. The epithelium of all parts of the uterine tube is simple columnar and is composed of ciliated cells and peg-shaped secreting cells. These cell types may be different functional states of the same cell. The relative numbers of these cell types vary, depending on the estrogenic or progesteronic influences. The muscularis consists of inner circular and outer longitudinal smooth muscle layers. These layers are relatively thicker in passing from the infundibulum to the interstitial portion. The serosa is lined by mesothelium and is a continuation of the peritoneal covering of the broad ligament.

## UTERUS

The uterus is comprised of a **body**, **fundus**, and **cervix**. The body and fundus are histologically similar, and their walls consist of three layers: perimetrium (serosa), myometrium (muscularis), and endometrium (mucosa). The **perimetrium** is the serosal continuation of the broad ligament. The **myometrium** is composed of an inner layer of longitudinal smooth muscle, a thick middle layer of circular smooth muscle and large blood vessels, and an outer layer of longitudinal and circular smooth muscle. During pregnancy, these smooth muscle cells undergo hyperplasia and hypertrophy.

The basic constituents of the **endometrium** are simple columnar epithelium that is partly ciliated; a lamina propria stroma containing mesenchyme-like cells, reticular fibers, and varying amounts of leukocytes; simple tubular glands; and two sets of arteries. One set of arteries (basal arteries) supplies the glands and stroma of the deepest part of the lamina propria. The other set (coiled, spiral arteries) supplies the rest of the endometrium.

The **endometrium** is under the **influence of ovarian progesterone and estrogen**. It reflects this influence in the marked structural changes characteristic of the menstrual cycle. Four uterine stages of the cycle are recognized: menstrual, proliferative (follicular, estrogenic), secretory (luteal, progesteronic), and premenstrual (ischemic). The **menstrual stage** takes place from *day 1 to day 5* of the cycle. The **proliferative stage** occurs from *day 5 to day 14*; the **secretory stage** from *day 14 to day 27*; and the **premenstrual stage** from *day 27 to day 28*. These are approximate times.

During the **proliferative stage**, the **endometrium grows from a height of 0.5 mm to 2 mm to 3 mm**. In this process, epithelial cells of the gland remnants form a new epithelial lining and straight glands. The stroma develops from the deep (basal) layer. Spiral arteries grow into this new functional layer.

In the **secretory stage**, under the influence of estrogen and progesterone, the **endometrium grows another 2 mm in height**. An increase in interstitial fluid in part of the functional layer divides it into an inner edematous spongy layer and an outer compact layer. The uterine glands grow, become corkscrew-shaped, and produce a glycogen-rich mucoid secretion. Coiled (spiral) arteries elongate and empty into venous sinusoids by way of capillaries.

The **premenstrual stage** is the result of a *decrease of progesterone and estrogen production* by the corpus luteum. In this stage, the coiled arteries kink, there is a drop in the blood supply to the functional layer, the edema decreases, and the glands begin to fragment.

During the **menstrual stage**, the *functional layer becomes anemic and ischemic.* Arteries and veins break down, and blood oozes into the uterine cavity through the degenerating glands and surface epithelium. The entire functional layer is sloughed off during this stage. The basal arteries are not affected, so the basal layer is retained.

The **cervix** consists of a mucosa, muscularis, and adventitia. It does not undergo the extensive cyclic changes of the endometrium, although some changes in structure are seen in pregnancy. The portio vaginalis of the cervix projects into the vagina. The mucosa of the cervix is thrown into folds (plicae palmatae). It consists of simple columnar epithelium with some cilia, extensive forked mucus-secreting glands, and a firm stroma that is rich in collagenous and elastic fibers. The epithelium changes to stratified squamous at the portio vaginalis. During **pregnancy**, the cervical glands become larger and secrete a mucus plug that seals the cervical canal. At the time of **parturition**, the lamina propria becomes more edematous, looser, and cellular. The muscularis layers are similar to those in the uterine body and fundus, except that there is no inner longitudinal muscular layer. The adventitia contains collagenous fibers that are continuous with surrounding structures.

## VAGINA

The vagina also is composed of a mucosa, muscularis, and adventitia. The lining epithelium of the mucosa is stratified, squamous, and nonkeratinized, but keratohyaline granules may be found in some of the cells. Under the influence of **estrogen**, the epithelium accumulates glycogen, and many pyknotic surface cells appear. When estrogen levels are low, a basal layer of cells is prominent. The lamina propria contains many elastic fibers and some large blood vessels. No glands are present. The muscularis consists of a thin inner circular layer and thicker outer longitudinal layer of smooth muscle. A sphincter of skeletal muscle is found at the lower end of the vagina.

## HYMEN

The hymen has the same structure as the vaginal mucosa. It is a thin fold at the opening of the vagina into the vestibule.

## CLITORIS

The clitoris corresponds to the dorsal penis in the male. It has two small cavernous bodies of erectile tissue that end in a rudimentary glans clitoridis. It is covered by stratified squamous epithelium. Specialized nerve endings, such as Meissner's and pacinian corpuscles, are located in the subepithelial stroma.

## LABIA MINORA

The labia minora flank the vestibule. They have a vascularized connective tissue core that is covered by stratified squamous epithelium possessing a thin keratinized layer. Sebaceous glands, not associated with hairs, are located in the stroma.

## LABIA MAJORA

The labia majora are folds of skin that cover the labia minora. The inner surface is like that of the labia minora. The outer surface is covered by skin containing hairs, sweat glands, and sebaceous glands. The interior of these folds contains much adipose tissue.

## VESTIBULE

The vestibule is lined by partially keratinized stratified squamous epithelium. Minor **vestibular glands**, placed chiefly near the clitoris and opening of the urethra, secrete mucus. The longer major vestibular glands are analogous to the bulbourethral glands of the male. They are located in the lateral wall of the vestibule. Their ducts open close to the attachment of the hymen.

## MAMMARY GLANDS

The mammary glands are **integumentary glands** located from the level of the second to sixth or seventh rib on the anterior of the thorax. In fetal development, the glands first appear as ectodermal thickenings along a milk line extending between the upper and lower extremities. As development proceeds, 15 to 20 ectodermal invaginations branch and hollow out to give rise to the 15 to 20 lobes of the mammary gland, which are arranged in a radial fashion deep to the nipple. Thus, each lobe has a single excretory duct opening on the nipple. Each of these ducts diverges at the base of the nipple and increases in size to form an ampulla (lactiferous sinus). Deep to the ampulla, the ducts branch into intralobular ducts.

In the **male and nonpregnant gland**, a few ducts are present, and the epithelium changes from stratified to simple cuboidal epithelium in proceeding from larger to smaller ducts. In the **lactating gland**, the ducts have proliferated, and their terminal portions develop into secretory alveoli lined by a simple pyramidal epithelium invested by myoepithelial cells. The lining epithelium secretes milk proteins by the exocytosis of secretion granules (merocrine type of secretion). Milk lipids in the membrane-bound lipid vacuoles are externalized by the apocrine type of secretion.

The **lobes of the gland** are supported by a dense connective tissue sheath, between the interstices of which are large accumulations of fat. Suspensory ligaments (of Cooper) run through the gland, attaching the deep layer of the superficial fascia to the dermis. The areola is covered by a thin, delicate pigmented skin. Underlying glands (of Montgomery) open on its surface.

## PLACENTA

The placenta consists of a **maternal component (decidua basalis)** and a **fetal component (chorion frondosum)**. Because the embryo implants into the compacta layer of the endometrium, the decidua basalis is that portion of the functional layer that lies deep to the embryo. Glands and blood vessels of this layer empty into intervillous spaces. The stromal cells swell markedly, accumulate glycogen, and are called **decidual cells**.

# Chapter 32

# Male Reproductive System

## TESTIS

The testis is ovoid and surrounded by a thick connective tissue capsule, the **tunica albuginea**. This capsule penetrates the testis at the mediastinum and sends radiating septula into it, dividing the testis into lobules. Within the lobules are seminiferous tubules and a loose fibrous stroma. The stroma contains **interstitial cells of Leydig**, which are characterized by rod-shaped crystalloids (of Reinke), much smooth endoplasmic reticulum (SER), and mitochondria with tubular cristae. Under the influence of **luteinizing hormone (LH)**, also known as **interstitial cell–stimulating hormone (ICSH)**, these cells produce testosterone.

The SER and tubular cristae are characteristic of steroid-secreting cells. The seminiferous tubules consist of contorted loops that join by straight tubules with the rete testis. The convoluted portions in the viable male are lined by a germinal (seminiferous) epithelium resting on a basal lamina bound by peritubular myoid cells that probably produce a peristaltic action. The seminiferous epithelium contains supportive Sertoli cells and sex cells in various stages of spermatogenesis.

The **Sertoli cells** are connected to each other by occluding and gap junctions. These cells support, protect, and nurture the sex cells. They also phagocytize excess cytoplasm in spermatozoan production, and, under the influence of **follicle-stimulating hormone (FSH)**, they secrete **androgen-binding protein (ABP)**, which serves to concentrate **testosterone** needed for spermatogenesis. They also secrete transferrin and inhibin and produce lactate needed by the germ cells.

The **developing sex cells** are located between the Sertoli cells. Spermatogonia are located next to the basal lamina in a basal (extratubular) compartment formed by the tight junctions of the Sertoli cells. The primary spermatocytes and secondary spermatocytes lie nearer the lumen in the adluminal (intratubular) compartment between Sertoli cells. **Spermatids** are embedded in the apices of the Sertoli cells where they transform into spermatozoa. A cross section of a tubule may show several of six stages of development. This is because of different timing in the proliferation and division of stem cells.

Spermatogonia are the only sex cells present until the time of puberty when two different types are present. These are the A, or stem, cells, and the B, or derivative, cells. **A cells** may divide into two A cells or two B cells. **B cells** grow and differentiate into primary spermatocytes.

Each **primary spermatocyte** undergoes the reduction division of meiosis and gives rise to two secondary spermatocytes containing the **haploid number (23) of chromosomes**. Each **secondary spermatocyte** divides quickly into **two spermatids**. When each spermatid undergoes transformation into a spermatozoa, the nucleus condenses, an acrosome vesicle is formed by the Golgi complex, the acrosome vesicle collapses as a lysosomal head cap over the nucleus, an axial filament (flagellum) grows from the proximal centriole, and mitochondria form a helix around the proximal flagellum of the middle

**TABLE 32-1. Summary of the Major Ducts of the Male Reproductive System**

| Component | Epithelium | Other Features |
|---|---|---|
| Testis | Complex stratified epithelium | Demonstrates many stages of spermatogenesis |
| Tubuli recti (straight tubules) | Primarily lined by Sertoli cells | Connect seminiferous tubules to the rete testis |
| Rete testis | Cuboidal epithelium | Located in the mediastinum testis, labyrinthine channels connecting straight tubules to efferent ductules |
| Efferent ductules | Clumps of ciliated epithelial cells alternating with nonciliated cells | Produces a characteristic scalloped appearance to lumen |
| Epididymis | Pseudostratified columnar epithelium | Epithelial cells demonstrate many long, branched irregular microvilli, stereocilia |
| Ductus deferens | Pseudostratified columnar epithelium | Thick layer of smooth muscle in the wall |
| Ejaculatory duct | Pseudostratified columnar epithelium | Courses through the substance of the prostate gland |

piece. Most of the cytoplasm is cast off, leaving only a thin cytoplasmic investment to the head, neck, middle piece, and tail of the spermatozoon. In the principal piece of the tail, the cytoplasm forms a fibrous sheath that does not extend into the end piece of the tail.

**Sperm** are transported through straight tubules, rete testis, and efferent ductules to be stored in the tail of the epididymis, where they mature and become motile. At ejaculation, sperm pass from the head of the epididymis into the ductus deferens, ejaculatory ducts, and urethra.

Table 32-1 presents a summary of the major ducts of the male reproductive system.

### Straight Tubules and Rete Testis

Straight seminiferous tubules and rete testis are lined by simple cuboidal to columnar epithelium. They are located in the mediastinum testis.

## EFFERENT DUCTULES

Rete testis empty into 10 to 15 efferent ductules lined by alternating cuboidal and tall ciliated columnar epithelium, giving cross sections of these tubules a scalloped appearance. A basal layer of rounded cells is surrounded by a lamina propria and some circular smooth muscle fibers. The cilia of the efferent ductules move spermatozoa into the ductus epididymidis (epididymis).

## EPIDIDYMIS

The epididymis contains pseudostratified columnar epithelium containing basal cells and principal cells with long microvilli (stereocilia). It supplies nutritive substances to the sperm and absorbs excess fluid accompanying the sperm. The epithelium is invested by a circular smooth muscle layer.

## DUCTUS DEFERENS

The **vas (ductus) deferens** has a mucosa similar to that of the ductus epididymidis. The muscular layer is highly developed into inner longitudinal, middle circular, and outer longitudinal layers. This duct is invested in the spermatic cord by the cremasteric muscle and the pampiniform plexus of veins. The ductus deferens dilates into an ampulla before terminating as the short slender ejaculatory duct that pierces the prostate and opens into the urethra at the urethral crest.

The mucosa of these structures is folded, and the epithelium is not as tall as in the rest of the ductus deferens. The supporting wall of the ejaculatory duct is made up of fibrous tissue. Muscular contraction of the ductus deferens and the ductus epididymidis propel the spermatozoa to the urethra during the ejaculatory process.

## SEMINAL VESICLE

The seminal vesicle consists of a mucosa folded into a complex system of elevations, a prominent muscularis, and an adventitia. The epithelium is low pseudostratified columnar epithelium, and it secretes a viscid alkaline fluid rich in fructose.

## PROSTATE

The prostate is an aggregation of **30 to 50 tubulosaccular glands**. The glandular epithelium is simple cuboidal to columnar, with patches of pseudostratified columnar. The cells have apical secretion granules that contribute to the formation of the faintly acid secretion that is rich in citric acid and acid phosphatase. The lumina may contain lamellated prostatic concretions (corpora amylacea). The stroma between the tubules contains smooth muscle fibers.

## URETHRA

The male urethra has three parts: the prostatic, membranous, and cavernous (penile) portions. The **prostatic urethra** is lined mostly with transitional epithelium. The **membranous and penile portions** are lined with pseudostratified and stratified columnar epithelium, except at the meatus where the epithelium is stratified squamous. The prostatic urethra has a crest on its posterior wall. The paired ejaculatory ducts and prostatic utricle open on this crest. The ducts of the prostatic gland open into the prostatic urethra. The membranous urethra is encircled by a sphincter of skeletal muscle fibers from the deep transverse perineal muscle. The cavernous urethra extends throughout the penis. It occupies the corpus spongiosum and receives the ducts of the bulbourethral glands and the branching mucous urethral glands (of Littre).

# BULBOURETHRAL GLANDS

The bulbourethral glands (**Cowper's**) are variably tubular, alveolar, or saccular mucous glands enclosed in the membranous urethral sphincter. Their ducts, containing mucous areas of epithelium, open into the cavernous urethra about 2.5 cm in front of the urogenital diaphragm.

# PENIS

The penis consists of three cylinders of erectile tissue: two dorsal corpora cavernosa and one ventral corpus spongiosum. The ventral corpus spongiosum terminates distally as the glans penis, and it contains the cavernous urethra. A dense fibrous **tunica albuginea** surrounds the cavernous bodies and separates the corpora from each other by an incomplete median (pectiniform) septum. The three corpora are enclosed in a common loose, irregularly arranged connective tissue layer that underlies the investing skin. The skin folds over the glans as the prepuce. The skin of the glans adheres firmly to the erectile tissue because the loose connective tissue layer is lacking. The epithelium of the inner surface of the prepuce and that of the glans is stratified squamous and is characteristic of that lining a moist surface. It is continuous at the urethral orifice with the epithelium lining the urethra.

The erectile tissue of the **corpora cavernosa** consists of endothelium-lined lacunae, which are separated by fibrous trabeculae containing smooth muscle. The lacunae are large centrally but narrow peripherally, where they communicate with a venous plexus underlying the tunica albuginea. In the corpus spongiosum, the arrangement is similar, except for an elastic tunica albuginea, thinner trabeculae, and uniform lacunae.

The corpora cavernosa are supplied by two branches of the **penile artery**: the dorsal artery and the deep arteries. The branches of the **dorsal arteries** supply capillaries of the trabeculae that drain through the lacunae to the venous plexus. The **deep arteries** are the chief vessels for filling the lacunae during erection. These vessels run in the cavernous bodies and give off trabecular branches that empty directly to the lacunae through helicine arteries. The helicine arteries have a thick circular muscle layer and an intima with longitudinal thickened cushions.

During erection, the **smooth muscle** in the cavernous trabeculae and helicine arteries **relaxes** under parasympathetic influence, the helicine arteries become patent, and the lacunae are engorged with more blood than can be rapidly drained by the compressed peripheral lacunae and venous plexus. At the end of erection, **smooth muscle contraction** caused by sympathetic control shuts off the blood supplied by helicine arteries and forces blood into the peripheral venous plexus.

# Chapter 33

# Organs of Special Sense

## EYE

The eyeball consists of three layers: (1) an outer fibrous tunic composed of the sclera and cornea; (2) a vascular coat (uvea) of choroid, ciliary body, and iris; and (3) the retina formed of pigment and sensory (nervous) layers. The anterior chamber lies between the cornea anteriorly and the iris and pupil posteriorly; the posterior chamber lies between the iris anteriorly and the ciliary processes, zonular fibers, and lens posteriorly. Both chambers possess aqueous humor, which is produced in the region of the ciliary processes and exits through the uveal meshwork and canal of Schlemm at the lateral iris angle of the anterior chamber. The **canal of Schlemm** drains into the anterior ciliary veins. The **vitreous body** occupies the space between the lens and the retina (Fig. 33-1).

### Cornea

The cornea constitutes the anterior one-sixth of the eye. Its free surface is covered by **stratified nonkeratinized epithelium**, and its posterior surface is lined with **endothelium** that is continuous with the spaces of the uveal meshwork. Underlying the endothelium is a prominent basement membrane called **Descemet's membrane**; below the anterior epithelium is a thin connective tissue membrane (Bowman's membrane). Between both membranes is the **substantia propria**, comprising the bulk of the cornea. This layer consists of many lamellae of collagenous fibrils held together by a glycoprotein ground substance. The collagen fibrils of adjacent layers run perpendicular to each other and may interweave from layer to layer.

### Sclera

The sclera forms the posterior five-sixths of the fibrous tunic and is composed of **dense fibrous connective tissue**. Although continuous with the cornea, it is delimited from the cornea by internal and external scleral sulci. Nerve fibers of the nerve perforate it posteromedially at the **optic disc**, forming the lamina cribrosa. Extrinsic eye muscles insert into the sclera, and the loose outer scleral layer is continuous with the loose tissue of Tenon's space investing the eyeball.

### Choroid

The choroid layer consists of **vascular loose connective tissue**. It is separated externally from the sclera by a potential perichoroidal space, and is firmly attached internally to the pigment layer of the retina. The **vessel** and **capillary layers** of the choroid are the most prominent layers. The capillary layer supplies the outer layers of the retina and is the only portion of the choroid not continued forward into the ciliary body.

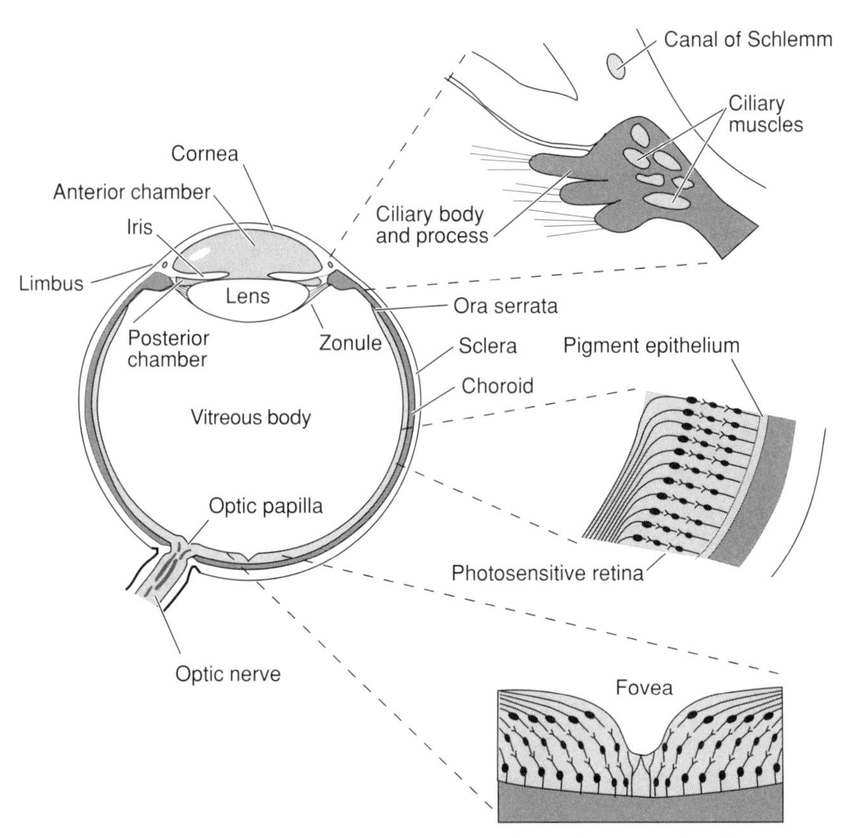


**Figure 33-1.** Diagram of the right eye, as seen from above. Insets depict the structure of the retina and the ciliary apparatus.

## Ciliary Body

The ciliary body is bound posteriorly at the ora serrata by the **retina** and the **choroid**; laterally by the **sclera**; medially by the **posterior chamber**, **vitreous body**, and **lens**; and anteriorly by the **iris**. The posterior two-thirds of the ciliary body is smooth on its inner surface, whereas the anterior one-third bears radially arranged ciliary processes. The forward continuation of the choroid forms the ciliary muscle layer, vessel layer, and lamina vitrea. The forward continuation of the retina gives rise to the outer pigment and inner ciliary epithelia layers and to the internal limiting membrane. The smooth muscle of the **ciliary body** is oriented in meridional, radial, and circular directions. Its action is to relax the tension on the suspensory zonular ligaments, thus allowing the **lens to become more convex because of its elasticity**. It is supplied by parasympathetic fibers of the oculomotor nerve.

In the ciliary processes, the **vascular layer** is thick, contains fenestrated capillaries, and is covered by the pigment epithelia. The **ciliary epithelium**, over the summits of the processes, is modified by basal infoldings of the plasmalemma for transport. This epithelium is involved in **producing aqueous humor**. Occluding junctions between ciliary epithelial cells may be a major site of a blood-aqueous barrier.

## Iris

The iris is attached peripherally to the anterior end of the ciliary body. The anterior surface of the iris demonstrates an inner pupillary zone separated from an outer ciliary zone by a collarette (iris frill). The iris is lined anteriorly by a discontinuous layer of **fibroblasts** and **melanocytes** and posteriorly by **pigment epithelium**. Underlying the anterior surface layer is an anterior border layer formed principally of chromatophores; deep to this is a vascular stromal layer containing the sphincter pupillae muscle. These layers are an anterior continuation of the uvea. The avascular stoma is bordered posteriorly by the pigment epithelium and the more deeply located dilator pupillae muscle; the two layers are the anterior continuation of the retina.

The **color of the iris** depends on the thickness of the anterior border layer and on the pigmentation of its cells. If the layer is thick and heavily pigmented, the eyes are seen as brown; if the layer is small and little pigment is present, the light passes through the vascular stoma and is reflected off of the pigment epithelium as blue.

## Retina

The retina is divisible into **ten layers**: (1) pigment epithelium; (2) layer of rod and cone outer and inner segments; (3) external limiting membrane; (4) outer nuclear layer; (5) outer plexiform layer; (6) inner nuclear layer; (7) inner plexiform layer; (8) ganglion cell layer; (9) nerve fiber layer; and (10) internal limiting membrane.

The simple cuboidal cells of the pigment epithelium have melanin-containing cytoplasmic processes that interdigitate with rod and cone outer segments. Layers two to five contain the **rod and cone cells** of the light pathway. The outer segments of rods and cones contain numerous stacked membranous discs derived from the plasma membrane and containing visual pigments. The outer discs of the rods differ from cones in that they lose their plasma membrane continuity and are discharged from the cell. Pigment cells phagocytize these extruded discs and also supply vitamin A to the receptor cells. The outer segments of cones and rods are connected to the inner segment by a connecting stalk containing an ilium. The inner segment contains the protein-producing endoplasmic reticulum and the Golgi complex necessary for the replacement of rod discs and the nurturing of cone discs. The cell bodies and nuclei of rods and cones constitute the outer nuclear layer. The axons (pedicles) of these cells pass into the outer plexiform layer, where they synapse with the dendrites of bipolar cells and processes of horizontal cells.

**Bipolar cells** are the second-order neuron in the visual pathway. Their nuclei are in the inner nuclear layer, and their axons synapse with dendrites of third-order neuron **ganglion cells** in the inner plexiform layer. The cell bodies of midget and diffuse ganglion cells constitute the ganglion cell layer, and their axons form the nerve fiber layer. By these arrangements of cells, one cone may synapse with one bipolar cell, which in turn synapses with one midget ganglion cell, or several rods or cones may synapse with one bipolar cell, which synapses with a diffuse ganglion cell.

**Horizontal interconnections** are accomplished between rods and cones by horizontal cells and between **ganglion cells** by **amacrine cells**. The outer and inner limiting membranes are formed by the ends of processes of supporting cells of Müller, whose nuclei, like those of bipolar, horizontal, and amacrine cells, are located in the inner nuclear layer.

## Lens

The lens is a **biconvex body** whose posterior surface has a greater convexity. It consists of a capsule, anterior epithelium, and lens substance. The capsule ensheathes the lens and is the site where zonular fibers insert. It consists of basal and reticular laminae. The anterior epithelium contains simple cuboidal dells that elongate at the equator of the lens where they give rise to new lens fibers. The lens substance consists of prismatic lens fibers that are meridionally arranged with older fibers more centrally located than the newer ones. **Desmosomal junctions** are present between the newer cells; **sutures** mark the junction of fibers in the central part of the lens.

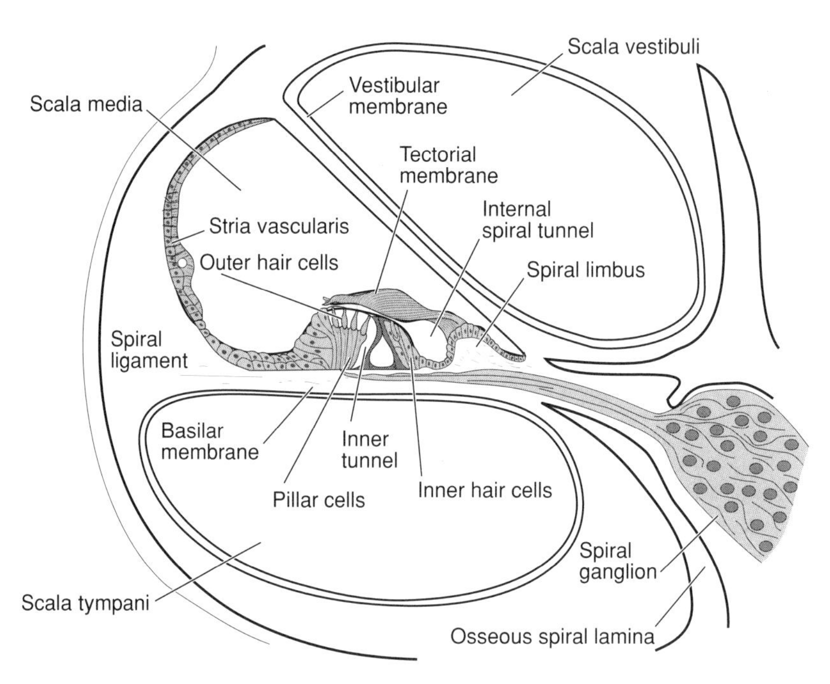


**Figure 33-2.** Diagram depicting the parts of the membranous labyrinth of the cochlea.

# EAR

The **membranous labyrinth** consists of an interconnected series of fibrous sacs lined by **simple squamous epithelium**. The epithelium is derived from an otic vesicle that develops from an otic placode of general surface ectoderm. With the exception of the vestibule, the membranous labyrinth conforms generally to the contour of the osseous labyrinth (Fig. 33-2).

The larger membranous portion in the upper posterior part of the vestibule is the utricle; that in front of the utricle is the saccule. The **utricle** receives the openings of the membranous semicircular canals. The **saccule** communicates with the membranous cochlear duct by the ductus reuniens. An utriculosaccular duct interconnects the utricle and saccule and continues backward through the vestibular aqueduct as the endolymphatic duct. This duct terminates as an endolymphatic sac under the dura lining the posterior surface of the petrous portion of the temporal bone.

There are **six neuroepithelial receptor areas** in each labyrinth: (1) macula utriculi; (2) macula sacculi; (3–5) one crista ampullaris in each ampulla; and (6) organ of Corti in the cochlear duct. The organ of Corti is supplied by the cochlear division of the eighth nerve; the maculae and cristae are innervated by the vestibular division. The cristae ampullares are thickened ridges of epithelium and connective tissue placed transversely to the long axis of each semicircular canal. The epithelium of each crista consists of sustentacular cells and two configurations (types I and II) of neuroepithelial cells called hair cells. Each hair cell contains one kinocilium and many **stereocilia** that project into an overlying gelatinous mass called the cupula. In the horizontal canals, the **kinocilia** are on the utricular side of the hair cells; in the superior (anterior) and posterior canals, the kinocilia are located away from the utricle.

Displacement of the stereocilia toward the kinocilia increases the rate of discharge from the hair cells, whereas movement in the opposite direction decreases the rate of discharge in the vestibular nerve; thus, **movement of endolymph** toward the utricle in the ampullary end of the horizontal semicircular canal causes an increased rate of discharge in that crista. When an individual is first rotated while the lateral canals are in a **horizontal position**, the endolymph flow in the horizontal canal on the side of direction of rotation would be essentially ampullipetal, resulting in increased rate of discharge, whereas the endolymph in the horizontal canal of the opposite ear is ampullifugal, and there is a decreased rate of discharge. In **postrotation**, the opposite occurs. If the head is positioned so that the horizontal canals are vertically oriented and warm water is added to one ear, then convection currents produce an ampullipetal flow in that ear, resulting in a rate of discharge that exceeds that from the unstimulated ear. The use of cold water produces an opposite direction of endolymph flow; thus, the results are opposite those for warm water.

Both the **type I and type II hair cells** are innervated by the dendritic terminals of bipolar cells of the vestibular ganglion. Some cells receive efferent neurons. The vestibular ganglion lies in the upper part of the outer end of the internal auditory meatus. Axons from the vestibular ganglion pass medially in the internal auditory canal as the vestibular portion of the vestibulocochlear nerve and enter the brainstem at the pons-medulla junction.

The **maculae** are similar to the cristae in that they are local thickenings of the membrane, they contain hair cells and sustentacular cells, and their hair cells penetrate a gelatinous membrane. The macular gelatinous membrane contains calcium carbonate crystals called **otoliths** (otoconia) and is called the **otolithic membrane**. The hair cells in various regions of the macula utriculi have their kinocilia placed on different sides so that the macula can detect linear acceleration and the position of the head in respect to gravitational forces. The macula of the saccule also is involved with equilibratory action.

The **organ of Corti** is located on the basilar membrane of the membranous cochlear duct. The **cochlear duct (scala media)** is filled with endolymph, runs throughout most of the cochlea, and is separated from the upper scala vestibuli by the vestibular membrane and from the lower scala tympani by the basilar membrane. The basilar membrane is suspended between the centrally located osseous spiral lamina of the modiolus and the peripherally located periosteal thickening called the spiral ligament. On the lateral wall of the cochlear duct is the stria vascularis. It is lined with pseudostratified columnar epithelium, which is highly vascularized and involved in endolymph production.

The **organ of Corti** is an arrangement of supportive and hair cells on the upper border of the basilar membrane. The neuroepithelial hair cells are arranged into inner and outer hair cells by their relationship to an inner tunnel (of Corti) formed by inner and outer pillar cells. The hair cells are supported by outer and inner phalangeal cells whose phalangeal processes form a firm reticular lamina at the peripheral surfaces of the hair cells. The microvillous hairs of the hair cells are in contact with the overlying gelatinous tectorial membrane, thus establishing a mechanism wherein vibrating movement of the basilar membrane causes stimulation of hair cells by bending of the microvilli.

The mechanical stimulus is transduced into electrical energy by the **hair cells** and transmitted to the terminal dendritic endings of the special somatic afferent (SSA) cells of the spiral cochlear ganglion. Axons of these bipolar nerve cells pass into the axis of the modiolus, course upward into the internal acoustic meatus, become the cochlear portion of the vestibulocochlear nerve, and synapse in the dorsal and ventral cochlear nuclei at the pons-medulla junction of the brainstem.

The blood supply of the labyrinth is by way of the internal **auditory (labyrinthine)** and **stylomastoid arteries**. The stylomastoid is a branch of the posterior auricular artery. The internal auditory artery arises from the basilar artery, or in common with the anterior inferior cerebellar artery, and traverses the internal acoustic meatus before branching into cochlear and vestibular branches. The veins accompany the arteries and drain as internal auditory veins into the superior petrosal or transverse sinuses.

# PART IV

# Neuroanatomy

# Chapter 34

## Spinal Cord

The spinal cord consists of a central canal lined with ependymal cells. The cord is surrounded by central gray matter and peripheral white matter. The junctional region between the gray and white matter, the **fasciculus proprius** (spinal-spinal system), consists of interneurons that interconnect adjacent segments of the spinal cord.

## GRAY MATTER

The H-shaped gray matter has anterior, posterior, and lateral horns, which consist of groups of nerve cell bodies (nuclei, cell columns), axons, dendrites, and glial cells that form a meshwork called **neuropil** (Fig. 34-1). An architectural lamination permits classification of the gray matter into nine **Rexed's layers**, of which laminae I–VI are in the posterior horn, laminae VIII and IX are in the anterior horn, and lamina VII is intermediate in position. These laminae are summarized in Table 34-1.

### Anterior Horn

The anterior horn contains the cell bodies of alpha and gamma motor neurons whose axons innervate extrafusal and intrafusal skeletal muscle fibers, respectively. These nerve cell bodies constitute the **general somatic efferent (GSE) cell column** and are grouped into nuclei that supply axons to specific regions; for example, those most medial in the anterior horn go to the more axial musculature, and those most lateral innervate the extremities and the lateral muscles of the trunk. The alpha and gamma motor neurons are in Rexed's layer IX; lamina VIII contains predominantly commissural neurons. Damage to alpha motor neurons results in a **lower motor neuron (LMN) syndrome** or flaccid paralysis, which causes muscle atrophy, atonia, and loss of deep and superficial reflexes. If only some GSE neurons to a muscle are damaged, then paresis, hypotonia, and hyporeflexia result.

### Lateral Horn

The lateral horn, formed by Rexed's layer VII, is found at T1–L2 and S2–S4 spinal cord levels. Segments **T1–L2** contain the cell bodies of **preganglionic sympathetic neurons**; **S2–S4** contain those of **preganglionic parasympathetic neurons**. Axons of the preganglionic neurons leave the cord by the ventral root and become part of the spinal nerve. **Sympathetic preganglionics** leave the nerve via the **white rami communicantes** and either enter the sympathetic chain of ganglia or become components of **splanchnic nerves**. The preganglionic fibers in the sympathetic chain synapse with postganglionic neurons in those ganglia; those in splanchnic nerves synapse in prevertebral ganglia. Unmyelinated postganglionic axons passing back into the spinal

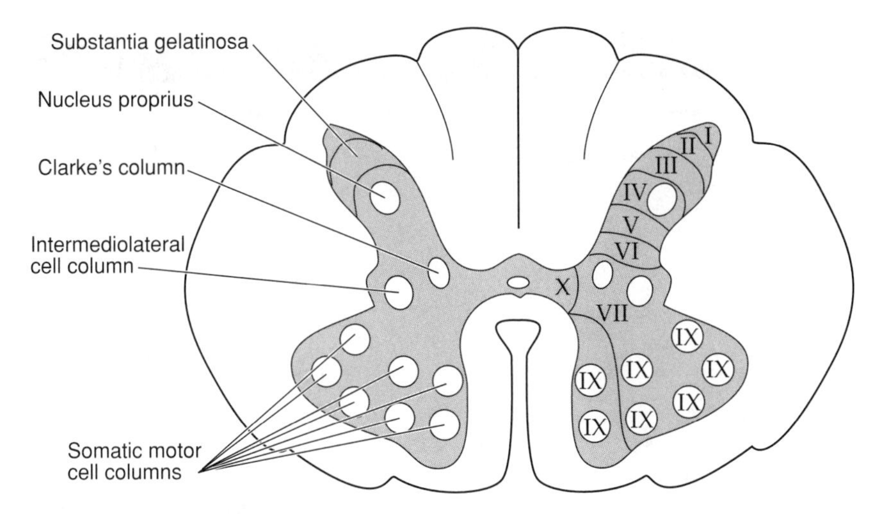


**Figure 34-1.** Cross-sectional diagram of the spinal cord illustrating the location of the cell columns (*left*) and Rexed's lamina (*right*).

nerve constitute the **gray rami communicantes. Sacral parasympathetic preganglionics** form **pelvic splanchnic nerves**, which branch from the spinal nerves and terminate on ganglia in or near the organs innervated.

## Posterior Horn

The posterior horn consists of several nuclear groups that constitute the **general somatic afferent (GSA) cell column**. Most prominent of these nuclei are the posteromarginal nucleus of Rexed's layer I, substantia gelatinosa in layer II, nucleus proprius in layer IV, and the nucleus dorsalis of Clarke in layer VII. These nuclei are "nuclei of termination" for incoming somatic afferents that pass through the dorsal roots; they also are involved in processing sensory information. Pain and temperature first-order afferent neurons terminate on second-order neuron cell bodies in the posteromarginal nucleus, nucleus proprius, and deeper layers of the posterior horn. Sensory information from intrafusal fibers of neuromuscular spindles and tendon spindles is transmitted via IA and IB myelinated first-order neurons, respectively. These synapse on cells of the nucleus dorsalis (Clarke) at cord levels C8–L3. Incoming fibers for crude (light) touch synapse in cells of the posterior horn. Visceral sensation [general visceral afferent (GVA)] is likely received by nuclei of termination in the lateral portion of the posterior horn. Its transmission to higher centers is likely through multisynaptic ascending paths in the **fasciculus proprius**, which surrounds the gray matter. Collaterals (branches) of incoming neurons of the dorsal root enter the gray matter and synapse on internuncials and alpha motor neurons for reflex purposes. Those coming from muscle spindle receptors and ending directly on alpha motor neurons are part of the monosynaptic stretch (myotatic) reflex. Terminals of association neurons interconnecting different segmental levels, as well as terminals of descending axons from suprasegmental levels, end on internuncials that synapse with gamma and alpha motor neurons.

# WHITE MATTER

The white matter is organized into posterior, lateral, and white funiculi, each of which contains both ascending and descending pathways (Fig. 34-2).

**TABLE 34-1. Rexed's Laminae of the Spinal Cord**

| Lamina | Name | Cross-sectional Location | Levels Present | Function |
|---|---|---|---|---|
| I | Posterior marginal nucleus | Cap of posterior horn | All | Receives small unmyelinated GSA DRFs; afferents cross → brainstem and thalamus |
| II | Substantia gelatinosa | Posterior horn | All | Receives small unmyelinated GSA DRFs; afferents to fasciculus proprius → all levels of cord |
| III | Nucleus proprius | Posterior horn | All | Same as II |
| IV | Nucleus proprius | Posterior horn | All | Receives large myelinated GSA DRFs; afferents crossed and uncrossed → thalamus |
| V | | Posterior horn | All | Receives GSA (cutaneous) and GVA DRFs → V and VI, GSA (proprioception) → VI, some corticospinal and rubrospinal fibers terminate here; afferents cross → thalamus |
| VI | | Base of posterior horn | Large at C5–T1 and L2–S3; absent T4–L2 | Same as V |
| VII | Zona intermedia | Intermediate gray | | |
| | Dorsal nucleus of Clarke | | C8–L2, L3 | Cell bodies of posterior spinocerebellar tract (GSA) |
| | Intermediolateral nucleus | | T1–L2, L3 | Cell bodies of preganglionic sympathetic neurons (GVE) |
| | | | S2, S3, S4 | Cell bodies of GVE |
| | Intermediomedial nucleus | | All levels | Receives GVA (?) DRF |
| VIII | GSE cell column | Base of anterior horn | All levels; large at C5–T1, L2–S3 | Receives descending fibers from cerebral cortex, vestibular nuclei, reticular formation; afferents to lamina IX |
| IX | GSE cell column | Most of anterior horn | | |
| | | Posterolateral anterior horn | All levels; large at C5–T1, L2–S3 | Alpha and gamma motor neurons to appendicular (distal) skeletal muscle fibers |
| | | Ventromedial anterior horn | All levels | Alpha and gamma motor neurons to axial (truncal) skeletal muscle fibers |

*DRF, dorsal root fiber; GSA, general somatic afferent; GSE, general somatic efferent; GVA, general visceral afferent; GVE, general visceral efferent.*

## Posterior White Funiculus

The **posterior white column** occupies most of the posterior funiculus and is concerned with two-point touch, vibratory and joint senses, and stereognosis. Spindle information by way of IA fibers from neuromuscular spindles and IB fibers from Golgi tendon

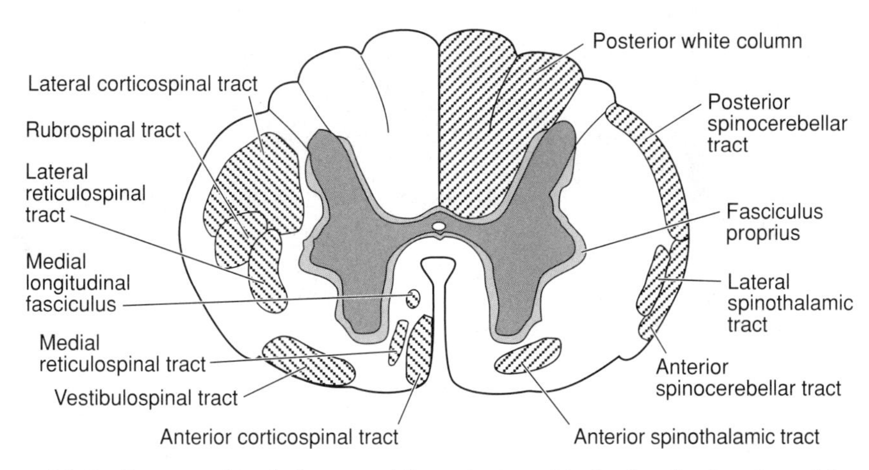


**Figure 34-2.** Cross-sectional diagram of the spinal cord indicating the location of the major pathways. The descending pathways are on the left, the ascending are on the right. The fasciculus proprius (spinal-spinal system) is stippled and consists of interneurons that both ascend and descend.

organs also are transmitted in the posterior white column. Axons of these pathways in the posterior white column arise from cell bodies in the dorsal root ganglia, enter the cord in the medial bundle of the dorsal root, and ascend to the medulla.

## Lateral White Funiculus

The most clinically prominent tracts of the lateral funiculus are the posterior spinocerebellar tract, the lateral spinothalamic tract, the lateral corticospinal tract, and the lateral reticulospinal tract. The **posterior spinocerebellar tract** is just deep to the surface in the posterolateral aspect of the lateral funiculus at all levels of the spinal cord above L4. The **lateral spinothalamic tract** is located ventrolaterally in the lateral funiculus and just deep to the anterior spinocerebellar tract. The **lateral corticospinal tract** is located just medial to the posterior spinocerebellar tract, predominantly in the dorsal half of the lateral funiculus. This tract constitutes the upper motor neurons of the pyramidal motor system. It is involved primarily in fine voluntary movements involving chiefly the more distal phylogenetically "newer" musculature and is facilitatory to the antagonists of the antigravity muscles. The **lateral reticulospinal tract** arises from large cells (nucleus gigantocellularis) in the medial reticular formation of the medulla. Axons from these cells descend ipsilaterally in the lateral funiculus and are interspersed with axons of the lateral corticospinal tract. Stimulation of cells of the lateral reticulospinal tract inhibits alpha and gamma neurons innervating antigravity muscles (i.e., extensors of the lower extremity and flexors of the upper extremity). Additional pathways in the lateral funiculus include the **rubrospinal pathway** from the red nucleus, **descending central autonomics** from the hypothalamus, the **spinotectal pathway** to the superior colliculus of the mesencephalon, **spinoreticular fibers** to the brainstem reticular formation, and interconnections between the spinal cord and inferior olivary nucleus.

**Unilateral damage to the lateral funiculus** results in (1) a loss of pain and temperature sensation contralaterally starting one segment below the lesion; (2) spastic paralysis, exaggerated deep reflexes, loss of superficial reflexes, and loss of fine distal motor activity ipsilaterally below the lesion; and (3) if the lesion is above the upper thoracic intermediolateral cell column level, Horner's syndrome, which consists of ptosis, pupillary constriction, and blanched and dry facial skin ipsilaterally.

## Anterior White Funiculus

The most prominent structures in the anterior funiculus are the anterior spinothalamic tract, anterior corticospinal tract, lateral vestibulospinal tract, medial reticulospinal tract, and medial longitudinal fasciculus (MLF). The **anterior spinothalamic tract** is located just anterior to the ventral horn. The **anterior corticospinal tract** has an origin and path similar to that of the lateral corticospinal tract. It differs, however, in that it is located near the anterior median fissure, it usually descends only to upper thoracic levels, and its axons do not cross in the pyramidal decussation. These axons do cross at the levels where they terminate on internuncials that synapse with alpha and gamma motor neurons, which supply muscles of the upper extremity and neck. The **vestibulospinal tract** is interspersed with the anterior spinothalamic tract. The **medial reticulospinal tract** lies lateral to the anterior corticospinal tract in the anterior funiculus. The **medial longitudinal fasciculus** is in the most dorsal portion of the anterior funiculus next to the anterior median fissure. This composite tract contains axons arising from the mesencephalic tectum (tectospinal tract), vestibular nuclei (medial vestibulospinal tract), and the reticular formation of the brainstem (medial reticulospinal tract). Unlike descending fibers of the lateral funiculus, which are facilitatory to the antagonists of antigravity muscles and inhibitory to antigravity muscles, the descending fibers of the anterior funiculus are facilitatory to antigravity muscles.

# BLOOD SUPPLY

The spinal cord is supplied by descending branches from the vertebral arteries and by radicular branches of segmental arteries. From these vessels, paired **posterior spinal arteries** arise that descend dorsal to the posterior funiculi and a single midline **anterior spinal artery** arises from the paired anterior spinal arterial branches of the vertebra. An arterial vasocorona plexus, lying in the pia adjacent to the lateral funiculus, interconnects the anterior and posterior radicular branches. The posterior spinal arteries supply the posterior funiculus, dorsal part of the dorsal horn of gray matter, and the posterolateral fasciculus (Lissauer). Sulcal branches of the anterior spinal artery supply all other parts of the spinal cord except the most peripheral portion of the lateral funiculus, which is supplied by the arterial vasocorona. Spinal **veins** have a distribution generally similar to that of the arteries. Sulcal and posterior veins empty into anteromedial, anterolateral, posteromedial, and posterolateral veins. These drain, in turn, to radicular veins that enter the epidural venous plexus.

# Chapter 35

# Brainstem

The structure of the brainstem reflects its phylogenetic development. Generally the older systems involved with crude sensibilities and gross axial motion are positioned centrally, and phylogenetically newer structures involving discriminatory sensation and fine and precise motion are located ventrolaterally. Structures associated with cranial nerves are found at various levels. Finally, the ventricular system, which consists of the small central canal in the spinal cord, is enlarged by the formation of the ventricles at certain levels.

## CELL COLUMNS OF THE BRAINSTEM

There are seven different types of fibers in the cranial nerves and thus seven different cell columns. No single cranial nerve, however, contains all seven types. These types of fibers (functional components) are as follows:

- **General somatic efferent (GSE)**—motor neurons to skeletal muscles of occipital and eye somite origin
- **Special visceral efferent (SVE)**—motor neurons to skeletal muscles of branchial arch origin
- **General visceral efferent (GVE)**—motor neurons to postganglionic parasympathetic ganglia
- **General visceral afferent (GVA)**—sensory neurons from visceral structures
- **Special visceral afferent (SVA)**—sensory neurons from taste and olfactory receptors
- **General somatic afferent (GSA)**—sensory neurons from exteroceptive and proprioceptive receptors
- **Special somatic afferent (SSA)**—sensory neurons from the eye and ear

The cell columns of the brainstem are less continuous from level to level than are those of the spinal cord. Their localization generally corresponds to the emergence of the cranial nerves (CNs); a notable exception is the GSA cell column, which continues upward from the dorsal horn of the spinal cord (as the spinal and chief nuclei of CN V) to the midpons level and receives trigeminal nerve fibers throughout its extent. The mesencephalic nucleus of CN V represents an upward extension of the GSA column but involves only proprioceptive neurons. There are no cell columns for the olfactory and optic nerves; thus, the cell columns are limited to the mesencephalon, pons, and medulla. The cross-sectional localization of brainstem cell columns differs markedly from those of the spinal cord. Most of this difference is a result of the development of the ventricles. In the spinal cord, the gray matter, as seen in cross section, is oriented vertically into a dorsal horn (alar plate) and ventral horn (basal plate) positioned lateral to the central canal. With the expansion of the central canal into the ventricles of the brain, the roof plate becomes stretched out as the tela choroidea (ependyma plus pia), and the alar plate is displaced lateral to the basal plate. Consequently, in the pons and medulla, the nuclei of termination of sensory fibers are located lateral to the motor nuclei in the pons and medulla. The basic sequential localization from the midline laterally is GSE, SVE, GVE, GVA, SVA, GSA, and SSA. All of the nuclei of these columns in the pons and medulla lie in the floor of the fourth ventricle or near the central canal with the exception of the SVE column, which migrates ventrolaterally into the reticular formation.

# MEDULLA

## Gross Topography

Based on the enlargement of the fourth ventricle, the medulla can be separated into a lower closed part and an upper open part.

**Dorsal aspect**. The upper (open) part of the medulla contains a portion of the **fourth ventricle**. The **tela choroidea** of pia and ependyma forms the roof of this part of the **fourth ventricle**, and the floor is divisible into a **medial eminence** and a more **lateral vestibular area**. The fourth ventricle is open laterally to the subarachnoid space through the **foramen of Luschka** at the junction of the pons and medulla. A midline opening, the **foramen of Magendie**, is located at the most caudal tip of the tela choroidea. The dorsal surface of the inferior closed part of the medulla represents an enlarged upward continuation of the fasciculus cuneatus and fasciculus gracilis of the spinal cord. These areas are referred to as the **tuberculum cuneatum** and **tuberculum gracilis** (clava). They lie caudal and lateral to the fourth ventricle.

**Ventral aspect**. The ventral surface of the medulla, from the midline laterally, consists of **pyramids** and **pyramidal decussation**, **preolivary sulcus**, olivary eminence, and postolivary sulcus. The **abducens nerve** exits at the pons-medulla junction in line with the hypoglossal nerve rootlets, which emerge from the preolivary sulcus. The **glossopharyngeal**, **accessory**, and **vagus nerves** exit from the postolivary sulcus. The facial and vestibulocochlear nerves emerge below the lateral recess of the fourth ventricle at the pons-medulla junction. They are in close anatomical relationship to the **inferior cerebellar peduncle** (restiform body), which courses anterior to the lateral recess in passing from the medulla to the cerebellum.

## Internal Structure

**Closed medulla**. The lower part of the closed medulla is similar to the spinal cord in that it has a central canal, fasciculus gracilis, fasciculus cuneatus, lateral and anterior spinothalamic tracts, posterior and anterior spinocerebellar tracts, lateral reticulospinal tracts, and anterior corticospinal tracts in essentially the same locations as in the spinal cord (Fig. 35-1). The substantia gelatinosa, the posteromarginal nuclei, and the posterolateral fasciculus continue upward as the spinal nucleus and tract of CN V, respectively. The major difference between the lower medulla and spinal cord is that the lateral corticospinal tract now forms the ventrally positioned pyramids. At this and higher levels, there is more of an admixture of gray matter and fibers in that area that corresponds to the gray matter of the spinal cord.

**Open medulla**. At higher regions of the closed medulla, the ascending axons of the posterior funiculus synapse on cell bodies in the **nucleus gracilis** and **nucleus cuneatus** (Fig. 35-2). Because these nuclei are displaced at higher levels by the expanding fourth

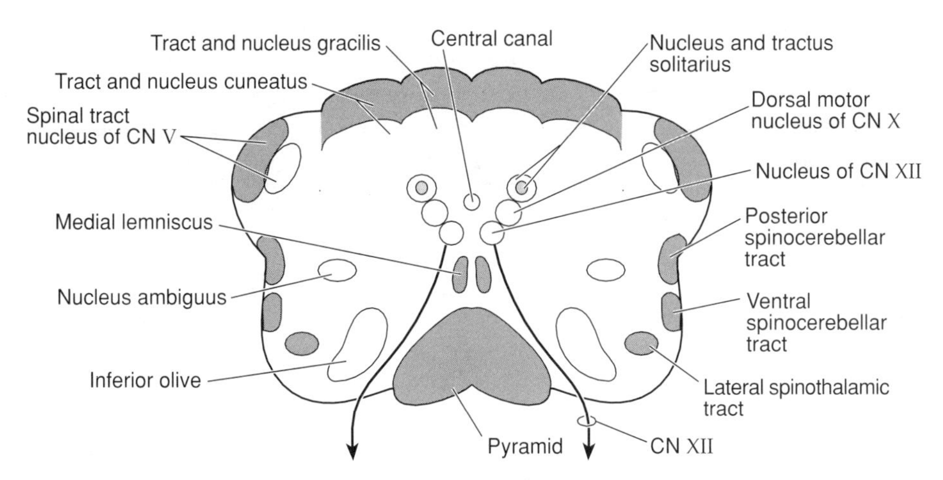


**Figure 35-1.** Cross-sectional diagram of the low or closed medulla. Tracts or bundles of fibers are *cross-hatched*; nuclei or cell columns are *clear*. CN, cranial nerve.

ventricle, axons from these nuclei arch ventrally around the central canal as the internal arcuate fibers and decussate to form the **medial lemniscus**. In the medial lemniscus, fibers carrying impulses from sacral levels are localized just above the pyramid; fibers representing cervical levels are located more dorsally just below the medial longitudinal fasciculus (MLF). The **inferior olivary nucleus** occupies the ventral part of the medulla lateral to the medial lemniscus and pyramids. The **vestibulospinal (lateral) tract** passes dorsal to this nucleus. The **spinothalamic** and **spinocerebellar tracts** lie along the lateral margin of the medulla in the postolivary sulcus region.

**Cranial nerve nuclei**. The **hypoglossal nucleus** (GSE) extends throughout most of the medulla and lies dorsal to the MLF. Lateral to it is the **dorsal motor nucleus** (GVE) of the vagus nerve. Axons from the hypoglossal nucleus pass ventrally and exit in the preolivary sulcus. In the closed medulla, the **fasciculus** and **nucleus soli-**

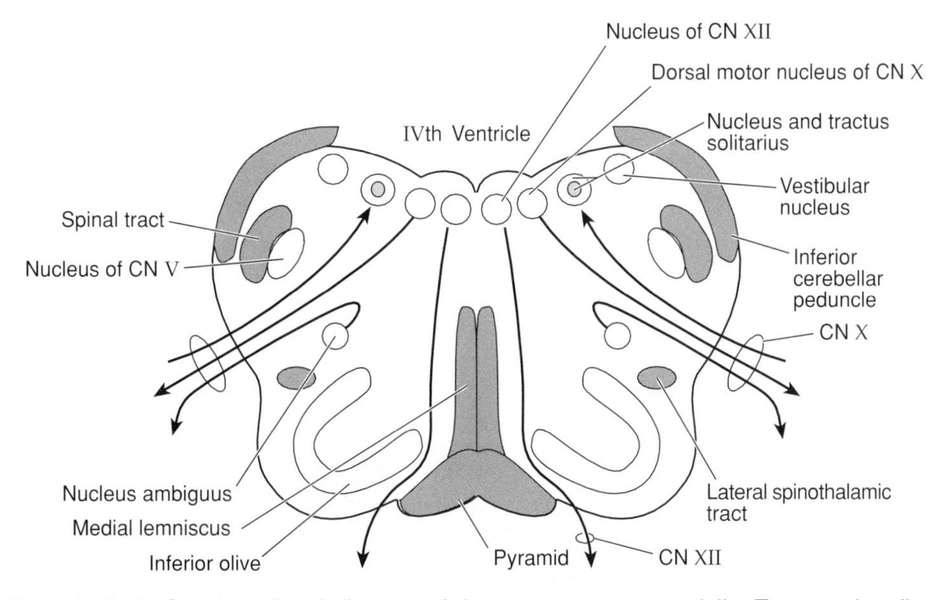


**Figure 35-2.** Cross-sectional diagram of the upper or open medulla. Tracts or bundles of fibers are *cross-hatched*; nuclei or cell columns are *clear*. CN, cranial nerve.

**tarius** lie dorsal to the dorsal motor nucleus of CN X, and at open-medulla levels, they lie lateral to the motor nuclei. In the open medulla, the **hypoglossal** and **vagal nuclei** form prominences in the floor of the fourth ventricle medial to the sulcus limitans. The sensory nuclei, the **nucleus solitarius** (GVA and SVA), **vestibular nuclei** (SSA), and **spinal nucleus of CN V** (GSA), lie lateral to the sulcus limitans. At high-medulla levels, the **vestibular nuclei** (SSA) lie dorsomedial to the spinal nucleus and form a vestibular area in the floor of the fourth ventricle lateral to the sulcus limitans. At the level of the lateral recess, **cochlear nuclei** (SSA) form the most lateral prominence in the floor of the fourth ventricle. The **nucleus ambiguus** (SVE) is located in the reticular formation halfway between the spinal nucleus of CN V and the inferior olivary nucleus.

Lateral to the spinal tract of CN V is the **inferior cerebellar peduncle**. It consists of the posterior spinocerebellar tract, olivocerebellar fibers from the contralateral inferior olivary nucleus, cuneocerebellar fibers from the accessory cuneate nucleus, and other ascending and reticular connections. The reticular formation is divisible into medial and lateral groups of nuclei. The lateral groups are sensory; they receive ascending sensory information. The lateral small-celled area is associated with respiratory responses and the location of the pressor center. The medial **gigantocellular nucleus** gives rise to the **lateral reticulospinal tract** and is inhibitory to antigravity muscles. This nucleus and adjacent medial nuclei constitute centers involved in respiratory and depressor circulatory responses. An ascending **reticular activating system** of multisynaptic connections arises primarily from the medial nuclei of the reticular formation.

## Blood Supply

The blood supply to the medulla, both the lower closed and the upper open parts, is via the branches of the spinal and vertebral arteries as well as the most proximal portion of the basilar artery. These vessels supply three longitudinal areas of the medulla: anteromedial, lateral, and posterior areas.

**Anterior area**. The **anterior spinal artery** supplies the anterior and medial aspects, specifically the pyramid, medial lemniscus, medial MLF, and hypoglossal nucleus.

**Lateral area**. The lateral area is larger and supplied by the **vertebral artery** throughout the medulla; it is joined by the **posterior inferior cerebellar artery** in the open medulla. The major structures in this area are the spinal tract and nucleus of CN V, nucleus ambiguus, dorsal motor nucleus of CN X, nucleus and tractus solitarius, spinothalamic system, dorsal spinocerebellar tract, and most of the medullary reticular system.

**Posterior area**. The posterior area is supplied predominantly by the **posterior spinal artery**. This area includes the cuneate and gracile nuclei and their tracts at the closed level and the inferior cerebellar peduncle more rostrally.

# PONS

## Gross Topography

The pons is bounded dorsally by the attached cerebellum. The lateral surface is composed of the middle cerebellar peduncle with its emerging root of the trigeminal nerve (CN V). Ventrally, the basis pontis forms a bridge between the middle cerebellar peduncles. On the superior aspect, the superior medullary velum, superior cerebellar peduncles, and cerebellum form the roof of the fourth ventricle. The tegmentum forms the floor of the ventricle and has a facial colliculus in the medial eminence located medial to the sulcus limitans.

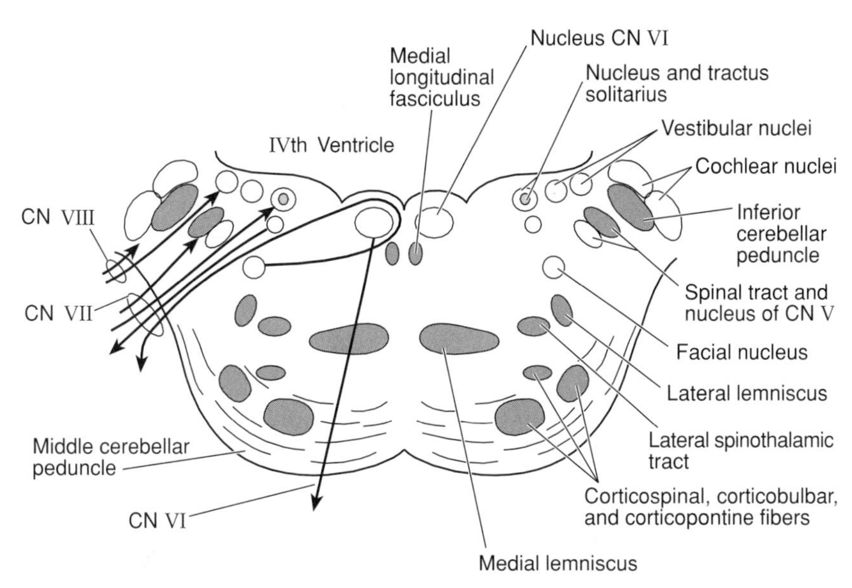


**Figure 35-3.** Cross-sectional diagram of the lower pons. Tracts or bundles of fibers are *cross-hatched*; nuclei or cell columns are *clear*. CN, cranial nerve.

## Internal Structure

The pons is divisible into the dorsal tegmentum and the ventral basis pontis, which is phylogenetically newer. The **tegmentum** is an upward continuation of the medulla. It differs from the medulla in that the inferior olivary nucleus disappears and a central tegmental tract and superior olivary nucleus are located in its place; the medial lemniscus is oriented horizontally in the basal part of the tegmentum; the spinothalamic tracts are located at the lateral tip of the medial lemniscus; the corticospinal tracts are in the basis pontis and not in pyramids; and cell columns and pathways relating to CN V to CN VIII are present.

**Pons-medulla junction.** At the pons-medulla junction, the inferior cerebellar peduncle passes ventral to the lateral recess of the fourth ventricle before entering the cerebellum. At this location, **dorsal** and **ventral cochlear nuclei** lie dorsolaterally and ventrally, the **spinal tract and nucleus of CN V** are medial, and the **vestibular nuclei** are dorsomedial to the inferior cerebellar peduncle. The **superior olivary nucleus** lies dorsal to the spinothalamic tracts in the ventrolateral tegmentum.

**Basis pontis at all levels.** The basis pontis contains the **corticospinal (pyramidal) tract**, **corticobulbar (corticonuclear) fibers** to cranial nerve motor nuclei, and corticopontine fibers that terminate on pontine nuclei (Figs. 35-3 through 35-5). Axons of the pontine nuclei cross the midline and pass laterally and dorsally into the cerebellum as the **middle cerebellar peduncle**.

**Low pons tegmentum.** In lower pontine levels, the facial (SVE) nucleus occupies the ventrolateral tegmentum (see Fig. 35-3). The internal genu of the facial nerve and the adjacent abducens nucleus form an **abducens (facial) colliculus** in the floor of the fourth ventricle. A **parabducens** group of cell bodies, a center for lateral gaze, lies inferior to the MLF in the paramedian pontine reticular formation. **Superior** and **inferior salivatory nuclei** containing the cell bodies of parasympathetic preganglionic nerves of the facial and glossopharyngeal nerves, respectively, occupy the tegmentum but are not sharply localized.

**Midpons tegmentum.** GSA fibers of the trigeminal nerve, whose cell bodies are located in the trigeminal ganglion, terminate in the **principal** (chief, main) **sensory nucleus** and **spinal nucleus of CN V** (see Fig. 35-4). The principal sensory nucleus is

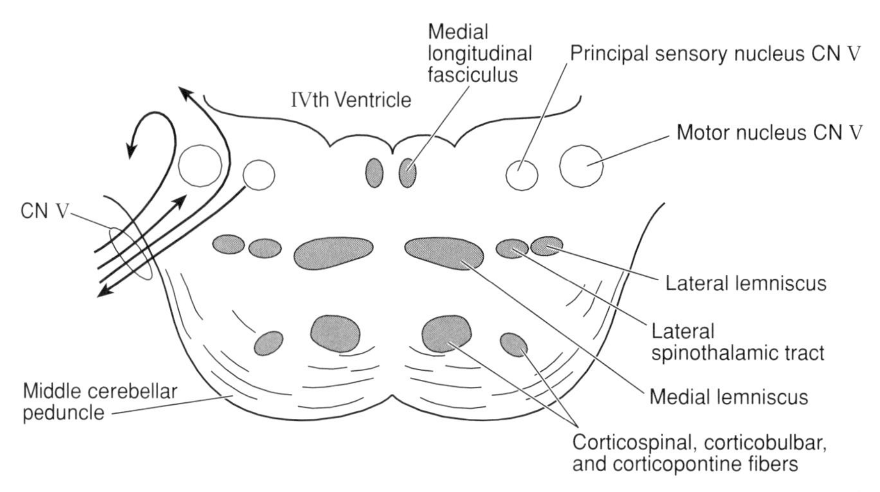


**Figure 35-4.** Cross-sectional diagram of the midpons. Tracts or bundles of fibers are *cross-hatched*; nuclei or cell columns are *clear*. CN, cranial nerve.

the upward continuation of the spinal nucleus and receives touch fibers. The **motor** (SVE) **nucleus of CN V** lies medial to the chief sensory nucleus of CN V. The **mesencephalic nucleus of CN V** extends into the mesencephalon from midpons levels and lies next to the lateral portion of the central gray material.

**Upper pons tegmentum**. At upper pons levels, the fourth ventricle narrows as the isthmus region of the pons-mesencephalon junction is approached. The superior medullary velum and superior cerebellar peduncles form the roof and lateral walls of the ventricle. In the region of the isthmus, the superior cerebellar peduncles move ventrally into the tegmentum. The fibers of these peduncles decussate in the tegmentum of the mesencephalon and ascend to the red nucleus and ventral lateral nucleus of the thalamus.

The **reticular formation** of the pons is an upward continuation of that in the medulla. The caudal and oral pontine reticular nuclei give rise to the **pontine (medial) reticulospinal tract**, which is facilitatory to antigravity muscles. The main ascending pathway of the reticular formation is the **central tegmental tract**.

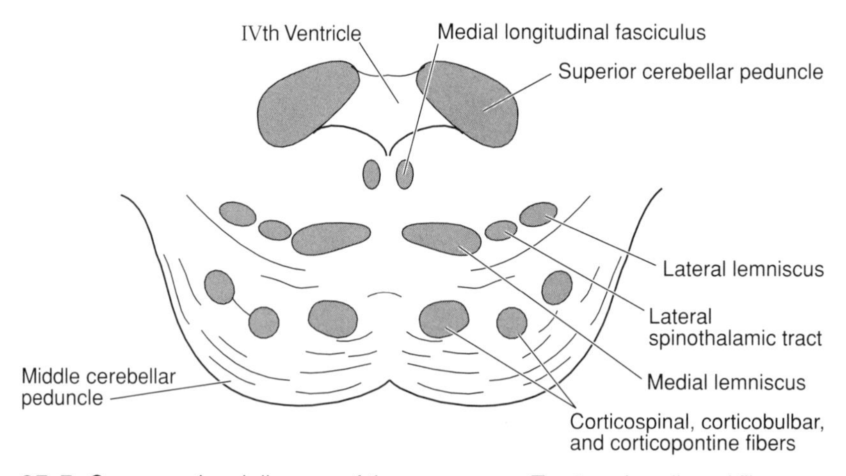


**Figure 35-5.** Cross-sectional diagram of the upper pons. Tracts or bundles of fibers are *cross-hatched*; nuclei or cell columns are *clear*.

## Blood Supply

The blood supply to the pons is provided by branches of the basilar artery. Like the medulla, the pons can be divided into longitudinal sections; however, in the pons, there are only two sections, and they are positioned anteriorly and posteriorly.

**Anterior area**. The **paramedian** and **short circumferential branches of the basilar artery** supply this section of the pons at low, mid, and high levels. The specific structures involved are the corticospinal and corticobulbar tracts, medial lemniscus, lateral spinothalamic tract, lateral lemniscus, pontine nuclei, and middle cerebellar peduncle; and at the low pons level, the fibers of the abducens nerve (CN VI).

**Posterior area**. This area is supplied by different arteries at each pontine level. It should be noted that, at all levels, the junctional region between the posterior and anterior areas is variable; thus, the medial and lateral lemnisci and the lateral spinothalamic tract may be supplied by either the anterior or posterior arteries or both. The **anterior inferior cerebellar artery** supplies the posterior section of the low pons; the structures involved are the MLF, cochlear and vestibular nuclei, inferior cerebellar peduncle, spinal tract and nucleus of CN V, nucleus and tractus solitarius, salivatory nucleus, and the motor nuclei of the facial and CN VI. The **long circumferential branches of the basilar artery** supply the posterior aspect of the midpons. The major structures supplied are the MLF, primary sensory and motor nuclei of the CN V, and the fibers of the exiting CN V. The high pons posteriorly is supplied by the superior cerebellar artery. The primary structures supplied are the MLF and the superior cerebellar peduncle.

# MESENCEPHALON

## Gross Topography

The mesencephalon or **midbrain** lies between the diencephalon and pons. Its dorsal surface consists of two superior colliculi, two inferior colliculi, the brachia of these colliculi, which connect to the lateral and medial geniculate nuclei of the diencephalon, respectively, and the emerging trochlear nerve. Ventrally, the mesencephalon has the cerebral peduncles, interpeduncular fossa, and emerging oculomotor nerves. The iter (cerebral aqueduct of Sylvius) connects the third ventricle of the diencephalon with the fourth ventricle in the pons.

## Internal Structure

The mesencephalon is divided at the level of the cerebral aqueduct into a **dorsal tectum** and two ventral **cerebral peduncles**. The tectum consists of two superior and two inferior colliculi. Each cerebral peduncle is subdivided by the substantia nigra into a dorsal tegmentum and a ventral crus cerebri.

**Level of the inferior colliculi**. At this level, the axons of the lateral lemniscus terminate in the inferior colliculus (Fig. 35-6). Fibers from this relay nucleus of the hearing pathway pass laterally into the brachium of the inferior colliculus and synapse on cells of the medial geniculate body, which, in turn, send axons to the transverse temporal gyri (of Heschl). The **mesencephalic nucleus** and root and the **nucleus pigmentosus** (locus ceruleus) lie ventral to the inferior colliculi in the lateral extent of the central gray. In the dorsomedial limits of the tegmentum, the **trochlear nucleus** indents the dorsal surface of the MLF. Below the MLF, fibers of the **superior cerebellar peduncle decussate** before passing upward toward the red nucleus; the **medial lemniscus**, secondary **tracts of CN V**, and **spinothalamics** lie ventral and lateral to this decussation. The **crus cerebri** contains the corticopontine,

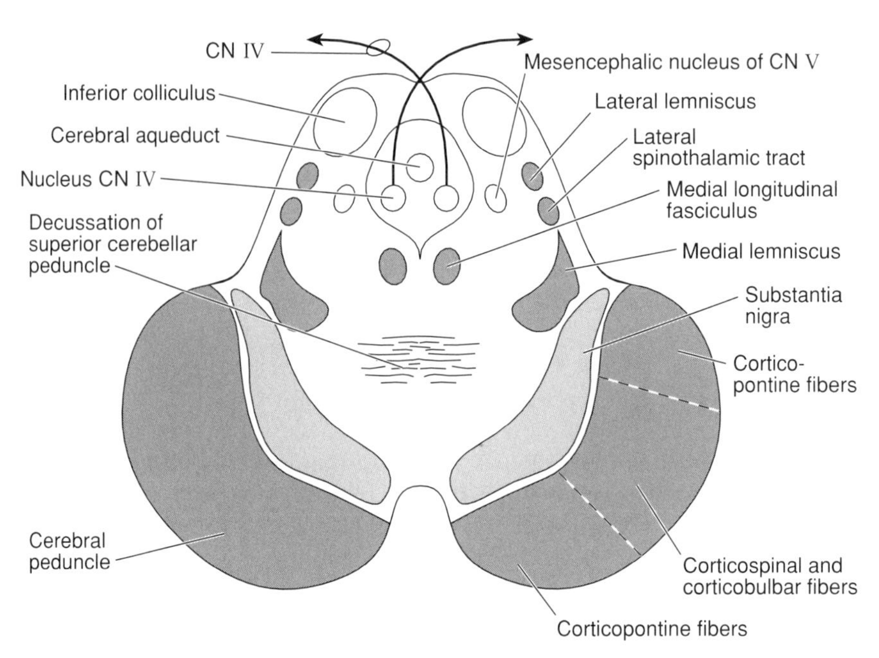


**Figure 35-6.** Cross-sectional diagram of the midbrain at the level of the inferior colliculus. Tracts or bundles of fibers are *cross-hatched*; nuclei or cell columns are *clear*. CN, cranial nerve.

corticobulbar, and corticospinal pathways; corticobulbar and corticospinal fibers occupy the middle third of the crus.

**Level of the superior colliculi**. The most significant features of this level are the superior colliculi, oculomotor nuclear complex and nerves, and the red nucleus (Fig. 35-7). The **superior colliculus** receives optic fibers from the optic tract and occipital cortex by way of the brachium of the superior colliculus, and is involved in aligning the fovea on a visual target (fixation reflex). The **oculomotor nuclear complex** consists of the oculomotor (GSE) and Edinger-Westphal (GVE) nuclei, which are located medial to the MLF. The **red nucleus** lies in the ventromedial tegmentum and is surrounded by dentatothalamic axons of the superior cerebellar peduncle that go to the ventral lateral nucleus of the thalamus. Some axons of the superior cerebellar peduncle synapse on cells of the red nucleus. The red nucleus also receives corticorubral fibers. Major efferent paths from the red nucleus are the **rubroreticular** and the **rubrospinal tracts**.

**Pretectal area**. The pretectal area at the junction with the diencephalon is considered part of the mesencephalon. Features of this area are the pretectal nuclei, posterior commissure, and subcommissural organ. **Pretectal nuclei** lie rostral to the superior colliculi, receive optic tract axons by way of the brachium of the superior colliculi, and send crossed and uncrossed axons to the Edinger-Westphal nucleus. The pretectal nuclei constitute the association limb of the pupillary light reflex. The crossing fibers of this reflex either pass through the posterior commissure or decussate in the central gray below the cerebral aqueduct. The **posterior commissure** lies dorsal to the cerebral aqueduct at the level where it joins the third ventricle. The **subcommissural organ** is modified ependyma located beneath the posterior commissure. The columnar ciliated cells of this organ may secrete aldosterone and may serve as a volume receptor.

The **reticular formation** consists of many nuclei whose cells release a variety of neurotransmitters. A median (raphe) group of nuclei contains serotonergic neurons. A medial group, which includes the cells of origin of the medial reticulospinal (nucleus pontis oralis and caudalis) and lateral reticulospinal (nucleus gigantocellularis) tracts, produces serotonin and possibly substance P. A lateral group of nuclei (e.g., the locus ceruleus) constitutes a norepinephrine and epinephrine system. A dopaminergic system of neurons includes cells of the substantia nigra.

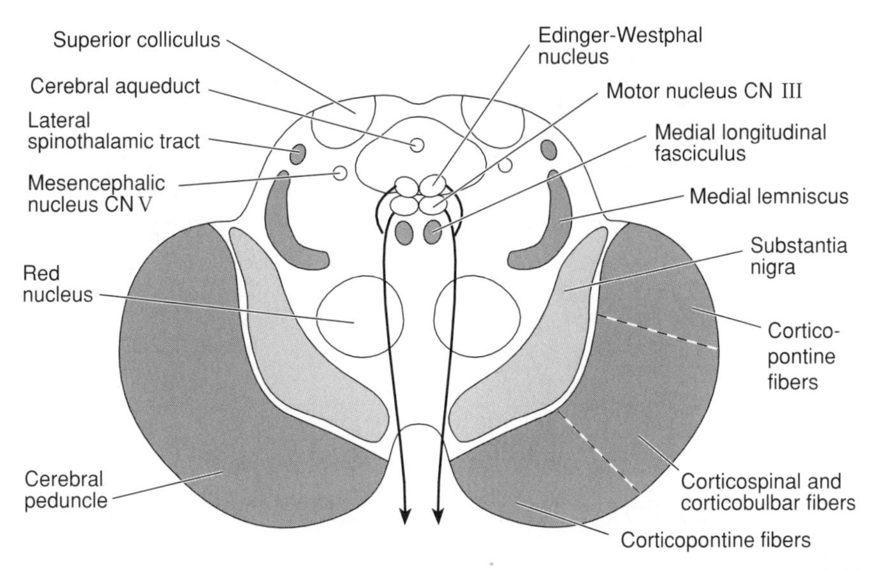


**Figure 35-7.** Cross-sectional diagram of the midbrain at the level of the superior colliculus. Tracts or bundles of fibers are *cross-hatched*; nuclei or cell columns are *clear*. CN, cranial nerve.

## Blood Supply

The primary blood supply is provided by branches of the terminal part of the basilar and of the posterior cerebral arteries; however, branches of the anterior choroidal arteries are also involved. Like the rest of the brainstem, there are longitudinal sections involving both the superior and inferior collicular levels that are supplied by similar branches of specific arteries.

**Ventromedial area.** This medial area extends from the cerebral aqueduct ventrally into the cerebral peduncle. It is supplied by **paramedian branches** of the basilar, posterior cerebral, posterior communicating, and anterior choroidal arteries. The major structures supplied are the red, interpeduncular, trochlear, and oculomotor nuclei; fibers of the oculomotor nerve; decussation of the superior cerebellar peduncle; corticopontine fibers from the frontal lobe; and MLF.

**Lateral area.** This section includes most of the cerebral peduncle and extends dorsally to the level of the cerebral aqueduct. It is supplied by **short circumferential branches** of the superior cerebellar, posterior cerebral, and anterior choroidal arteries. The structures in this section are the corticospinal and bulbar tracts; corticopontine fibers from the temporal, occipital, and parietal lobes; lateral aspect of the substantia nigra; medial lemniscus; mesencephalic nucleus of the trigeminal nerve; and the lateral spinothalamic tract.

**Dorsal area.** The roof consists of the superior and inferior colliculi and, at the level of the inferior colliculus, the fibers of the trochlear nerve. It is supplied by the long circumferential branches of the posterior cerebral, posterior choroidal, and superior cerebellar arteries.

# DIENCEPHALON

The diencephalon contains the **third ventricle** and is divisible into four parts: (1) roof or **epithalamus**, (2) dorsolaterally located **thalamus**, (3) floor and ventromedially positioned **hypothalamus**, and (4) ventrolateral **subthalamus**. The diencephalon is bound above by the transverse cerebral fissure. Although this fissure extends caudally between the occipital lobe and cerebellum and is occupied by the tentorium cerebelli, rostrally it lies between the corpus callosum and fornix above and the epithalamus (tela choroidea, pineal body, habenula) and thalamus below. It extends laterally and ventrally as the choroid fissures adjacent to the choroid plexus of the lateral and third ventricles, respectively. Rostrally the transverse cerebral fissure ends blindly behind the interventricular foramen (of Monro), where the choroid plexus of the third ventricle is continuous with that of the lateral ventricle. The diencephalon is bounded laterally by the internal capsule and the cerebral peduncles. Basally, from anterior to posterior, it includes the optic chiasm, infundibular stalk and pituitary gland, and mammillary bodies.

## Epithalamus

The epithalamus is in the roof of the third ventricle and is composed of the **tela choroidea**, **striae medullaris**, **habenular trigones**, and **pineal gland**. The stria medullaris conveys fibers from the septal and preoptic areas to the habenular nuclei. Axons from the habenular nuclei pass to the mesencephalon through the fasciculus retroflexus.

## Thalamus

The thalamus is bound laterally by the posterior limb of the internal capsule and medially by the third ventricle. It extends anteroposteriorly from the interventricular foramen to the pretectal area, and is positioned above the hypothalamus and subthalamus. The thalamus is separated into medial, lateral, and anterior nuclear groups by the **internal medullary lamina**.

**Medial nuclear group**. This group consists of the midline and dorsomedial nuclei. The midline nuclei make connections with the hypothalamus. The dorsomedial nucleus receives afferents from the amygdaloid nucleus, hypothalamus, and temporal cortex, and sends axons to the prefrontal cortex. Intralaminar nuclei receive spinothalamic fibers and reticulothalamic fibers from the reticular activating system. The centromedian nucleus is an intralaminar nucleus that receives fibers from the motor cortex (area four) and globus pallidus and sends axons to the putamen.

**Lateral nuclear group**. The lateral group of thalamic nuclei are divisible into ventral and dorsal tiers. The **ventral tier nuclei** consist of the ventral anterior, ventral lateral, and ventral posterior [ventroposterolateral (VPL) and ventroposteromedial (VPM)] nuclei. All of these ventral tier nuclei are relay nuclei. The VPL nucleus receives the medial lemniscus and spinothalamics. The VPM receives secondary ascending trigeminal pathways. Axons from both the VPM and VPL nuclei pass into the posterior limb of the internal capsule and terminate in the great somesthetic area (postcentral gyrus, areas three, one, and two). Those fibers from the VPM terminate closer to the more ventral portion of the postcentral gyrus than those from the VPL. Some axons conveying nociceptive information from the VPL and VPM nuclei terminate in somatic sensory area two, located in the parietal lobe above the lateral fissure. The ventral lateral nucleus receives dentatothalamic fibers and pallidothalamic fibers and sends axons to the precentral gyrus (motor cortex). The latter fibers arise from cells in the medial aspect of the globus pallidus and pass through the fasciculus lenticularis to reach the prerubral field. They then project to the thalamus via the thalamic fasciculus. The ventral anterior nucleus receives pallidothalamic fibers by this same pathway and projects axons to the premotor cortex. The **dorsal tier nuclei** are the lateral dorsal, lateral posterior, and pulvinar nuclei, which are **association nuclei** and have interconnections with the parietal cortex. The pulvinar receives fibers from the metathalamus (medial and lateral geniculate bodies) and sends axons to the parastriate (area 18) and peristriate (area 19) areas in the occipital cortex and also to the inferior parietal cortex.

**Anterior nucleus**. The anterior nucleus of the thalamus relays mammillothalamic impulses from the mammillary bodies to the cingulate gyrus.

## Subthalamus

The subthalamus lies ventral to the thalamus between the hypothalamus and posterior limb of the internal capsule. It consists of the **subthalamic nucleus**, **zona incerta**, **fasciculus lenticularis**, **prerubral field** (of Forel), and the fasciculus thalamicus. The subthalamic nucleus lies on the internal capsule and substantia nigra, separated from the zona incerta above by the fasciculus lenticularis. It has interconnections with the globus pallidus. **Lesions** of this nucleus produce **hemiballism contralaterally**.

## Hypothalamus

The hypothalamus consists of groups of nuclei in the floor of the third ventricle. The nuclei are divided into medial and lateral groups by the fornix as it passes through the hypothalamus on its way to the mammillary body. The lateral group includes the lateral and tuberal nuclei. The medial group is subdivided anatomically into three groups: (1) anterior group, which includes the preoptic, anterior, supraoptic, and paraventricular nuclei; (2) middle group, containing the dorsomedial and ventromedial nuclei; and (3) posterior group of posterior and mammillary nuclei. The hypothalamic nuclei can also be grouped on the basis of their functional connections into autonomic, neuroendocrine, and olfactory groups.

**Autonomic nuclei**. The **anteromedial** hypothalamic group of nuclei generally are concerned with **parasympathetic** regulation, whereas the **posterolateral** group is more involved in **sympathetic** regulation. Both nuclear groups receive ascending GVA, GSA, and SVA (taste) input from the reticular formation and send out descending central autonomics to preganglionic neurons by way of reticulospinal and reticuloreticular pathways. These regions also receive olfactory input from the amygdaloid and septal areas and have interconnections with the thalamus, prefrontal cortex, and limbic lobe.

**Neuroendocrine nuclei**. The neuroendocrine nuclei are the supraoptic and paraventricular group, which produce posterior pituitary hormones, and a hypophysiotrophic group, which produce adenohypophyseal releasing and inhibiting factors.

**Olfactory nuclei**. The **mammillary**, **preoptic**, and **lateral hypothalamic nuclei** receive input from the olfactory cortex. These nuclei have reciprocal connections with the limbic lobe and also project to other hypothalamic and brainstem nuclei.

## Blood Supply

The blood supply to the thalamus is variable in terms of the specific areas supplied by various vessels, and is provided by deep branches of the circle of Willis, proximal aspects of the middle and posterior cerebral arteries, and the terminal part of the basilar artery. In addition to the variability within the thalamus, vessels that supply the thalamus also supply portions of the basal ganglia and internal capsule.

**Medial region**. This area is supplied by paramedian branches of the terminal part of the **basilar artery** and the proximal aspect of the **posterior cerebral artery**. The primary nuclei supplied are the medial (midline) and dorsal medial groups.

**Anterolateral region**. The ventral tier of lateral nuclei, as well as portions of the dorsomedial nucleus, are supplied by deep branches of the **posterior communicating artery**.

**Posterolateral region**. This region is supplied by long branches of the **posterior cerebral arteries** that curve around the posterior aspect of the thalamus and penetrate its superior surface. These vessels supply primarily the dorsal tier of lateral nuclei, but may include the centromedian nucleus as well.

# Chapter 36

# Cranial Nerves

The cranial nerves (CNs) are presented in this chapter and in Chapter 4. The course of each nerve within the cranial cavity and in the periphery is reviewed in Chapter 4 within the discussion on cranial nerves. The functional components and structures supplied by each nerve are summarized in Table 4-2, and the cranial parasympathetics are summarized in Table 4-3. The level and nuclei of origin or termination of the functional components of each nerve are summarized in Table 36-1. The internal (within the brainstem) courses of the fibers of each nerve are discussed here.

## OLFACTORY NERVE (CN I)

The olfactory nerve is composed of the central processes of **bipolar olfactory cells** [special visceral afferent **(SVA)**] whose cell bodies are located in the olfactory epithelium on the upper part of the nasal septum and the lateral nasal wall. These unmyelinated fibers (fila olfactoria) pass into the anterior cranial fossa through the lamina cribrosa of the ethmoid bone and enter the olfactory bulb. After synapsing with the mitral cells of the bulb, the impulses travel in the **olfactory tract** to the **lateral olfactory stria** and terminate in the **pyriform lobe**. The cortical receptive area for olfaction, the pyriform lobe, consists of the **lateral olfactory stria**, the **uncus** (periamygdaloid area), and the anterior part of the **parahippocampal gyrus**.

## OPTIC NERVE (CN II)

The optic nerve is formed by the central processes of the **retinal ganglion cells [special somatic afferent (SSA)]**, which converge at the optic papilla. After piercing the sclera, the optic nerve passes through the orbital fat, traverses the optic canal, and joins with the optic nerve of the opposite to form the optic chiasma. The visual pathway from retina to occipital cortex consists of four neurons. The **first-order neurons** are the **rods** and **cones**. The **second-order neurons** are **bipolar cells**, and the **third-order neurons** are **ganglion cells**. The cell bodies of all three of these neurons are located in the retina. Axons of the ganglion cells comprise the optic nerve, chiasm, and tract and terminate in the lateral geniculate body. Twenty percent of the axons continue directly to the pretectal area of the mesencephalon. Those axons from the nasal half of the retina decussate in the optic chiasm; those from the temporal half of the retina remain uncrossed. Thus, the optic tract consists of axons from the temporal half of the ipsilateral eye and the nasal half of the contralateral eye. The **fourth-order neurons** are located in the **lateral geniculate nucleus**. Axons of these cells pass into the sublenticular and retrolenticular portions of the internal capsule, loop over the roof of the inferior horn of the lateral ventricle, and

**TABLE 36-1. Function, Components, and Nuclei of Cranial Nerves**

| Nerve | Functional Components (Afferent Neuron Cell Body Location) | Nucleus of Origin or Termination |
|---|---|---|
| Olfactory | SVA (olfactory epithelium) | Mitral cells in the olfactory bulb |
| Optic | SSA (retina, rod and cone cells) | Retina (bipolar cells) |
| Oculomotor | GSE | Motor nucleus of CN III |
| | GVE: parasympathetic | Edinger-Westphal |
| Trochlear | GSE | Motor nucleus of CN IV |
| Trigeminal | SVE | Motor nucleus of CN V |
| | GSA: pain and temperature (trigeminal ganglion) | Spinal nucleus of CN V |
| | GSA: touch (trigeminal ganglion) | Principal sensory nucleus of CN V |
| | GSA: proprioception (mesencephalic nucleus of CN V) | Mesencephalic nucleus of CN V |
| Abducens | GSE | Abducens nucleus |
| Facial | GSA (geniculate ganglion) | Spinal nucleus of CN V |
| | SVA: taste (geniculate ganglion) | Solitary (gustatory) nucleus |
| | GVA (geniculate ganglion) | Solitary nucleus |
| | GVE: parasympathetic | Superior salivatory nucleus |
| | SVE | Motor nucleus of CN VII |
| Vestibulocochlear | SSA: cochlear [cochlear (spiral) ganglion] | Dorsal and ventral cochlear nuclei |
| | SSA: vestibular (vestibular ganglion) | Vestibular nuclei |
| Glossopharyngeal | GSA (superior ganglion of CN IX) | Spinal nucleus of CN V |
| | SVA: taste (inferior ganglion of CN IX) | Solitary nucleus |
| | GVA (inferior ganglion of CN IX) | Solitary nucleus |
| | SVE | Nucleus ambiguus |
| | GVE: parasympathetic | Inferior salivatory nucleus |
| Vagus | GSA [superior (jugular) ganglion of CN X] | Spinal nucleus of CN V |
| | SVA [inferior (nodose) ganglion of CN X] | Nucleus solitarius |
| | GVA (inferior of CN X) | Nucleus solitarius |
| | SVE | Nucleus ambiguus |
| | GVE | Nucleus ambiguus |
| | GVE: parasympathetic | Dorsal motor nucleus of CN X |
| Accessory | SVE | Nucleus ambiguus |
| Hypoglossal | GSE | Hypoglossal nucleus |

*CN, cranial nerve; GSA, general somatic afferent; GSE, general somatic efferent; GVA, general visceral afferent; GVE, general visceral efferent; SSA, special somatic afferent; SVA, somatic visceral afferent; SVE, special visceral efferent.*

course posteriorly as the optic radiations to the striate cortex (area 17) located above and below the calcarine fissure in the occipital lobe. This pathway is also referred to as the **geniculocalcarine tract**. The pathway from the upper retina projects to the cuneus; that from the lower retina projects to the lingual gyrus below the calcarine fissure.

## Visual Pathway Lesions

Lesions in the visual pathway give rise to **deficits in the visual fields and some loss of visual reflexes**. Because light rays from specific portions of the visual field project to opposite parts of the retinal fields, lesions in the nasal retinal field result in **loss of vision**

**in the temporal visual field**, whereas damage to the upper retinal field results in **blindness in the lower visual field**. The following results are caused by the distribution of fibers in the pathway:

- **Lesions in front of the chiasma**—blindness (anopsia), or partial blindness, on the side of the damaged optic nerve or retina
- **Midline lesions of the optic chiasm**—bitemporal heteronymous hemianopsia (loss of vision in the temporal visual fields of both eyes)
- **Lesions posterior to the chiasm**—homonymous hemianopsias on the side opposite the lesion (e.g., right lesion results in loss of left temporal and right nasal visual fields)

Because of the topography of optic radiations and terminations in the striate cortex, lesions in the rostral portion of the temporal lobe or below the calcarine fissure can give upper quadrantic anopsias to the opposite side.

## Visual Reflexes

**Pupillary light reflex**. When light is shone into one eye, the ipsilateral pupil constricts (**direct response**) and the contralateral pupil constricts (**consensual response**). The **afferent limb** of this pupillary light reflex consists of rods and cones, bipolar cells, and ganglion cells whose axons pass through the brachium of the superior colliculus to the pretectal area. The **association limb** is composed of neurons of the pretectal nucleus that send axons ipsilaterally and contralaterally, by way of the posterior commissure or central gray, to the Edinger-Westphal nucleus. The **efferent limb** of the pupillary light reflex involves innervation of the sphincter pupillae muscles by a pathway involving the preganglionic neurons of both Edinger-Westphal nuclei and oculomotor nerves and the postganglionic neurons of both ciliary ganglia. Section of one optic nerve results in blindness in that eye and loss of direct and consensual pupillary light response when light is shone into the blind eye. When light is shown into the other eye, the direct and consensual responses are intact. Damage to an oculomotor nerve results in loss of pupillary constriction in that eye regardless of which eye is stimulated.

**Near reflex**. When a person focuses on a near object after focusing on a far object, three responses take place: (1) convergence of the eyes by contraction of the medial rectus muscles, (2) pupillary constriction by contraction of the sphincter pupillae muscles, and (3) rounding of the lens because of constriction of the ciliary muscle. This reflex is referred to as the **near reflex** (accommodation-convergence-pupillary reflex). Two notable distinctions exist between this reflex and the pupillary light reflex: (1) The pathway involves cortical connections, and (2) the efferent limb involves both general somatic efferent (GSE) and general visceral efferent (GVE) parasympathetic components. The **afferent limb** involves rods and cones, bipolar cells, ganglion cells, the lateral geniculate nucleus, and area 17. The **association limb** consists of connections from area 17, to area 18 (parastriate), to area 19 (peristriate) and its corticomesencephalic axons to the oculomotor complex. The **efferent limb** involves the oculomotor and Edinger-Westphal nuclei, the oculomotor nerves, and the ciliary ganglia and nerves.

**Vision and head position**. In the reflexes involving the turning of the head and eyes in response to visual impulses, cortical connections project to a center for convergence in the midbrain tegmentum, the rostral interstitial nucleus of the medial longitudinal fasciculus (MLF) for vertical gaze in the midbrain and diencephalon, and the paramedian pontine reticular formation for lateral gaze in the pons. The association limb for reflex turning of the head probably relays through the superior colliculus and tectospinal tract to anterior horn cells and spinal accessory neurons in the cervical cord.

## OCULOMOTOR NERVE (CN III)

The oculomotor nerve (CN III) contains both GSE and GVE parasympathetic fibers, which arise from the **oculomotor** (GSE) and **Edinger-Westphal (GVE) nuclei** of the midbrain at the level of the superior colliculus. These nuclei are positioned ventral to the central gray and just lateral to the midline. From these nuclei, the fibers pass ventrally through the tegmentum and emerge on the medial aspect of the cerebral peduncles in the interpeduncular fossa.

**Oculomotor lesions**. Damage to CN III results in several characteristic symptoms. **Ptosis** (drooping of the upper eyelid) occurs as a result of denervation of the levator palpebrae superioris; the **pupil is dilated** because the dilator pupillae is unopposed by the sphincter pupillae; and the direct and consensual responses of the **pupillary light reflex** are lost. The **ciliary muscle is paralyzed** and **accommodation is impaired**. Denervation of the extrinsic muscles results in external strabismus because of unopposed action of the lateral rectus and superior oblique muscles.

The upper motor neurons (UMNs) to the nuclei of CN III, CN IV, and CN VI arise from three primary areas: (1) occipital lobe, areas 18 and 19, for smooth pursuit movements; (2) frontal eye fields, area 8; and (3) caudal portion of middle frontal gyrus for saccadic (jerky) eye movements. Jerky eye movements are used in reading or "searching out" an object. Both UMNs (corticobulbar) for smooth pursuit and saccades work through lower motor neurons via the conjugate gaze centers.

## TROCHLEAR NERVE (CN IV)

Caudal to the nucleus of the oculomotor nerve and in the same general location (ventral to the central gray), the trochlear nerve (CN IV) arises from the trochlear nucleus (GSE) at the level of the inferior colliculus. Axons of these cells pass dorsally around the cerebral aqueduct, decussate with fibers of the opposite side, and emerge as the fourth nerve at the superior medullary velum.

**Trochlear lesions**. A lesion of CN IV, and thus **paralysis** of the superior oblique muscle, causes **extorsion** of the involved eye because of the unopposed action of the inferior oblique muscle. Double vision (**diplopia**) and **weakness of downward gaze** also occur.

## TRIGEMINAL NERVE (CN V)

The trigeminal nerve (CN V) contains both general somatic afferent (GSA) fibers and somatic visceral efferent (SVE) fibers. The motor fibers arise from the motor nucleus of CN V in the midpons, located just medial to the chief sensory nucleus of CN V. The fibers then pass ventrolaterally and emerge laterally from the middle cerebral peduncle as the **motor root** of the trigeminal nerve. The GSA fibers of the ophthalmic, maxillary, and mandibular nerves enter the pons as the **sensory root** of CN V. On entering the pons, the fibers from **pain** and **temperature** receptors descend as the spinal tract of CN V to terminate in its spinal nucleus. **Touch fibers** end in the principal sensory nucleus and in the upper part of the spinal nucleus of the fifth nerve. Crossed secondary fibers for pain, temperature, and touch ascend as the ventral secondary tract of the CN V to the ventroposteromedial (VPM) nucleus. Some touch fibers ascend ipsilaterally in the dorsal secondary tract to the VPM nucleus. Third-order neurons of the VPM nucleus send axons to the postcentral gyrus. **Proprioceptive fibers** in CN V ascend to the mesencephalic nucleus of CN V. The location of secondary pathways to the cerebral cortex and cerebellum from this nucleus is not yet fully understood.

**Trigeminal lesions.** Lesions of CN V result in **exteroceptive deficits** of **pain**, **temperature**, and **touch** in the areas supplied by the damaged components. Lesions of the ophthalmic division result in **loss** of the afferent limb of the **corneal blink reflex**. Damage to one ophthalmic nerve also produces **loss** of the **direct and consensual blink**. If the efferent limb, mediated by the facial nerve, is damaged and the ophthalmic is intact, then stimulation of the cornea results in blink of only that eye that has an intact seventh nerve innervating the orbicularis oculi muscle. **Lesion of the mandibular nerves** can result in **deviation of the jaw** to the affected side when the jaw opens, caused by the unopposed action of the contralateral lateral pterygoid muscle.

## ABDUCENS NERVE (CN VI)

The abducens nerve (CN VI) arises from GSE cell bodies in the floor of the fourth ventricle at the facial colliculus level of the pons. From the nucleus, the fibers pass ventrally through the tegmentum and exit from the brainstem as the most medial nerve at the pons-medulla junction. The abducens nucleus receives afferents from the ipsilateral **lateral gaze center** [parabducens nucleus, paramedian pontine reticular formation (PPRF)] located near the abducens nucleus below the MLF. Some axons from the abducens nucleus cross the midline and ascend in the contralateral MLF to reach cells of the oculomotor nucleus that innervate the contralateral medial rectus muscle; thus, stimulation of neurons in the PPRF results in conjugate deviation of the eyes to the side stimulated because the ipsilateral abducens, lateral rectus, contralateral oculomotor, and medial rectus are activated. The abducens and oculomotor nuclei also are supplied by fibers that course through the MLF from vestibular nuclei and superior colliculi.

**Abducens lesions.** Section of CN VI may lead to medial deviation (**strabismus**) of the affected eye because the medial rectus is unopposed. Strabismus is accompanied by **diplopia** and **inability to turn the eye laterally**. Lesion of the abducens nucleus area often involves both the abducens and lateral gaze center and causes an **inability of both eyes to look laterally** toward the side of the lesion as well as a tendency for persistent conjugate **deviation toward the opposite side**.

## FACIAL NERVE (CN VII)

The facial nerve (CN VII) contains GSA, SVA (taste), general visceral afferent (GVA), SVE, and GVE functional components. **GSA fibers** arise from receptors in a small area near the external ear, pass to pseudounipolar cell bodies in the geniculate ganglion, and course into the pons at the pons-medulla junction to terminate in the **spinal nucleus of CN V**. SVA fibers from cell bodies in the geniculate ganglion enter the pons and descend in the fasciculus solitarius to end in the upper part of the **solitary (gustatory) nucleus**. **GVA fibers** arise from the submandibular, sublingual, lacrimal, nasal, and minor salivary glands. Their cell bodies are in the geniculate ganglion, and their axons terminate in the **nucleus solitarius**. **General visceral efferent** (GVE) **preganglionic parasympathetic neurons** arise in the **superior salivatory nucleus** of the pons. **SVE neurons** arise in the **facial motor nucleus**, which is located dorsal to the superior olivary nucleus in the ventrolateral tectum of the pons. From this nucleus, the fibers pass dorsally and loop around the abducens nucleus (internal genu); they then pass ventrolaterally and enter the facial nerve, which emerges through the lateral aspect of the pons-medulla junction (cerebellopontine angle).

**Facial nerve lesions**. Lesions of CN VII (such as **Bell's palsy**) result in weakness of the facial muscles ipsilateral to the lesion. This is distinct from lesions of the UMNs (corticobulbar tract), which bilaterally innervate that part of the facial nucleus supplying the upper facial muscles but only contralaterally supply cells to the lower face; thus, in UMN lesions, there is **weakness** to the **contralateral lower face**. Peripheral nerve lesions also may result in (1) **loss of taste** in the anterior two-thirds of the tongue, (2) **impaired lacrimation**, and (3) **hyperacusis** caused by loss of the dampening effect of the stapedius muscle.

# VESTIBULOCOCHLEAR NERVE (CN VIII)

The vestibulocochlear nerve (CN VIII) contains SSA fibers arising from receptors for hearing in the cochlear duct and from the maculae and cristae of the vestibular apparatus. It is composed of the processes of bipolar cells whose cell bodies lie in the **spiral cochlear and vestibular ganglia**. Axons of these cells pass through the internal acoustic meatus and enter the pons at the lateral aspect of the pons-medulla junction (cerebellopontine angle).

## Cochlear Part of the Nerve

Cochlear fibers synapse in the dorsal and ventral cochlear nuclei. From these nuclei, axons pass bilaterally, synapsing in the superior olivary nucleus, nucleus of the lateral lemniscus, and inferior colliculus. From the inferior colliculus, axons pass to the medial geniculate nucleus, where impulses are relayed to the transverse temporal gyrus (of Heschl). **Lesion** of one nerve results in **deafness** to that ear. Unilateral lesions of the central pathway lead only to a **diminution in hearing** caused by the bilateral representation.

## Vestibular Part of the Nerve

Entering vestibular fibers pass directly to the flocculonodular lobe of the cerebellum or synapse on **vestibular nuclei**. There are reflex connections from vestibular nuclei to the spinal cord (by way of the MLF and vestibulospinal tract), to the center for lateral conjugate gaze, and to motor nuclei of the brainstem reticular formation. Some of these connections are better appreciated when the pathways involving reflex activities that accompany angular rotation of the head are considered.

When the head is inclined 30 degrees forward and is first rotated to the right, the endolymph of both horizontal semicircular canals does not move initially as fast as the cristae ampullares. This gives a relative displacement of endolymph to the left, although endolymph is actually moving to the right. After rotation the cristae stop moving, but there is continued brief movement of endolymph to the right in the direction of rotation. Because stimulation of cristae of the horizontal canal toward the utricle results in depolarization of the hair cells, an imbalance occurs during rotation and postrotation between the two canals. In postrotation to the right, the left vestibular nerve is stimulated while the right vestibular nerve is inhibited, resulting in rapid alternating eye movements (**nystagmus**). The pathway for the slow component of nystagmus to the right in postrotation to the right involves the left vestibular nerve, left vestibular nuclei, right lateral gaze center, right abducens nucleus and nerve, left MLF, and oculomotor nucleus and nerve. In this pathway, axons from vestibular nuclei either ascend in the MLF to reach the oculomotor nucleus and cross to the opposite abducens nucleus or pass to the opposite lateral gaze center, which, in turn, makes connections with the abducens and oculomotor nuclei. **Past-pointing** and a tendency to **fall to the right** are mediated through connections of the left vestibular nucleus with the cerebellum and of the left vestibulospinal tract with anterior horn cells of antigravity muscles, thus causing a thrust to the right. **Nausea**, **increased salivation**, and **vomiting** are mediated through connections of the vestibular nuclei with such motor nuclei as the dorsal motor nucleus of CN X, the nucleus ambiguus, the salivatory nuclei, and other reticular nuclei.

## GLOSSOPHARYNGEAL NERVE (CN IX)

The glossopharyngeal nerve (CN IX) contains GSA, SVA (taste), GVA, SVE, and GVE functional components. **GSA fibers** pass through the auricular branch of the vagus to the glossopharyngeal nerve near the jugular foramen. Their cell bodies are located in the superior ganglion in the jugular foramen. Axons of these cells enter the upper medulla, in the postolivary sulcus, and synapse on cells of the **spinal nucleus of CN V**.

**SVA fibers** arise from taste buds in the posterior one-third of the tongue. They pass through lingual branches, and their cell bodies are in the inferior ganglion. Axons terminate in the solitary (gustatory) nucleus. **GVA fibers** arise in the mucosa and glands of the posterior tongue, fauces, and pharynx. Their cell bodies are located in the inferior ganglion, and their axons terminate in the solitary nucleus. The carotid sinus nerve of the glossopharyngeal conveys GVA fibers from the carotid sinus to the nucleus solitarius. Axons from this nucleus course to the dorsal motor nucleus of CN X, where they synapse on neurons whose fibers pass into the vagus nerve and constitute the efferent limb of the carotid sinus reflex. **SVE fibers** arise in the nucleus ambiguus, and **GVE preganglionic parasympathetics** have their cell bodies in the inferior salivatory nucleus.

The afferent limb of the **gag reflex** is via GVA fibers from the oropharynx to the association limb in nucleus solitarius. Axons of the nucleus solitarius are both crossed and uncrossed to the hypoglossal nucleus for tongue movements and to the nucleus ambiguus for movements of the pharynx, larynx, and soft palate.

**Lesions** of CN IX can result in some **loss of taste** in the posterior one-third of the tongue and in a **loss of the gag reflex** when the affected side is stimulated.

## VAGUS NERVE (CN X)

The vagus nerve (CN X) contains the same five functional components as the facial and glossopharyngeal nerves. The small **GSA component** travels from the external ear in the auricular branch of the vagus and has its cell bodies located in the superior (jugular) ganglion lying in the posterior part of the jugular foramen. Axons from this ganglion enter the medulla in the postolivary sulcus and end in the spinal nucleus of CN V. The **SVA fibers** arise from taste buds in the region of the epiglottis. SVA cell bodies are in the inferior (nodose) ganglion, located inferior to the jugular foramen, and their axons terminate in the gustatory part of the nucleus solitarius. **GVA fibers** arise in the mucosa and walls of the intestine (as far as the splenic flexure), the stomach, esophagus, pharynx, larynx, trachea, lungs, heart, carotid body, and kidney. Their cell bodies are located in the inferior ganglion, and their axons terminate in the nucleus solitarius.

**SVE nerve fibers** arise from the nucleus ambiguus and are distributed to the skeletal muscles of the pharynx, soft palate, larynx, and esophagus. The **GVE fibers** (**parasympathetic**) arise from the dorsal motor nucleus of CN X. These preganglionics are distributed through numerous branches of the vagus to terminal parasympathetic ganglia in or on the heart, larynx, trachea, lungs, and digestive system from the pharynx to the splenic flexure. The terminal ganglia of most of the gastrointestinal tract are the myenteric and submucosal nerve ganglia and plexuses. Postganglionic axons from terminal ganglia supply smooth and cardiac muscle and glands.

**Vagal lesions**. A unilateral lesion of CN X or its nuclei produces paresis of the ipsilateral vocal cord, resulting in **abnormal phonation** and **hoarseness**. In addition, the affected side of the soft palate is lower than the normal side. On phonation, the uvula points toward the normal side. Lesions of the vagus nerve also demonstrate **abnormal gag and swallowing reflexes** because the vagus constitutes the efferent limbs of those reflexes.

## ACCESSORY NERVE (CN XI)

The accessory nerve (CN XI) has cranial and spinal parts. The **cranial part** arises in the nucleus ambiguus (SVE), exits from the postolivary sulcus of the medulla, passes through the jugular foramen, and merges with the vagus near the inferior vagal ganglion. The **spinal part** arises from an SVE cell column in the anterior horn of the upper five cervical segments. Axons of these cells pass laterally through the lateral funiculus, ascend between the dentate ligament and the dorsal roots, and pass as a nerve trunk through the foramen magnum into the cranial cavity. This nerve trunk communicates with the cranial part of the accessory and the vagus and exists from the jugular foramen as the accessory nerve.

## HYPOGLOSSAL NERVE (CN XII)

The hypoglossal nerve (CN XII) contains **GSE fibers** that arise from cell bodies in the hypoglossal nucleus, which is located dorsal to the MLF and just lateral to the midline in the dorsal tegmentum of the medulla. From this nucleus, the fibers pass ventrally between the pyramid and inferior olive before emerging in the preolivary sulcus.

**Hypoglossal lesions**. Unilateral lesions of the hypoglossal nerve or nucleus result in **deviation of the tongue** to the affected side upon protrusion. This is caused by the action of the unopposed genioglossus of the normal side. There may be fasciculations and wasting on the affected side. Because corticobulbar fibers are crossed, damage to these fibers causes the tongue to deviate away from the lesion on protrusion.

# Chapter 37

# Major Ascending and Descending Pathways

The major ascending and descending pathways of the spinal cord and brainstem are summarized in this chapter. Each system is traced by following the courses of its primary neurons. Loss of function can be predicted by localizing the level and side of a lesion; loss of motor function or sensation occurs ipsilaterally if the lesion is below the level of decussation, and contralaterally if the lesion is above.

## ASCENDING PATHWAYS OF THE SPINAL CORD

### Lateral Spinothalamic Tract, the Pain and Temperature Pathway (Fig. 37-1)

**Neuron I.** The cell bodies are in the dorsal root ganglion, and their central processes enter the spinal cord via the dorsal root and end in the dorsal gray horn.

**Neuron II.** These cell bodies are in the posteromarginal nucleus and nucleus proprius. Their axons cross to the opposite side in the anterior white commissure and ascend to the ventroposterolateral (VPL) nucleus of the thalamus as the lateral spinothalamic tract, located in the lateral white funiculus.

**Neuron III.** These cell bodies are in the VPL nucleus of the thalamus. Their axons ascend in the posterior limb of the internal capsule to the postcentral gyrus and to a region of the parietal lobe bordering the lateral fissure (somesthetic area II).

**Lesion.** Unilateral destruction of the lateral spinothalamic tract results in loss of pain and temperature on the contralateral side, beginning approximately one segment below the level of the lesion.

### Posterior White Columns, the Two-point Touch, Stereognosis, and Vibratory Sense Pathway (Fig. 37-2)

**Neuron I.** The cell bodies are in the dorsal root ganglion. The central processes enter the spinal cord via the dorsal root; they do not synapse at that level, but ascend in the posterior white funiculus as the more medial fasciculus gracilis (those entering below T6) and the lateral fasciculus cuneatus (those entering above T6). These fibers ascend to the nuclei gracilis and cuneatus in the low medulla.

**Neuron II.** These cell bodies are in the nucleus gracilis and cuneatus. Their axons sweep ventrally as the internal arcuate fibers and then cross the midline in the decussation of the medial lemniscus. The medial lemniscus ascends through the brainstem and ends in the VPL nucleus of the thalamus.

**Neuron III.** From cell bodies in the VPL nucleus of the thalamus, axons pass to the postcentral gyrus by way of the posterior limb of the internal capsule.

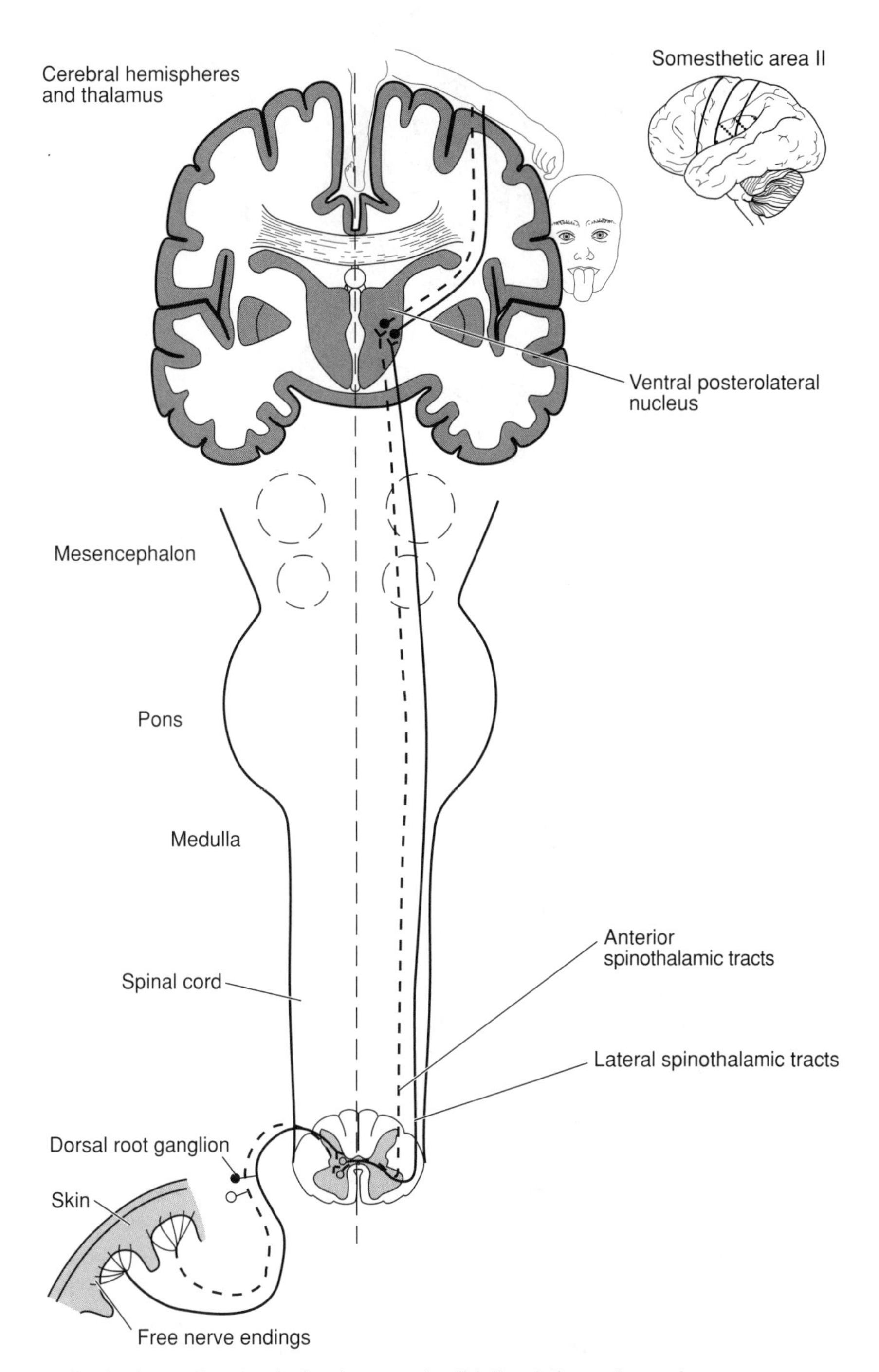


**Figure 37-1.** Lateral spinothalamic tract (*solid lines*) for pain and temperature; anterior spinothalamic tract (*broken lines*) for light touch.

**Lesion**. Unilateral destruction of the posterior white column results in astereognosis and a loss of two-point touch and vibratory sense on the ipsilateral side below the level of the lesion.

## Anterior Spinothalamic Tract, Crude (Light) Touch Pathway (see Fig. 37-1)

**Neuron I**. The cell bodies are in the dorsal root ganglion and their central processes pass through the dorsal root to end in lamina V at the level of entry.

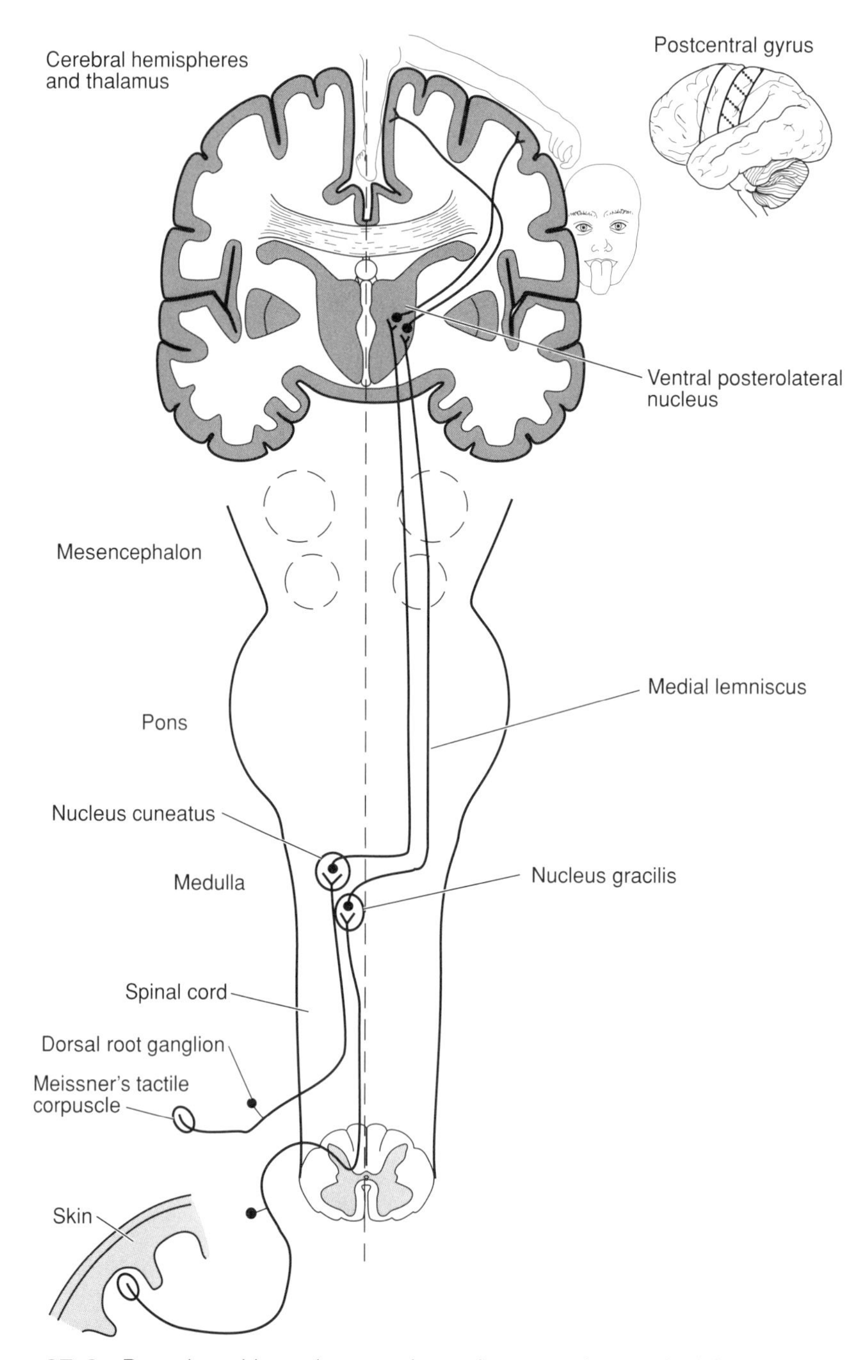


**Figure 37-2.** Posterior white column pathway for two-point touch, joint sense, vibratory sense, and stereognosis. That part of the pathway using the nucleus gracilis begins in the lower limb; that using the nucleus cuneatus begins in the upper limb.

**Neuron II.** These cell bodies are in laminae V and VI in the dorsal horn; the axons ascend and over the next several segments cross to the contralateral side via the anterior white commissure. The tract is located in the anterior white funiculus just anterior to the anterior horn and the axons terminate in the VPL nucleus of the thalamus.

**Neuron III.** These cell bodies are in the VPL nucleus of the thalamus; their axons pass to the postcentral gyrus of the cerebral cortex via the posterior limb of the internal capsule.

**Lesions**. A lesion affecting the anterior spinothalamic likely produces minimal or no discernible loss because touch is also carried by the posterior white columns.

## Posterior Spinocerebellar and Cuneocerebellar Pathways, the Position-sense Pathway

**Neuron I**. The peripheral processes of these neurons are associated with neuromuscular spindles and Golgi tendon organs. The cell bodies are in the dorsal root ganglion, and the central processes enter the spinal cord via the dorsal roots. These fibers take one of three courses: Those below L3 ascend in the posterior white column to the nucleus dorsalis; those above C8 ascend to the accessory cuneate nucleus of the medulla; and the rest enter the nucleus dorsalis near the level of entry.

**Neuron II**. These cell bodies in the nucleus dorsalis (Clarke's column), where first-order neurons below C8 terminate, have axons that ascend ipsilaterally in the posterior spinocerebellar tract to the cerebellum. This tract is superficially positioned in the posterolateral aspect of the lateral white funiculus. From cell bodies in the accessory cuneate nucleus, which receives first-order neurons from levels above C8, axons form the cuneocerebellar tract, which ascends through the cervical cord and then becomes part of the inferior cerebellar peduncle.

**Lesion**. Because proprioceptive information is conveyed by the posterior white columns and the posterior spinocerebellar tract projects to the cerebellum that is subcortical, it is difficult to characterize the effects of a lesion affecting this tract. Even so, ataxia may result, and position and vibratory sense may be diminished.

# ASCENDING PATHWAYS ARISING AT BRAINSTEM LEVELS

## Pain and Temperature Pathway (Fig. 37-3)

**Neuron I**. These neurons are those of the ophthalmic, maxillary, and mandibular divisions of the trigeminal nerve; their cell bodies are in the trigeminal ganglion. The central processes of these cell bodies enter the pons and descend to the low medulla as the spinal tract of CN V. Additional first-order neurons of CN VII, CN IX, and CN X are from the posterior ear region. Their axons also enter the spinal nucleus of CN V.

**Neuron II**. These cell bodies are in the spinal nucleus of CN V. Their axons cross the midline and ascend to the thalamus as the ventral ascending secondary tract of CN V.

**Neuron III**. From cell bodies in the ventroposteromedial (VPM) nucleus, axons pass to the postcentral gyrus and somesthetic area II via the posterior limb of the internal capsule.

## Touch Pathways (see Fig. 37-3)

**Neuron I.** The cell bodies are in the trigeminal ganglion. Central processes enter the pons and terminate in the principal sensory nucleus of CN V and in the upper part of the spinal nucleus of CN V.

**Neuron II**. From cell bodies in the principal sensory and spinal nuclei of CN V, axons cross the midline and ascend to the thalamus as the ventral secondary ascending tract of CN V. Other axons from the principal sensory nucleus of CN V ascend ipsilaterally as the dorsal secondary ascending tract of CN V.

**Neuron III**. These cell bodies are in the VPM nucleus of the thalamus; their axons project to the postcentral gyrus.

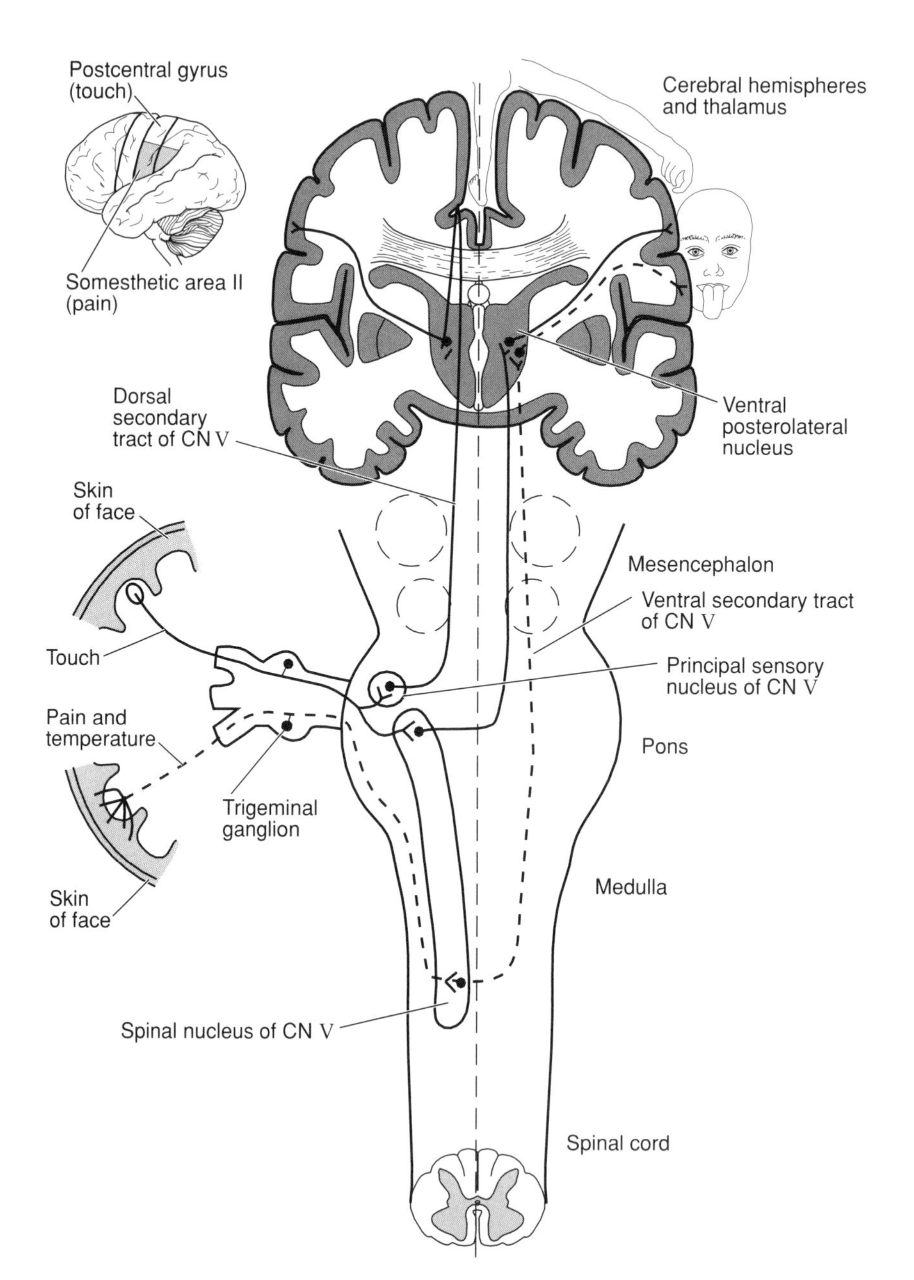


**Figure 37-3.** The trigeminal nerve pathways for pain and temperature (*broken lines*) and touch (*solid lines*) from the face. CN, cranial nerve.

## Hearing Pathway

**Neuron I.** These neurons are the bipolar cells in the spiral cochlear ganglion. Their dendrites extend from the hair cells of the organ of Corti to the cell bodies in the ganglion; their axons project to cochlear nuclei at the pons-medulla junction.

**Neuron II.** From cell bodies in the dorsal and ventral cochlear nuclei, both crossed and uncrossed axons ascend as the lateral lemniscus to the inferior colliculus. Some of these axons pass through the superior olivary nucleus, nucleus of the lateral lemniscus, and nucleus of the trapezoid body in this ascent.

**Neuron III.** These cell bodies are in the inferior colliculus. Their axons pass into the brachium of the inferior colliculus on their way to the medial geniculate body.

**Neuron IV.** From cell bodies in the medial geniculate body, axons pass in the sublenticular limb of the internal capsule to the transverse temporal gyrus of Heschl.

## Visual Pathway

**Neuron I**. These neurons are the rods and cones of the retina.

**Neuron II**. These neurons are the bipolar cells of the retina.

**Neuron III**. From cell bodies in the ganglion cell layer of the retina, axons pass centrally to the lateral geniculate body via the optic nerve, optic chiasm, and optic tract. Axons from the nasal half of the retina cross the midline in the optic chiasm.

**Neuron IV**. These cell bodies are in the lateral geniculate body. Their axons pass through the sublenticular and retrolenticular limbs of the internal capsule and the geniculocalcarine tract (optic radiations) to the calcarine striate cortex (area 17).

# DESCENDING PATHWAYS OF THE BRAIN AND SPINAL CORD

## Pyramidal System (Fig. 37-4)

### *Corticospinal Tract*

**Upper motor neuron (UMN)**. From cell bodies in the precentral (area four) and postcentral gyri and the premotor cortex, axons descend through the posterior limb of the internal capsule, crus cerebri, basis pontis, and pyramids. Most fibers cross the midline in the pyramidal decussation and descend in the lateral corticospinal tract, which occupies most of the posterior aspect of the lateral funiculus. These fibers synapse either directly on lower motor neurons (LMNs) or on internuncials, which in turn synapse on LMNs. This pathway tonically facilitates the antagonists of antigravity muscles and phasically controls the distal muscles in fine movements.

**LMN**. These neurons are the alpha motor neurons to extrafusal muscle fibers and the gamma motor neurons to intrafusal muscle fibers.

**Lesion**. Damage to the lateral corticospinal and reticulospinal tracts results in a UMN syndrome (see Chapter 41).

### *Corticobulbar Pathways*

**UMN**. The cell bodies are located near the lateral fissure in the precentral and postcentral gyri and in the premotor cortex. The axons descend through the genu and posterior limb of the internal capsule and the crus cerebri. The crossed fibers are from the frontal eye fields to the lateral gaze center (paramedian pontine reticular formation) for saccadic eye movements, to that portion of the hypoglossal nucleus that supplies the genioglossus nuclei, and to that portion of the facial motor nucleus that supplies muscles of the lower face. Those that supply the lateral gaze center for smooth pursuit movements and those that terminate on the spinal accessory nucleus are uncrossed. All other corticobulbar fibers are both crossed and uncrossed.

**LMN**. These cell bodies are in the somatic efferent (SE) and somatic visceral efferent nuclei of CN III through CN VII and CN IX through CN XII.

## Major Extrapyramidal Tracts (see Fig. 37-2)

### *Vestibulospinal Tract*

**UMN**. The cell bodies of the vestibulospinal tract are located in the lateral vestibular nucleus. The axons are uncrossed and located in the anterior white funiculus; they end on internuncial neurons.

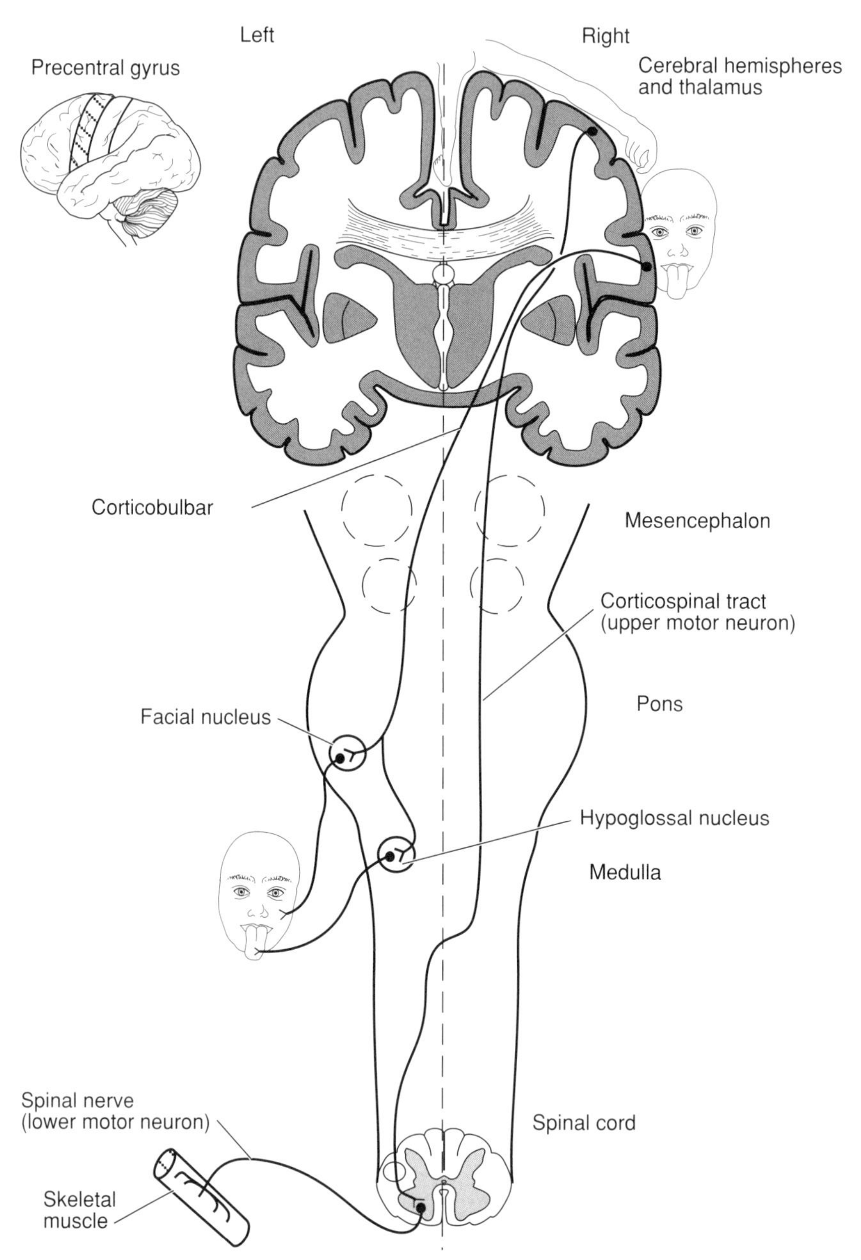


**Figure 37-4.** Corticospinal and corticobulbar tracts. (Only the crossed corticospinal pathway to the upper extremity muscles and the crossed corticobulbar pathways to the lower face and genioglossus muscles are shown.)

**LMN**. The alpha and gamma motor neurons to antigravity muscles are facilitated by this tract.

### *Lateral Reticulospinal Tract*

**UMN**. From cell bodies in the gigantocellular nucleus in the reticular formation of the medulla, axons are uncrossed and located in the lateral white funiculus.

**LMN**. The alpha and gamma motor neurons to antigravity muscles are inhibited by the lateral reticulospinal tract.

### *Medial Reticulospinal Tract*

**UMN**. The cell bodies are in the oral and caudal pontine reticular nuclei. The axons of the medial reticulospinal tract are primarily uncrossed and descend in the anterior white funiculus.

**LMN**. The alpha and gamma motor neurons of antigravity muscles are facilitated by this tract.

### *Rubrospinal Tract*

**UMN**. These cell bodies are in the red nucleus. The axons cross the midline in the ventral tegmental decussation and descend in the rubrospinal tract, which is located in the lateral reticular formation of the brainstem and in the lateral white funiculus of the spinal cord. The axons end on internuncials.

**LMN**. The alpha and gamma motor neurons to the antagonists of antigravity muscles appear to be facilitated by this tract.

### *Central Descending Autonomics*

**UMN**. From cell bodies in the hypothalamus and the reticular formation, axons descend through the lateral reticular formation of the brainstem and in the lateral white funiculus of the spinal cord.

**LMN**. These neurons consist of the preganglionic sympathetic neurons in the intermediolateral cell column of the lateral horn at thoracic and upper lumbar levels of the spinal cord and of preganglionic parasympathetic neurons in the brainstem and sacral levels of the spinal cord.

# Chapter 38

# Cerebellum

## GROSS TOPOGRAPHY

The cerebellum is positioned posterior to the pons and upper medulla, and occupies the majority of the posterior cranial fossa. It is formed by an outer gray cortex, four pairs of deep nuclei (dentate, globose, emboliform, and fastigial), and an intervening medullary core of white matter called the **arbor vitae**. The surface of the cerebellum consists of small elevations or ridges (**folia**), which are separated by indentations called **sulci**. The cerebellum is connected to the midbrain, pons, and medulla by the **superior** (brachium conjunctivum), **middle** (brachium pontis), and **inferior** (restiform body) **cerebellar peduncles**, respectively.

The phylogenetic development and function of the various parts of the cerebellum can be described on the basis of two different anatomical classifications. First, there are three transverse divisions that are based on embryologic development. On the superior surface, the primary fissure separates the **anterior lobe** (paleocerebellum) from the **posterior lobe** (neocerebellum); on the inferior surface, the posterolateral fissure demarcates the posterior lobe from the **flocculonodular lobe** (archicerebellum). Second, based on its medial to lateral organization, the cerebellum can be divided into a medial **vermis** (archicerebellum), an intermediate **paravermis** (paleocerebellum), and the **lateral hemispheres** (neocerebellum).

The components of these two organizational patterns relative to function are similar but not exact. The size of the neocerebellum (posterior lobe or lateral hemispheres) corresponds to the size of the cerebral cortex and is best developed in mammals with large cerebral cortices. It is functionally related to skilled, intricate peripheral activities. The anterior and flocculonodular lobes, the vermal and paravermal regions, are larger in lower vertebrates and are functionally associated with balance and control of proximal and axial activities.

## CEREBELLAR CORTEX

The cerebellar cortex consists of outer molecular, inner granule cell, and intermediate Purkinje cell layers. The **Purkinje cell layer** consists of a single layer of Purkinje cell bodies, which are the largest neurons in the cerebellar cortex. Its dendrites form an extensive network that extends superficially into the molecular layer where they are interrelated to the axons (parallel fibers) of large numbers of granule cells. The Purkinje cell axons are the efferent fibers of the cerebellar cortex and project into the deeper cerebellar white matter as they pass toward the deep nuclei. The dendritic tree of each

Purkinje cell is also closely related to the terminal arborization of a **climbing fiber**, which arises from a cell body in the contralateral inferior olivary nucleus of the medulla.

The **granule cell layer** is the thickest layer of the cortex and houses the granule cell bodies. The axons of these cells project perpendicularly into the molecular layer where they branch into two processes (**parallel fibers**), which are long and parallel with the surfaces of the folia. The short dendrites and the cell bodies of the granule cells are influenced by incoming **mossy fibers** from the spinal cord, vestibular system, and pontine nuclei.

The outer **molecular layer** consists largely of parallel fibers and dendritic arborizations of the Purkinje cells. **Golgi**, **basket**, and **stellate cells** are found in all three layers of the cortex.

# CIRCUITRY OF THE CEREBELLUM

In general, the cerebellum functions as a feedback system. As a result, its circuits are somewhat circular because cerebellar efferents project to the source of the afferent fibers to each part of the cerebellum. The general scheme of the circuitry is also predictable: Afferent fibers terminate in the cerebellar cortex; Purkinje cells of the cortex project to the deep nuclei; and neurons of the deep nuclei project either directly or indirectly to the source of the afferent fibers.

## Vestibulocerebellum

The input to the cortex of the flocculonodular lobe is primarily from vestibular receptors as well as from cranial nerve nuclei controlling eye movements. There is no deep nucleus involved in this circuit; thus, the efferent fibers pass directly to vestibular nuclei and the medial longitudinal fasciculus. In this way, this oldest part of the cerebellum influences head position and eye movements as well as posture.

## Spinocerebellum

The vermis and anterior lobe, or median and intermediate zones, receive their input from auditory, visual, and vestibular receptors as well as axial (neck and trunk) and proximal limb musculature. Most of this information enters the cerebellum primarily through the inferior cerebellar peduncle, but some also enters through the superior cerebellar peduncle via the anterior spinocerebellar tract. The cortex of the median zone projects to the fastigial nucleus, which in turn projects to reticular and vestibular nuclei. The cortex of the intermediate zone projects to the interposed nuclei (globose and emboliform). Efferent fibers from these nuclei pass through the superior cerebellar peduncle to the red nucleus and thalamus. Through these connections, the medial and intermediate zones influence (via reticulospinal, vestibulospinal, rubrospinal, and corticospinal tracts) balance, posture, locomotion, and limb musculature.

## Pontocerebellum

The cortex of the lateral hemispheres receives a large input through the middle cerebellar peduncle from the contralateral pontine nuclei. These nuclei are part of the corticopontocerebellar system that connects the contralateral cerebral cortex with the neocerebellum. The lateral hemisphere cortex projects to the dentate nucleus, which in turn projects to the contralateral red nucleus and thalamus via the superior cerebellar peduncle. Because the thalamus projects to the cerebral cortex where the corticospinal system begins, the neocerebellum controls precise distal contralateral limb activity.

## BLOOD SUPPLY

The cerebellum is supplied by three pairs of arteries, all branches of the vertebral/basilar system. The **posterior inferior cerebellar arteries** branch from the terminal parts of the vertebrals; the **anterior inferior** and **superior cerebellar arteries** branch from the basilar artery. In general, the peripheral portions of these arteries supply the areas of the cerebellum indicated by the name of the artery; however, there is considerable overlap of these areas, making it difficult to specifically characterize the cerebellar deficits that result from lesions of each vessel. The proximal parts of these vessels are related to specific parts of the brainstem and supply consistent structures. As a result, blockage of an entire vessel can be accurately determined because of the specificity of the brainstem symptoms.

# Chapter 39

# Telencephalon

The **cerebral hemispheres** are interconnected by the corpus callosum, anterior commissure, and the lamina terminalis, which lies rostral to the third ventricle. Each hemisphere consists of an outer **gray cortex** (pallium), underlying **white matter**, a deeply located nuclear mass called the **basal ganglia**, and a **lateral ventricle**.

## CEREBRAL CORTEX

### Gross Topography

The cortex and its underlying white matter have a fairly consistent pattern of gyri and sulci that make it possible to define frontal, parietal, temporal, occipital, insular, and limbic lobes.

**Frontal lobe**. The more rostral frontal lobe is separated from the parietal lobe by the **central sulcus** (of Rolando). In the frontal lobe, the **precentral gyrus (motor area)** borders the central sulcus and is demarcated from the more rostral superior, middle, and inferior frontal gyri by the **precentral sulcus** (Fig. 39-1). On the basal surface, the frontal lobe is composed of an olfactory bulb and tract lying between the gyrus rectus and the orbital gyri.

**Parietal lobe**. The parietal lobe consists of the more rostral **postcentral gyrus** (somesthetic area) and the posteriorly located superior and inferior parietal gyri. On the medial aspect, the parietal and occipital lobes are divisible by the **parieto-occipital sulcus**; laterally the boundary is less precise. The **lateral fissure** (sulcus) helps to form an inferior boundary to the frontal and parietal lobes. It terminates caudally where the inferior parietal gyrus caps it as the supramarginal gyrus. The rest of the inferior parietal gyrus abuts the posterior extent of the superior temporal sulcus and is called the **angular gyrus**.

**Occipital lobe**. The lateral surface of the occipital lobe consists of lateral occipital gyri. The medial surface is divided by the **calcarine fissure** into a cuneus above and a lingual gyrus below. That portion of the cortex immediately bordering the calcarine fissure is the striate (visual) cortex.

**Temporal lobe**. The lateral surface of the temporal lobe is composed of superior, middle, and inferior **temporal gyri**. The transverse temporal gyri (of Heschl) lie medial to the superior temporal gyrus in the floor of the lateral fissure. It is the primary receptive area for hearing. The insula lies deep in the lateral fissure and constitutes a cortical cover to the lenticular nucleus. On the basal surface of the temporal lobe, the occipitotemporal gyrus and parahippocampal gyri lie medial to the inferior temporal gyri. The more medial parahippocampal gyrus ends rostrally as the uncus, is bound laterally by the collateral fissure and medially by the hippocampal fissure, and is continuous around the caudal end (splenium) of the corpus callosum with the cingulate gyrus. The rostral part of the parahippocampal gyrus, the uncus, and the lateral olfactory stria and gyrus (which project from the olfactory trigone and tract) compose the primary olfactory receptive area.

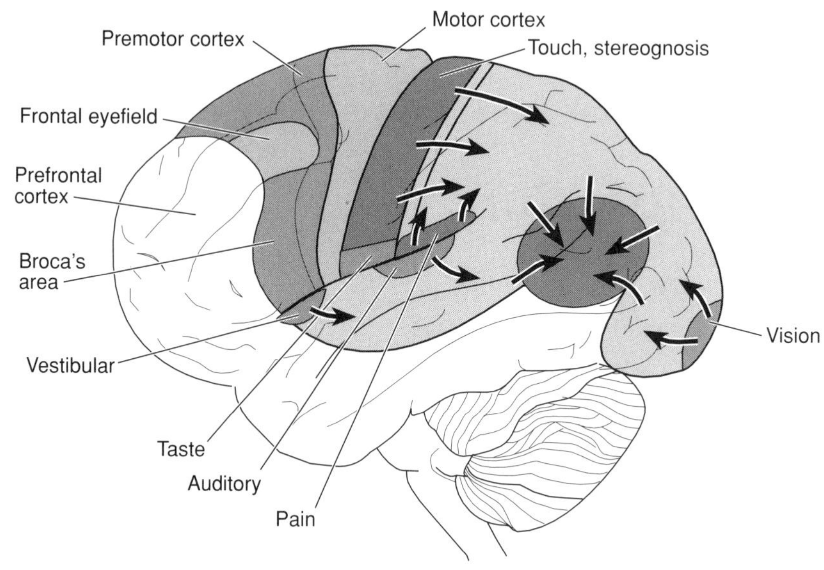


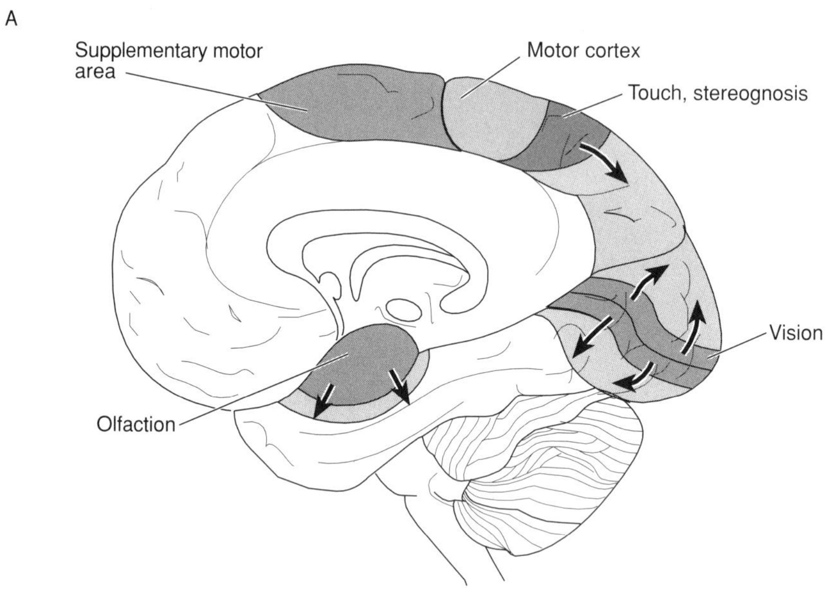


**Figure 39-1.** Functional areas of the cerebral cortex as shown in lateral **(A)** and medial **(B)** views. The primary receptive areas are *cross-hatched*; the unisensory association areas are *small stipples*; the multisensory association areas are *large stipples*; and the motor areas are *doubly cross-hatched*.

**Limbic lobe**. The limbic lobe includes the subcallosal, cingulate, and parahippocampal gyri and the dentate gyrus and hippocampus, which lies deep to the hippocampal fissure.

## Topographic Localization

**Sensory areas**. The **primary receptive areas** of each cerebral hemisphere are the postcentral gyrus (areas 3, 1, 2) for two-point touch, joint sense, and vibratory sense; the striate cortex (17) for vision; the transverse temporal gyri (41, 42) for hearing; the base of the postcentral gyrus (43) for taste; and the periamygdaloid region (34) for smell (see Fig. 39-1). The gyri adjacent to these areas constitute the **unisensory association**

**areas**, where sensory information is recognized as that perceived before (gnosis). These unisensory areas are the adjacent postcentral gyrus (5), superior parietal gyrus (7), and supramarginal gyrus (40) for areas 3, 1, 2; the peristriate (19) and parastriate cortex (18) for area 17; the adjacent superior temporal gyrus (42, 22) for areas 41, 42; areas 5 and 40 for area 43; and the parahippocampal gyrus (42, 22) for area 34. The angular gyrus (39) and adjacent areas 19 and 22 constitute **multisensory association areas**, where an individual can recognize an object through perception of one sense and can recall what the other sensations would be for that object (e.g., seeing and recognizing a chicken and recalling how it would feel, taste, sound, and smell). This multisensory area of the parietotemporal cortex is an area for language and the formulation of complex motor activities, especially in the dominant hemisphere (usually the left); thus, damage to the angular gyrus on the left may result in receptive aphasia and apraxia. **Lesions of unisensory areas** in the dominant hemisphere result in agnosia (e.g., visual agnosia, astereognosis). **Lesions of the right posterior and inferior parietal** lobe may lead to extinction and denial of one's contralateral body parts or environment.

**Motor areas**. Fibers from the parietal-temporal-occipital cortex connect with the prefrontal, premotor, and motor areas of the frontal lobe by way of the arcuate, superior longitudinal, and inferior occipitofrontal fasciculi. The **prefrontal cortex** is for initiative, judgment, and creativity. **Broca's area** in the inferior frontal gyrus of the dominant hemisphere is the motor speech area. **Lesions** of this area result in **expressive aphasias**. The frontal eye fields (8) and Exner's writing center (8) are located in the posterior part of the middle frontal gyrus. Destructive lesions of the former lead to transient conjugate gaze to the side of the lesion, whereas lesions of the latter result in a writing apraxia. The **motor cortex** (area 4) in the precentral gyrus and the **premotor cortex** (6, 8) in front of area 4 control motor activity, especially of the contralateral extremities.

## Blood Supply

The blood supply of the cerebral cortex is provided by the terminal portions of the three cerebral arteries. The branches of the proximal aspects of each of these arteries supply the deeper structures.

**Anterior cerebral artery**. On the medial aspect of the hemisphere in the medial longitudinal fissure, the anterior cerebral artery curves around the anterior and superior aspects of the corpus callosum. It supplies the medial aspects of the frontal and parietal lobes; its terminal branches also pass around the superior edge of the cortex onto its lateral aspect. Because this artery supplies primarily the paracentral lobule, interruption of its flow affects the contralateral lower limb.

**Middle cerebral artery**. This artery passes through the lateral fissure and branches across most of the lateral aspect of the cortical surface. It supplies most of the lateral aspects of the frontal, parietal, occipital, and temporal lobes as well as the **insula**. Occlusion of the cortical branches results in paralysis and loss of sensation of the contralateral face and upper limb. If the dominant hemisphere is affected, a motor aphasia, agraphia, and visual field deficit also occur.

**Posterior cerebral artery**. The posterior cerebral artery supplies the medial and inferior aspects of the temporal and occipital lobes. Its loss results primarily in cortical blindness (homonymous hemianopsia).

# INTERNAL CAPSULE

The cerebral cortex receives and projects fibers from and to lower centers by way of the internal capsule. The **anterior limb** of the internal capsule, positioned between the head of the caudate and lenticular nucleus, contains connections with the prefrontal cortex. The **posterior limb** of the internal capsule is located between the thalamus and lenticular nucleus and contains corticospinal fibers from the frontal lobe and projections to the parietal and temporal lobes. The **genu** of the internal capsule is at the level of the interventricular foramen and transports corticobulbar fibers. **Sublenticular fibers** of the internal capsule are auditory and optic radiations; **retrolenticular fibers** are also optic radiations.

# BASAL GANGLIA

Like the cerebellum, the basal ganglia influence motor activity through input to the cerebral cortex and through other descending systems. The primary components of these deeply positioned telencephalic nuclei are the caudate nucleus, putamen, and globus pallidus. Nomenclature of these ganglia can be confusing because other terms are used to indicate different combinations of the ganglia. These terms are summarized in Table 39-1. Two other nuclei, the substantia nigra of the midbrain and the subthalamic nucleus, are not considered part of the basal ganglia but are related functionally.

The **caudate nucleus** is long and curved and corresponds to the shape of the lateral ventricle. Its expanded head is related to the anterior horn of the ventricle. Its body follows the ventricle posteriorly and inferiorly, and its tail is related to the inferior horn. The **putamen** and **globus pallidus** are fused and located in the lateral angle formed by the internal capsule. The caudate and lenticular nucleus are separated by the internal capsule except cephalically, where the two structures are interconnected through the anterior limb of the internal capsule.

## Circuitry

The specific details regarding the sources and destinations of the afferent and efferent fibers are discussed in Chapter 35. Within the basal ganglia, most of the afferent fibers to the basal ganglia project to the caudate and putamen; the major efferent component is the globus pallidus. Within the basal ganglia, there are feedback connections between the globus pallidus and subthalamic nucleus and also between the caudate and putamen and the substantia nigra.

## Blood Supply

Because of the anatomical relationship between the basal ganglia and internal capsule, their blood supplies are provided by the same vessels. These vessels are deep vessels that arise from the internal carotid and proximal aspects of the anterior and middle cerebral arteries. The **medial striate arteries**, including the **artery of Heubner**, branch from both the anterior and middle cerebral arteries. These vessels ascend into the base of the brain and supply the anterior limb of the internal capsule and anterior

**TABLE 39-1. Nomenclature of the Basal Ganglia**

| Term | Components |
|---|---|
| Corpus striatum | Caudate nucleus, putamen, globus pallidus |
| Lenticular nucleus | Putamen, globus pallidus |
| Striatum (neostriatum) | Caudate nucleus, putamen |
| Pallidum (paleostriatum) | Globus pallidus |

components of the basal ganglia. The **lateral striate arteries** arise more posteriorly from the middle cerebral artery and supply both the anterior and posterior limbs of the internal capsule as well as the putamen and globus pallidus. The **anterior choroidal artery** is usually a branch of the internal carotid, which supplies the globus pallidus and putamen as well as the genu of the internal capsule.

# Chapter 40

# *Blood Supply of the Brain and Brainstem*

## VERTEBRAL ARTERIES

The brain is supplied by the vertebral and internal carotid arteries. The vertebral arteries enter the cranial cavity through the foramen magnum and give off **posterior inferior cerebellar** and **anterior and posterior spinal arteries**. The vertebral arteries unite at the pons-medulla junction to form the basilar artery. The **basilar artery** runs in the basilar sulcus of the pons and terminates at the level of the mesencephalon by branching into **posterior cerebral arteries**. The branches of the basilar artery are the anterior inferior cerebellar, labyrinthine, paramedian, short and long circumferential, and superior cerebellar arteries. Branches of the vertebral and basilar arteries supply the cerebellum, mesencephalon, pons, medulla, medial portion of the occipital lobe, and part of the temporal lobe and diencephalon.

## INTERNAL CAROTID ARTERIES

The internal carotid artery courses through the cavernous sinus and emerges from the sinus medial to the anterior clinoid process. It gives off the ophthalmic artery to the eye; the posterior communicating artery to the posterior cerebral artery; and the anterior choroidal artery, which supplies the optic tract, choroid plexus of the lateral ventricle, basal ganglia, posterior part of the internal capsule, and the hippocampus. Lateral to the optic chiasm, the internal carotid bifurcates into the middle cerebral and anterior cerebral arteries. The **anterior cerebral arteries**, connected by an anterior communicating artery, supply the medial surfaces of the frontal and parietal lobes. The **middle cerebral artery** passes into the lateral fissure and supplies the insula, lateral surface of the cerebral hemisphere, and part of the inferior surface of the temporal lobe. Basal ganglia and part of the internal capsule are supplied by the lenticulostriate branches of the middle cerebral artery. The internal carotid and vertebral arterial supplies are interconnected, forming the **circle of Willis**, which consists of the anterior communicating, anterior cerebral, posterior communicating, and posterior cerebral arteries. Central branches from the circle of Willis supply the basilar portion of the diencephalon and basal ganglia and give rise to the hypophyseal portal arterial system of the adenohypophysis.

## CEREBRAL VEINS

Superficial and deep cerebral veins drain into the dural venous sinuses. The **great cerebral vein (of Galen)** receives the internal cerebral veins and drains to the straight sinus. Superior, middle, and inferior **superficial veins** drain to the superior sagittal or basal sinuses. The midbrain, pons, and medulla drain by small veins into the sinuses at the base of the brain. The cerebellum is drained by superior and inferior cerebellar veins into adjacent dural venous sinuses. The dural venous sinuses are described in Chapter 4.

# Chapter 41

# Somatic Motor Control Mechanisms

## LEVELS OF CONTROL

The somatic motor anterior horn cells of the spinal cord and the general somatic efferent (GSE) [nuclei of cranial nerve (CN) III, CN IV, CN VI, and CN XII] along with special visceral efferent (SVE) (nucleus ambiguus and motor nuclei of CN V and CN VII) neurons of the brainstem are regulated by afferent and association fibers from all levels of the central nervous system (CNS).

**Segmental**. Afferent and internuncial neurons innervating anterior horn cells within one level of the spinal cord constitute a segmental level of motor control. For example, the myotatic stretch reflex involves IA fibers from the muscle spindle of a stretched muscle that synapse (monosynaptically) with an alpha motor neuron supplying the extrafusal muscle fibers of the muscle.

**Intersegmental**. Intersegmental connections of afferent and association neurons within the spinal cord constitute a mechanism for intersegmental regulation of motor activity. Such connections may involve collaterals from long ascending and descending pathways or shorter spinospinal fibers of the fasciculus proprius. For example, the pain withdrawal reflex involves the stimulation of pain fibers, the spread of these impulses to several levels of the cord via the fasciculus proprius, and the subsequent stimulation of motor neurons that mediate the withdrawal motor response.

**Suprasegmental**. Suprasegmental control pathways involve connections between the spinal cord and higher centers of the CNS. This constellation of numerous suprasegmental connections is best understood if it is separated into phylogenetically older and newer systems. Such a scheme includes older antigravity and vestibular regulation, the intermediate regulation of the more stereotyped grosser movements involving the more axial musculature, and the newer regulation of fine, discrete movements.

## SUPRASEGMENTAL CONTROL MECHANISMS

### Antigravity and Vestibular Regulation

Antigravity and vestibular connections involve input from **muscle spindles** and the **vestibular apparatus**. These afferents make connections through the spinocerebellar tracts and vestibular nuclei with the cerebellum; the vestibular fibers end in the flocculonodular lobe, while spindle information is conveyed to vermal and paravermal areas of the anterior and posterior lobes of the cerebellum. Cerebellar efferents, through dentatorubral fibers of the superior cerebellar peduncle, synapse in the red nucleus with cells of the **rubrospinal tract**, which bring about the **facilitation of the**

**antagonists of the antigravity muscles**. Cerebellar efferents from the flocculonodular lobe arise from the fastigial nuclei and terminate in the lateral vestibular nuclei. Axons from the lateral nucleus constitute the **lateral vestibulospinal tract**, which is **facilitatory to antigravity muscles**. The **medial** and **lateral reticulospinal tracts** receive input from all levels of the CNS and are **facilitatory** and **inhibitory**, respectively, to the **antigravity muscles**.

## Gross Stereotyped Movement Regulation

Gross stereotyped movements are regulated by the **basal ganglia** through their **ascending** connections with the cortex and their **descending connections** through the reticular formation. In these circuits, the premotor cortex and centromedian nucleus of the thalamus receive ascending afferent input and relay this information to the caudate and putamen. The striatum (putamen and caudate) also receives input from the substantia nigra. The neostriatal nuclei send axons to, and appear to regulate, the globus pallidus. The **efferent outflow** from the basal ganglia is from the globus pallidus to (1) the **motor and premotor cortices** by way of the ventral anterior and ventral lateral nuclei of the thalamus, and (2) the **motor neurons of the spinal cord and brainstem** by way of the subthalamus, prerubral field, red nucleus, and the reticulospinal, rubrospinal, and reticuloreticular pathways. **Pallidal efferents** to the thalamic nuclei course through the internal capsule and then pass between the subthalamic nucleus and zona incerta in the fasciculus lenticularis to reach the prerubral field (of Forel). From the prerubral field, axons loop laterally toward the thalamus in the fasciculus thalamicus, or they descend in the reticular formation to contralateral motor nuclei. Other pallidal efferents loop around the internal capsule in the ansa lenticularis on their way to the thalamus.

**Lesions of the basal ganglia**. Lesions of the basal ganglia, subthalamus, and substantia nigra produce disturbances in both involuntary movements (**dyskinesias**) and in muscle tone. Lesions of the subthalamic nucleus can lead to **hemiballism** on the opposite side because of release of the globus pallidus from the inhibitory control of the subthalamus. **Parkinsonism** (paralysis agitans) is characterized by a decrease of **dopamine** in the substantia nigra and striatum. The tremor at rest and rigidity of this disorder may be caused by an inability of the striatum to regulate the globus pallidus. Lesions producing **chorea** and **athetosis** are most likely in the striatum, but are not as clearly localized as the other basal ganglia disorders.

## Fine Coordinated Movement Regulation

The **pathways** involved in the initiation and performance of a fine coordinated movement involve (1) input to the cortex, (2) a feedback loop between the cerebral cortex and the cerebellum, (3) corticospinal and corticobulbar pathways, and (4) alpha and gamma motor neurons of the spinal cord and brainstem.

**Input to cortex**. In these pathways, ascending exteroceptive input is relayed through the thalamus to the **primary receptive areas** of the cortex (areas 3, 1, 2; 41 and 42; 43; 17). Projections from the primary receptive areas go to **unisensory association areas** lying adjacent to the primary receptive areas (see Fig. 39-1). The specific sensation for each receptive area is "recognized" in each unisensory association area. Projections from the unisensory association areas congregate in **multisensory association areas** in the junctional area of the inferior parietal gyrus (areas 39 and 40), superior temporal gyrus (area 22), and lateral occipital gyri (area 19). In these areas, gnosis from more than one sensation is used in the initiation and formulation of learned complex motor activity.

**Feedback between cerebral cortex and cerebellum**. Axons from the parietal, occipital, and temporal regions, from the prefrontal cortical center for initiative and

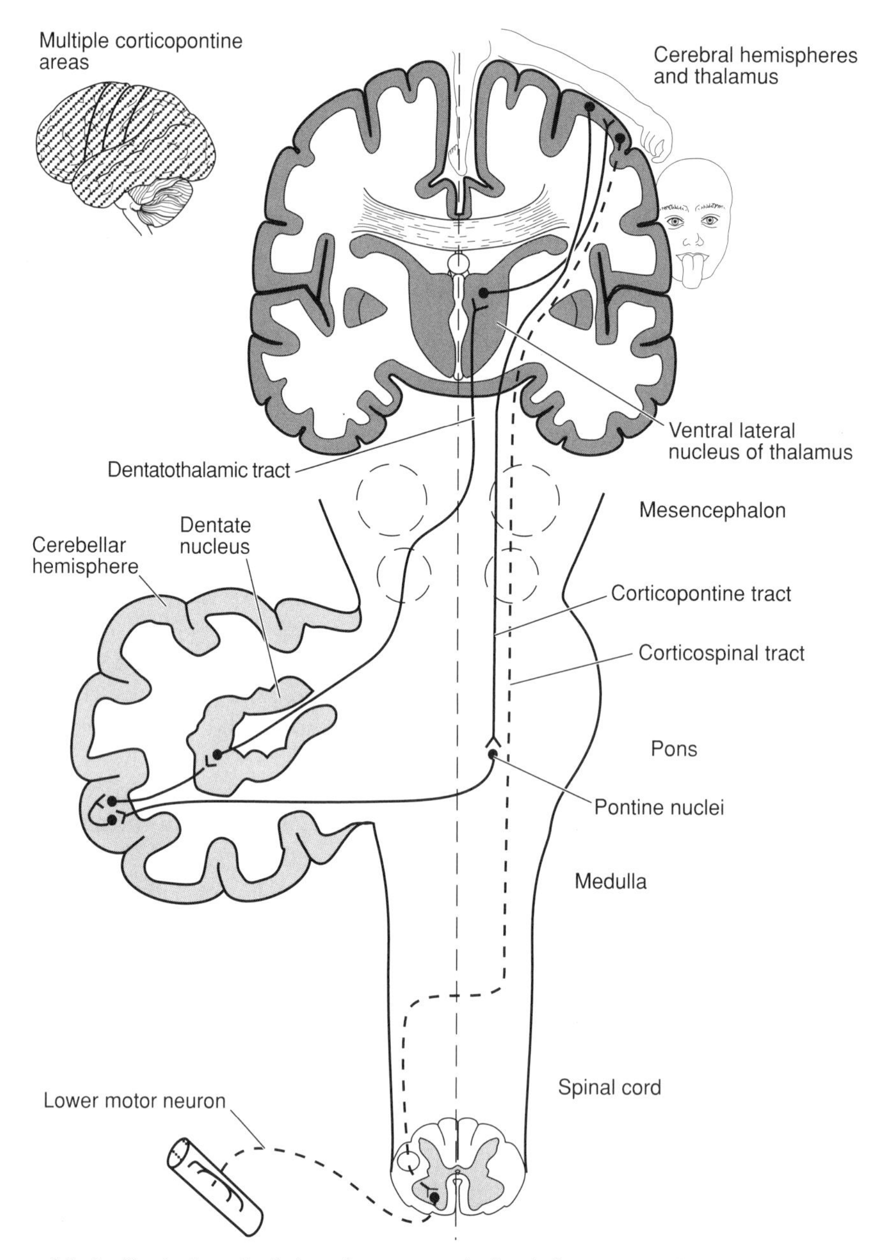


**Figure 41-1.** Illustration depicting the neocerebellar influence on the cerebral cortex. The neocerebellar circuit (*solid lines*) as well as the corticospinal tract and lower motor neurons (*broken lines*) are indicated.

judgment, and from premotor and motor regions, descend as **corticopontine pathways to pontine nuclei** (Fig. 41-1). Frontopontine fibers descend through the anterior limb of the internal capsule and medial portion of the crus cerebri. The other corticopontine fibers descend in the posterior limb of the internal capsule and lateral part of the crus cerebri. After they synapse on cells of the pontine nuclei, axons of these cells carry impulses across the midline to the **cerebellar cortex** in the middle cerebellar peduncle. These pontocerebellar fibers are mossy fibers that end in the granule cell layer of the neocerebellar cortex. Axons of Purkinje cells pass to the **dentate nucleus**, where they synapse on cells whose fibers cross the midline in the decussation of the superior cerebellar peduncle and ascend to the **ventral lateral nucleus** of the thalamus. From the ventral lateral nucleus axons, they ascend to the **motor cortex**.

**Corticospinal and corticobulbar pathways**. The ultimate descending pathway for fine discrete movements is the corticospinal and corticobulbar tracts to anterior horn cells and brainstem GSE and SVE neurons. This complex pathway involves a double crossing of fibers in the cerebral cortex–cerebellar loop and a single crossing of the corticospinal tract. The cerebellum acts as a computer coordinating the activity of numerous neurons.

**Cerebellar lesions**. Lesions of the cerebellum may cause **deficits in motor activity**. **Neocerebellar lesions** produce a lateral cerebellar syndrome in which the symptoms are on the same side of the body as the involved cerebellar hemisphere. This syndrome includes an asynergia of voluntary skilled activity characterized by hypotonia and postural fixation defects, particularly affecting the limbs. Some clinical signs and abnormal reflexes of neocerebellar disease are dysmetria, intention tremor, rebound phenomenon, dysdiadochokinesia, explosive speech, decomposition of movement, and a tendency to fall to the side of the lesion. **Flocculonodular syndrome** is primarily a disorder of locomotion and equilibrium, characterized by truncal ataxia, falling, and walking with a wide base or "drunken gait."

**Cortical lesions**. Lesions of various portions of the cerebral cortex lead to deficits in motor activity. Immediately after partial or complete lesions of the **precentral gyrus**, there is **flaccid paralysis** of the contralateral limbs and loss of superficial and deep reflexes; hypotonus and exaggerated deep reflexes may also occur. Involvement of the **premotor area** and **corticoreticular fibers** may lead to **spastic paralysis**. Lesions of **Broca's area** in the inferior frontal opercular and triangular areas of the dominant hemisphere may result in **expressive aphasia**. **Sensory aphasias** are more often associated with lesions of the posterior temporoparietal region. **Ideomotor** and **ideational apraxias** appear to be associated with lesions of the dominant parietal lobe in the region of the inferior parietal gyrus. **Ablative lesions** of the frontal eye fields in the posterior part of the middle frontal gyrus interfere with voluntary conjugate eye movements to the opposite side. **Corticobulbar lesions** can result in contralateral lower facial paralysis, a deviation of the tongue to the opposite side on protrusion, and ipsilateral weakness in shoulder shrugging. **Damage to the corticospinal tract** in the spinal cord will give an ipsilateral upper motor neuron complex of symptoms. This constellation of symptoms includes (1) increased segmental muscle tone (hypertonus), especially in extensors of the lower extremity and in flexors of the upper extremity caused by the loss of lateral reticulospinal fibers; (2) increased deep tendon reflexes (hyperreflexia); (3) absence of or diminished superficial reflexes; (4) presence of pathologic reflexes such as the Babinski reflex; and (5) loss of fine movements. **Lesions of lower motor neurons** produce flaccid paralysis or paresis, loss of deep and superficial reflexes, and denervation atrophy.

# Chapter 42

# *Visceral Motor Control Mechanisms*

The **hypothalamus** is the highest subcortical center for the regulation of visceral activity. It receives ascending information from the spinal and cranial nerves by way of the reticular formation. In addition to these spinoreticular and reticuloreticular pathways, the hypothalamus receives input from the thalamus, basal ganglia, and cerebral cortex. Efferents from the hypothalamus descend to the brainstem and spinal cord and also terminate in the thalamus, cortex, and hypophysis.

As previously indicated, hypothalamic nuclei are structurally divisible into medial and lateral groups by the columns of the fornix. The lateral group includes lateral and tuberal nuclei. The medial group is further subdivided into anterior, middle, and posterior groups. The anterior group includes the preoptic, anterior, supraoptic, paraventricular, and periventricular nuclei. The dorsomedial and ventromedial nuclei comprise the middle group; the posterior and mammillary nuclei make up the posterior group.

Functional classification of nuclei into autonomic and endocrine neurosecretory groups makes the hypothalamic region easier to understand. An anteromedial group of nuclei is involved in **parasympathetic regulation**. A posterolateral group of posterior, tuberal, and lateral nuclei regulates **sympathetic activity**. Outflow from the hypothalamus to preganglionic parasympathetic and preganglionic sympathetic cells of the brainstem and spinal cord occurs via periventricular fibers, a multisynaptic system that uses the dorsal longitudinal fasciculus or the reticuloreticular and reticulospinal pathways.

The **endocrine neurosecretory nuclei** are the supraoptic, paraventricular, and hypophysiotropic. The supraoptic and paraventricular nuclei produce antidiuretic (vasopressin) and oxytocin hormones, which are secreted in the pars nervosa. The **hypophysiotropic nuclei** include several hypothalamic nuclei (ventromedial, arcuate, preoptic) whose axons terminate in the infundibular stalk adjacent to capillary loops of the hypophyseal portal vascular system. Releasing and inhibiting factors produced in neurons of the hypophysiotropic area pass into these capillary loops and are transported by way of venous trunks of the pituitary stalk and capillaries of the pars distalis to the chromophobes and chromophiles of the anterior pituitary.

Visceral motor activity is influenced by descending pathways from the **olfactory cortex** and the **limbic system**. Olfactohypothalamic fibers course from the pyriform cortex to the hypothalamus. Impulses from the amygdala course in a ventral path and also in the stria terminalis to reach the hypothalamus and septal nuclei. Basal olfactory regions and the septal area are connected through the hypothalamus with the mesencephalic reticular formation by way of the medial forebrain bundle. The septal region also communicates with the mesencephalic reticular formation through a pathway that includes the stria medullaris, habenular nucleus, habenulopeduncular tract (fasciculus retroflexus), interpeduncular nucleus, and tegmental nuclei.

The **limbic cortex** is the phylogenetically "older" cortex that forms a ring around the corpus callosum and diencephalon. It includes the subcallosal area, cingulate gyrus, isthmus (retrosplenial area), parahippocampal gyrus, and hippocampal formation (hippocampus and dentate gyrus). All of these regions are connected through an association bundle, the **cingulum**. Axons from the hippocampal formation pass as the fimbria and fornix to the mammillary bodies and septal region. The mammillary bodies are linked to the cingulate gyrus through the mammillothalamic tract and its relay through the anterior nucleus of the thalamus. The mammillotegmental tract is a descending pathway from the mammillary bodies to the reticular formation of the brainstem.

In addition to the interconnections of the hypothalamus and amygdala with the phylogenetically older olfactory and limbic cortex, there are neocortical connections in the limbic lobe. An important circuit is between the prefrontal cortex and the hypothalamus through a relay in the dorsomedial nucleus of the thalamus.

**Lesions of the hypothalamus** affect visceral activity. Damage to the lateral hypothalamic nucleus may abolish appetite and lead to **weight loss**, whereas lesions of the satiety center in the ventromedial nucleus may lead to **obesity** through excessive eating. Lesions of the supraoptic nuclei can lead to **diabetes insipidus**. Bilateral temporal lobe lesions involving the pyriform cortex, amygdala, and hippocampal formation may produce disturbances in **olfaction**, **emotional behavior**, and **recent memory**.

# ? ANATOMICAL SCIENCES QUESTIONS

**DIRECTIONS: Each of the numbered items or incomplete statements in this section is followed by answers or by completions of the statement. Select the ONE lettered answer or completion that is BEST in each case.**

**1.** In the cubital fossa, the brachial artery is easily palpable as it passes

**(A)** between the tendons of the biceps brachii and brachioradialis muscles
**(B)** medial to the median nerve
**(C)** medial to the tendon of the biceps brachii muscle
**(D)** posterior to the medial epicondyle of the humerus
**(E)** superficial to the bicipital aponeurosis

**2.** In the region of the wrist, the radial artery is readily palpable where it passes

**(A)** around the medial aspect of the ulna
**(B)** between the tendons of the palmaris longus and flexor carpi ulnaris muscles
**(C)** lateral to the tendon of the flexor carpi radialis muscle
**(D)** superficial to the tendons of the extensor pollicis brevis and abductor pollicis longus muscles
**(E)** through the carpal tunnel

**3.** Fracture of the proximal shaft of the humerus (surgical neck) is commonly accompanied by significant and rapid swelling. This is most likely the result of injury to which of the most closely related vessels?

**(A)** Axillary vessels
**(B)** Brachial artery
**(C)** Cephalic vein
**(D)** Circumflex scapular artery
**(E)** Posterior circumflex humeral vessels

**4.** The structure best positioned to reinforce (support) the proximal radio-ulnar joint is the

**(A)** annular ligament
**(B)** biceps brachii muscle
**(C)** pronator teres muscle
**(D)** radial collateral ligament
**(E)** ulnar collateral ligament

**5.** The muscles of the rotator cuff of the shoulder

**(A)** are important in stabilization of the scapula
**(B)** are innervated by fibers from spinal cord segments C8 and T1
**(C)** extend from the humerus to the axial skeleton
**(D)** provide the major stabilization of the joint
**(E)** surround the shoulder joint, reinforcing it on its anterior, posterior, inferior, and superior aspects

6. Which of the following statements about the superficial palmar arterial arch is correct?

(A) It is formed by the ulnar and median arteries
(B) It is found between the palmar aponeurosis and long digital flexor tendons
(C) It is found distal to the metacarpophalangeal joints
(D) It is found in the subcutaneous tissue of the palm
(E) It is proximal to the carpal tunnel

7. Which statement about the carpal tunnel is most accurate?

(A) It contains all of the nerves that enter the hand
(B) It contains the tendon of the flexor carpi radialis and palmaris longus muscles
(C) It contains the tendons of the extrinsic finger and wrist flexors
(D) It contains the ulnar and median nerves
(E) It is formed ventrally by the transverse carpal ligament

8. A small area of tenderness just lateral to the acromion and pain on abduction of the humerus would most likely indicate inflammation in the tendon in that location. This is the tendon of the

(A) infraspinatus muscle
(B) pectoralis major muscle
(C) subscapularis muscle
(D) supraspinatus muscle
(E) trapezius muscle

9. A small boy falls on a piece of glass and sustains a laceration to the palm of his hand. On physical examination, his major neurologic deficit is an inability to oppose his thumb to the other digits. The most likely structure lacerated was the

(A) common digital branch of the median nerve
(B) recurrent (motor) branch of the median nerve
(C) tendon of the abductor pollicis longus
(D) tendon of the flexor pollicis longus muscle
(E) ulnar nerve

10. Contraction of the gluteus medius and minimus muscles can be palpated

(A) anterior to the anterior inferior spine of the ilium
(B) between the crest of the ilium and the greater trochanter of the femur
(C) between the ischial tuberosity and the lesser trochanter of the femur
(D) inferior to the greater trochanter of the femur
(E) just superior to the ischial tuberosity

11. The capsule of the hip joint is best described as

(A) attaching to the femur at the edge of the articular surface of the head of the femur
(B) being atypical for a synovial joint in that the synovial layer of the capsule is not present
(C) being relatively lax and containing no particularly strong reinforcing ligaments
(D) containing strong reinforcing bands of fibers (ligaments) that tighten as the femur is extended
(E) extending distally the entire length of the neck of the femur, both anteriorly and posteriorly

**12.** Loss of the quadriceps femoris reflex is indicative of a lesion involving predominantly spinal cord segment

**(A)** L1 (L1–L2)
**(B)** L3 (L2–L4)
**(C)** L5 (L4–S1)
**(D)** S1 (L5–S2)
**(E)** S3 (S2–S4)

**13.** Which statement about the anterior cruciate ligament is correct?

**(A)** It attaches to the posterior intercondylar region of the tibia
**(B)** It is obliquely oriented so its superior attachment is more posterior than its inferior attachment
**(C)** It protects against anterior displacement of the femur on the tibia
**(D)** It protects against posterior dislocation of the tibia on the femur
**(E)** It resists lateral displacement of the tibia on the femur

**14.** The posterior tibial pulse is best palpated

**(A)** about half-way between the lateral malleolus and the calcaneal tendon
**(B)** along the anterior border of the medial malleolus
**(C)** approximately 2 cm posterior to the medial malleolus
**(D)** between the peroneus longus and calcaneal tendons
**(E)** just posterior to the lateral malleolus

**15.** Forceful plantar flexion at the ankle and inversion of the foot places considerable tensile stress on the

**(A)** anterior talofibular ligament
**(B)** deltoid ligament
**(C)** medial collateral ligament of the ankle
**(D)** plantar calcaneonavicular ligament
**(E)** tendon of the tibialis posterior muscle

**16.** During quiet standing, weight is transferred from the foot to the floor through the

**(A)** bases of the metatarsals, navicular and cuboid
**(B)** distal phalanges and talus
**(C)** heads of the metatarsals and the tuberosity of the calcaneus
**(D)** sustentaculum tali and calcaneus
**(E)** talus

**17.** Which statement about the ligamentum flavum is correct?

**(A)** It blends with the posterior aspect of the intervertebral disk
**(B)** It interconnects the anterior aspects of the vertebral bodies
**(C)** It interconnects the lamina of adjacent vertebra
**(D)** It is a continuous ligament that extends from the sacrum to the occipital bone
**(E)** It is composed of hyaline cartilage

**18.** The anterior aspect of the intervertebral foramen is formed by the

**(A)** articular processes
**(B)** lamina
**(C)** pedicles
**(D)** vertebral bodies and intervertebral disk
**(E)** zygapophyseal joint

**19.** The dorsal rami of the spinal nerves

**(A)** contain primarily sensory or afferent fibers
**(B)** end as intercostal nerves in the thoracic region
**(C)** form the brachial and lumbosacral plexus
**(D)** supply some but not all of the extrinsic muscles of the shoulder
**(E)** supply the deep muscles and medial skin of the back

**20.** The aponeurosis of the external abdominal oblique muscle forms or contributes to the formation of the

**(A)** conjoined tendon (falx inguinalis)
**(B)** deep inguinal ring
**(C)** floor of the inguinal canal
**(D)** internal spermatic fascia
**(E)** posterior wall of the inguinal canal

**21.** Which of the following structures is partially peritonealized?

**(A)** Descending colon
**(B)** Ileum
**(C)** Kidney
**(D)** Pancreas
**(E)** Transverse colon

**22.** A radiograph taken after the ingestion of a barium meal reveals a small ulcer in the posterior wall of the second or descending part of the duodenum. If a perforation occurs at that point, what adjacent structure would be in danger?

**(A)** Abdominal aorta
**(B)** Gallbladder
**(C)** Right quadratus lumborum muscle
**(D)** Right renal pelvis
**(E)** Superior mesenteric artery

**23.** After a barium enema, the colon would be readily recognizable in a radiograph because of

**(A)** appendices epiploicae
**(B)** prominent haustra (sacculations)
**(C)** the large number of circular folds (valvulae conniventes)
**(D)** the number of vascular arcades
**(E)** three longitudinal muscle bands, the taeniae coli

**24.** The most posterior part of the stomach is the

**(A)** body
**(B)** cardiac orifice
**(C)** fundus
**(D)** lesser curvature
**(E)** pylorus

**25.** The fundus of a slightly enlarged gallbladder could best be palpated

**(A)** at the anterior end of the right ninth costal cartilage
**(B)** at the inferior angle of the right scapula
**(C)** at the junction of the xiphoid process and right subcostal margin
**(D)** at the most inferior aspect of the right side of the rib cage
**(E)** posterolaterally at the tip of the right eleventh costal cartilage

**26.** The head of the pancreas is most closely associated with the

**(A)** ascending colon
**(B)** left lobe of the liver
**(C)** second part of the duodenum
**(D)** spleen
**(E)** splenic artery

**27.** Most of the lymphatic fluid from the stomach drains through the

**(A)** celiac lymph nodes
**(B)** inferior mesenteric lymph nodes
**(C)** left renal lymph nodes
**(D)** right lumbar lymph nodes
**(E)** superior mesenteric lymph nodes

**28.** The innervation of the gut where the small and large intestines join is provided primarily by the

**(A)** celiac plexus
**(B)** ilioinguinal nerve
**(C)** inferior mesenteric plexus
**(D)** subcostal nerve
**(E)** superior mesenteric plexus

**29.** The parasympathetic innervation of the sigmoid colon is provided by the

**(A)** inferior mesenteric plexus
**(B)** lumbar plexus
**(C)** lumbar splanchnic nerves
**(D)** pelvic splanchnic nerves
**(E)** sacral plexus

**30.** By placing your stethoscope on the right midclavicular line at the level of the fifth intercostal space, you can best listen to sounds in

**(A)** both the upper and middle lobes of the right lung
**(B)** the horizontal fissure of the right lung
**(C)** the lower lobe of the right lung
**(D)** the middle lobe of the right lung
**(E)** the upper lobe of the right lung

**31.** Superiorly the pleural cavities extend

**(A)** approximately 2.5 cm above the clavicle
**(B)** approximately 2.5 cm below the clavicle
**(C)** to a level about midway between the first rib and clavicle
**(D)** to the level of the clavicle
**(E)** to the level of the costal cartilage of the first rib

**32.** To enter the costodiaphragmatic recess of the pleural cavity in the midaxillary line, a needle should be directed horizontally through the

**(A)** eleventh intercostal space
**(B)** ninth intercostal space
**(C)** seventh intercostal space
**(D)** fifth intercostal space
**(E)** fourth intercostal space

**33.** The posterior surface of the heart is formed primarily by the

**(A)** left atrium
**(B)** left ventricle
**(C)** right atrium
**(D)** right ventricle
**(E)** left and right ventricles

**34.** A major loss of the blood supply to the adjacent diaphragmatic surfaces of the left and right ventricles would most likely be caused by an occlusion of the

**(A)** anterior interventricular (descending) artery
**(B)** circumflex artery
**(C)** left coronary artery
**(D)** marginal branch of the right coronary artery
**(E)** posterior interventricular (descending) artery

**35.** Which of the following statements regarding the pericardium is correct?

**(A)** The parietal layer of serous pericardium is also called the epicardium
**(B)** The pericardial cavity is deep to the visceral layer of serous pericardium
**(C)** The pericardial cavity is located between the fibrous pericardium and parietal layer of serous pericardium
**(D)** The reflection of the serous pericardium around the large veins bounds the oblique sinus
**(E)** The transverse sinus is located on the diaphragmatic surface of the heart

**36.** Which statement regarding the azygos vein is correct?

**(A)** It drains most of the thoracic viscera
**(B)** It is located to the left of the thoracic duct
**(C)** It is positioned deep to the descending aorta
**(D)** It passes through the root of the right lung
**(E)** It receives venous blood either directly or indirectly from both right and left posterior intercostal veins

**37.** In the mediastinum

**(A)** the aortic arch passes posteriorly and to the right
**(B)** the left brachiocephalic vein passes posterior to the brachiocephalic artery
**(C)** the left recurrent laryngeal nerve passes around the arch of the aorta
**(D)** the ligamentum arteriosum interconnects the aortic arch and left pulmonary vein
**(E)** the thoracic duct is found between the esophagus and trachea

**38.** The smallest diameter of the pelvis is the distance between the

**(A)** arcuate lines of the ilia
**(B)** ischial spines
**(C)** ischial tuberosities
**(D)** pubic symphysis and the coccyx
**(E)** pubic symphysis and the sacral promontory

**39.** "Back-room" abortions occasionally result in peritonitis because at one point the vagina is separated from the peritoneal cavity by only the thin muscular wall of the vagina and a layer of the peritoneum. This point is at the

**(A)** anterior wall of the vagina as it passes through the urogenital diaphragm
**(B)** anterior wall of the vagina at the level of the pelvic sling
**(C)** lateral walls of the vagina where they are associated with the broad ligament
**(D)** posterior fornix of the vagina
**(E)** posterior wall of the vagina at the level of the pelvic sling

**40.** The pudendal nerve can be anesthetized before the delivery of an infant. This nerve is easily anesthetized where it is related to the

**(A)** ischial spine
**(B)** ischial tuberosity
**(C)** pubic tubercle
**(D)** sacral promontory
**(E)** symphysis pubis

**41.** After an injury in the perineum, fluid (blood, urine) can extravasate into the abdominal wall just deep to the membranous layer of superficial fascia. This can occur because of the continuity between this plane of the abdominal wall and the

**(A)** anal triangle
**(B)** anterior recess of the ischiorectal fossa
**(C)** deep perineal space
**(D)** posterior recess of the ischiorectal fossa
**(E)** superficial perineal space

**42.** In the female, the perineal body (central tendinous point) is located between the

**(A)** coccyx and the anus
**(B)** external urethral opening and the vaginal entrance
**(C)** symphysis pubis and the external urethral opening
**(D)** symphysis pubis and vaginal opening
**(E)** vaginal opening and the anus

**43.** The skin of the perineum is innervated primarily by

**(A)** branches of the pelvic splanchnic nerves
**(B)** branches of the posterior and lateral femoral cutaneous nerves
**(C)** the ilioinguinal and obturator nerves
**(D)** the pudendal and ilioinguinal nerves
**(E)** the pudendal and obturator nerves

**44.** The carotid sinus is located at the same level as the

**(A)** interval between the cricoid and thyroid cartilages
**(B)** interval between the thyroid prominence and hyoid bone
**(C)** lower border of the cricoid cartilage
**(D)** mastoid process
**(E)** prominence of the thyroid cartilage (Adam's apple)

**45.** The respective afferent and efferent limbs (nerves) of the gag reflex are the

**(A)** glossopharyngeal and hypoglossal
**(B)** glossopharyngeal and vagus
**(C)** trigeminal and glossopharyngeal
**(D)** trigeminal and vagus
**(E)** vagus and hypoglossal

**46.** After a severe blow to the side of the head, a patient's jaw deviates to the same side when he attempts to protrude his mandible or open his mouth against resistance. X-rays reveal a fracture across the base of the skull. The most likely location of the fracture is the

**(A)** frontal bone or anterior cranial fossa
**(B)** occipital bone or posterior cranial fossa
**(C)** petrous bone or ridge between the posterior and middle cranial fossae
**(D)** sphenoid bone or middle cranial fossa
**(E)** temporal bone or posterior cranial fossa

**47.** Which of the following muscles is essential to the process of mastication but is supplied by the facial nerve (CN VII)?

**(A)** Buccinator muscle
**(B)** Lateral pterygoid muscle
**(C)** Masseter muscle
**(D)** Medial pterygoid muscle
**(E)** Temporalis muscle

**48.** An angiogram reveals an aneurysm of the internal carotid artery just above where it emerges from the cavernous sinus. Based on the location of the following nerves, which would appear to be the most vulnerable?

**(A)** Abducens nerve
**(B)** Mandibular nerve
**(C)** Maxillary nerve
**(D)** Optic nerve (or chiasm)
**(E)** Trochlear nerve

**49.** While testing the eye movements of a patient, the physician determines that the right eye cannot be depressed when adducted. Such a finding would logically result from an injury involving the

**(A)** abducens nerve
**(B)** inferior division of the oculomotor nerve
**(C)** optic nerve
**(D)** superior division of the oculomotor nerve
**(E)** trochlear nerve

**50.** Nerve cell bodies mediating taste from the anterior two-thirds of the tongue are located in the

**(A)** geniculate ganglion
**(B)** inferior ganglion of the vagus nerve
**(C)** otic ganglion
**(D)** submandibular ganglion
**(E)** trigeminal (semilunar) ganglion

**51.** A lesion of the right facial nerve as it exits from the stylomastoid foramen would cause

**(A)** loss of taste on the right anterior tongue and loss of salivation from the right sublingual and submandibular glands
**(B)** paralysis of all of the muscles of facial expression on the right
**(C)** paralysis of all of the muscles of facial expression on the right except the buccinator
**(D)** paralysis of all of the muscles of facial expression on the right, loss of lacrimation on the right, and hyperacusis (increased sensitivity to sound) in the right ear
**(E)** paralysis of all of the muscles of facial expression on the right, loss of taste on the right anterior tongue, and loss of salivation from the right sublingual and submandibular glands

**52.** Because of the location of the ostium (opening) of the right maxillary sinus, the sinus is best drained

**(A)** in the prone position
**(B)** in the supine position
**(C)** in the upright position
**(D)** while lying on the left side
**(E)** while lying on the right side

**53.** Fertilization generally occurs in the

**(A)** ampulla of the uterine tube
**(B)** anterior wall of the body of the uterus
**(C)** internal os of the uterus
**(D)** ovary
**(E)** rectouterine cavity

**54.** The extraembryonic mesoderm that covers the yolk sac is termed the

**(A)** cytotrophoblast
**(B)** exocoelomic membrane
**(C)** somatopleuric mesoderm
**(D)** splanchnopleuric mesoderm
**(E)** syncytiotrophoblast

**55.** The neural tube is formed from the

**(A)** amniotic ectoderm
**(B)** intraembryonic paraxial mesoderm
**(C)** mesenchyme
**(D)** notochord
**(E)** surface ectoderm of the embryoblast

**56.** Which of the following is found only in tertiary chorionic villi?

**(A)** Central core of mesenchyme
**(B)** Cytotrophoblast cells
**(C)** Fetal capillaries
**(D)** Syncytiotrophoblast

**57.** Myoblasts are embryonic cells derived from

**(A)** dermatome
**(B)** epimere
**(C)** myotome
**(D)** sclerotome

58. During development of the heart, partitioning of the atria is the result of growth of the

(A) aorticopulmonary septum
(B) endocardial cushions
(C) septum primum
(D) septum secundum

59. The fourth left aortic arch is primarily responsible for the development of the

(A) aortic arch
(B) internal carotid artery
(C) pulmonary artery
(D) subclavian artery

60. The primary palate is formed by the fusion of the

(A) lateral nasal processes
(B) maxillary processes
(C) medial nasal processes
(D) medial nasal processes and the maxillary processes

61. The production of surfactant by type II pneumocytes begins during which of the following periods?

(A) Alveolar
(B) Canalicular
(C) Pseudoglandular
(D) Terminal sac

62. Which of the following statements about derivatives of the foregut is correct?

(A) They are connected to the anterior body wall by way of the ventral mesentery
(B) They include all parts of the gastrointestinal tract as far distal as the attachment of the yolk stalk
(C) They include the spleen
(D) They receive their blood supply from the superior mesenteric artery

63. The main duct of the pancreas is formed

(A) entirely from the duct of the dorsal pancreatic bud
(B) entirely from the duct of the ventral pancreatic bud
(C) proximally by the duct of the dorsal pancreatic bud and distally by the duct of the ventral pancreatic bud
(D) proximally by the duct of the ventral pancreatic bud and distally by the duct of the dorsal pancreatic bud

64. The organ that becomes secondarily retroperitoneal caused by rotation of the gastrointestinal tract during development is the

(A) liver
(B) pancreas
(C) sigmoid colon
(D) transverse colon

65. Separation of the cloaca into rectal and genitourinary passages depends on growth of the

(A) cloacal membrane
(B) metanephric blastema
(C) proctodeum
(D) urachus

**66.** In a cell with especially high energy [adenosine triphosphate (ATP)] requirements, which of the following organelles would you expect to be the most highly developed?

**(A)** Centrioles
**(B)** Mitochondria
**(C)** Peroxisomes
**(D)** Rough endoplasmic reticulum

**67.** Cells with large amounts of rough endoplasmic reticulum are most likely to

**(A)** line the lumen of blood vessels
**(B)** produce steroids
**(C)** produce structural proteins that remain in the cell
**(D)** synthesize a proteinaceous secretory product

**68.** Which statement about the nucleus is correct?

**(A)** Chromosomes and the nuclear envelop are most prominent during the metaphase stage of mitosis
**(B)** Chromosomes in areas of euchromatin are most likely less functionally active than in areas of heterochromatin
**(C)** The nuclear envelope consists of two membranes and is penetrated by pores to allow passage of material between the nucleus and the cytoplasm
**(D)** The nucleolar membrane separates the nucleolus from the chromosomes

**69.** Assume that a person has received sufficient exposure from x-rays to the whole body that cells are destroyed as they attempt to divide. In this person, which one of the following functions would survive best?

**(A)** Cardiac contraction
**(B)** Hair growth
**(C)** Intestinal absorption of fat
**(D)** Red blood corpuscle production

**70.** Which of the following statements about mitochondria is true?

**(A)** Mitochondria of pancreatic cells possess a preponderance of tubular cristae
**(B)** Pyruvate is converted to acetyl coenzyme A (acetyl-CoA) between the outer and inner membranes
**(C)** The electron transport system of enzymes is on the inner membrane of the cristae
**(D)** The energy that is produced is used to form adenosine diphosphate (ADP)

**71.** The Golgi apparatus is involved in which of the following functions?

**(A)** Adding the lipid component to the glycosaminoglycans of ground substance
**(B)** Packaging secretory products in membrane-bound vacuoles
**(C)** Producing proteins for intracellular use
**(D)** Synthesizing secretory products

**72.** Which statement regarding the free surface specializations of epithelial cells is correct?

**(A)** Cilia contain a core of nine peripheral and two central microfilaments
**(B)** Microvilli contain a core of microfilaments
**(C)** Microvilli insert into centrioles (basal bodies)
**(D)** Stereocilia contain a core of microtubules that inserts into centrioles (basal bodies)

**73.** The specific intercellular junctional mechanism through which cells are electrically coupled is the

**(A)** desmosome
**(B)** gap junction
**(C)** tight junction
**(D)** zonula adherens

**74.** Which of the following fibers or fibrils are most similar to collagen in chemical composition?

**(A)** Elastic fibers
**(B)** Muscle fibers
**(C)** Neurofibrils
**(D)** Reticular fibers

**75.** Tendons are composed of

**(A)** dense, irregularly arranged connective tissue
**(B)** dense, regularly arranged connective tissue
**(C)** large collagenous fibers with fibroblasts lying in lacunae between the fibers
**(D)** large elastic fibers with fibroblasts lying between the fibers

**76.** In which of the following organs would you be most likely to find elastic cartilage?

**(A)** Developing long bone
**(B)** Inner ear
**(C)** Larynx
**(D)** Lungs

**77.** The stiffness of cartilage is caused primarily by the presence of

**(A)** chondroitin sulfate
**(B)** collagen fibers
**(C)** hyaluronic acid
**(D)** the perichondrium

**78.** Which of the following statements about bone remodeling is true?

**(A)** It can occur in response to mechanical stress and fluctuations in the blood calcium level
**(B)** It involves deposition of calcified cartilage on existing trabeculae of bone
**(C)** It involves removal of existing bone by osteoblasts
**(D)** It occurs during the "growing years" but ceases by age 50

**79.** Which statement about muscle tissue is correct?

**(A)** Cardiac and smooth muscle tissue are more vascular than skeletal muscle tissue
**(B)** Cardiac and smooth muscle cells both branch and have central nuclei
**(C)** Cardiac and smooth muscle tissue both possess gap junctions
**(D)** Skeletal muscle satellite cells lie outside the basement membrane of the muscle and are connective tissue cells

**80.** Which of the following statements about neurons is true?

**(A)** They and glial cells in the gray matter of the spinal cord make up a meshwork called neuropil
**(B)** Those of the central nervous system are invested by myelin produced by Schwann cells
**(C)** Those of the central nervous system have Nissl material that extends into both dendrites and axons
**(D)** Those of sympathetic ganglia are unipolar or pseudounipolar

**81.** Which of the following statements about neuromuscular spindles is true?

**(A)** They are located in the myenteric plexus of the intestine
**(B)** They contain both afferent and efferent nerve fibers
**(C)** They contain intrafusal fibers
**(D)** They regulate the state of contraction of cardiac muscle

**82.** Which of the following cells is morphologically closest to the earlier blast stage?

**(A)** Neutrophilic metamyelocyte
**(B)** Platelet
**(C)** Promyelocyte
**(D)** Reticulocyte

**83.** Which of the following cell types is most numerous in a normal blood smear?

**(A)** Eosinophil
**(B)** Lymphocyte
**(C)** Monocyte
**(D)** Neutrophil

**84.** Which of the following best characterizes myelocytes?

**(A)** They are incapable of division
**(B)** They arise from metamyelocytes
**(C)** They have an indented nucleus
**(D)** They have specific granules

**85.** Which of the following statements is correct about the structure of the heart?

**(A)** Papillary muscles are involuntary smooth muscle tissue that inserts into chordae tendinea
**(B)** Purkinje fibers are modified nerve fibers constituting part of the cardiac conduction system
**(C)** The endocardium of the atria is thicker than that of the ventricles
**(D)** The sinoatrial node is modified connective tissue of the cardiac skeleton

**86.** Which of the following is supplied by afferent lymphatic vessels?

**(A)** Lymph nodes
**(B)** Peyer's patches
**(C)** Spleen
**(D)** Thymus

**87.** Which of the following statements about the liver is true?

**(A)** Blood flows from the central veins into the sinusoids
**(B)** Continuous capillaries line the hepatic sinusoids
**(C)** Hepatocytes project villous processes into the space of Disse
**(D)** Kupffer cells are mainly localized in the sinusoidal lumen

**88.** Which of the following statements about the kidney is correct?

**(A)** Antidiuretic hormone causes the collecting tubules to become more permeable to water, thus concentrating the urine
**(B)** As much as 80% of the amino acids and glucose of the ultrafiltrate are reabsorbed by the thin portion of the loop of Henle
**(C)** Interlobular arteries arise from arcuate arteries and pass into the cortex via the medullary rays
**(D)** Renin produced by the juxtaglomerular apparatus acts directly on arterial smooth muscle to cause vasodilation

**89.** During the differentiation of a spermatozoan (spermiogenesis), the acrosome arises by accumulation of material in the

**(A)** Golgi complex
**(B)** mitochondria
**(C)** nuclear envelope
**(D)** nucleus

**90.** Which of the following statements about the male reproductive system is true?

**(A)** Spermatozoa pass in sequence through efferent ductules, rete testis, and ductus epididymis
**(B)** The ductus deferens, seminal vesicle, and prostate all have smooth muscle in their walls
**(C)** The prostate gland is a site where spermatozoa are stored and become mature
**(D)** The seminal vesicle is the main source of acid phosphatase in the semen

**91.** Correct statements about the female reproductive system include which of the following?

**(A)** Ovulation occurs when progesterone has reached its highest level in the blood plasma
**(B)** The cytotrophoblast of the placenta is most prominent during the third trimester of pregnancy
**(C)** Theca externa cells of ovarian follicles secrete most of the ovarian estrogen
**(D)** Uterine glands secrete a carbohydrate-rich substance and also are necessary for regeneration of the surface epithelium of the uterus during the menstrual cycle

**92.** Which one of the following combinations of a structure-secretory product is correct?

**(A)** Acidophils of the pars distalis—follicle-stimulating hormone
**(B)** Beta cells of the islets of Langerhans—glucagon
**(C)** Syncytiotrophoblast—progesterone and estrogen
**(D)** Zona glomerulosa of the suprarenal gland—cortisone

**93.** Which of the following is characteristic of the twenty-seventh day of the menstrual cycle?

**(A)** Constriction of the coiled spiral arteries
**(B)** Increased edema of the functional spongy layer of the endometrium
**(C)** Increased proliferation of the endometrium
**(D)** Stasis of blood in the basal straight arteries of the endometrium

**94.** Which one of the following organs possesses all of the following characteristics: has both an exocrine and endocrine organ, is under the influence of hypophyseal hormones, and has cells completing the second meiotic division?

**(A)** Liver
**(B)** Ovary
**(C)** Pancreas
**(D)** Testis

**95.** Which one of the following organs possesses all of the following characteristics: serous acini, intercalated ducts, striated ducts, and without mucous alveoli?

**(A)** Pancreas
**(B)** Parotid gland
**(C)** Sebaceous gland
**(D)** Sublingual gland

**96.** Which one of the following organs possesses all of the following characteristics: stratified squamous nonkeratinized epithelium, serous and mucous glands, skeletal muscle, and special visceral afferent nerve fibers?

**(A)** Anal canal
**(B)** Esophagus
**(C)** Tongue
**(D)** Vagina

**97.** Three sites where substances readily pass between the vascular system and a surface lining epithelium are in the lung alveoli, renal corpuscles, and chorionic villi. Which of the following cells is found in all three structures?

**(A)** Angiotensin-producing cells
**(B)** Fenestrated endothelial cells
**(C)** Lining cells
**(D)** Macrophages

**98.** Which of the following statements about the ear is true?

**(A)** Bipolar neurons in the spiral cochlear ganglion each have a peripheral process that ends on hair cells
**(B)** Endolymph fills the scala tympani and scala vestibuli
**(C)** The organ of Corti contains hair cells, each of which possesses one true cilium
**(D)** The stapes is located in the round window of the scala tympani

**99.** Which of the following statements about the eye is true?

**(A)** Accommodation involves the relaxation of ciliary smooth muscle
**(B)** The blind spot produced by the optic disc is medial to the visual axis
**(C)** The pupil increases in diameter in response to parasympathetic nerve stimulation
**(D)** Visual pigments are located in discs of the outer segments of cells of the pigmented layer of the retina

**100.** Which of the following types of cells characteristically contain abundant smooth endoplasmic reticulum and mitochondria with tubular cristae?

**(A)** Cardiac muscle cells
**(B)** Cells of the zona fasciculata of the suprarenal gland
**(C)** Granulosa lutein cells
**(D)** Pancreatic acinar cells

**101.** Which of the following type of cell is considered to be phagocytic?

**(A)** Fibroblast
**(B)** Kupffer cell
**(C)** Lymphocyte
**(D)** Plasma cell

**102.** Which of the following statements regarding epithelia is correct?

**(A)** All cells of pseudostratified columnar epithelium reach the basement membrane, but not all reach the luminal surface
**(B)** Epithelial cells usually receive nutritive substances from capillaries located in the intercellular spaces of the epithelium
**(C)** Stratified squamous epithelium lines the interior of blood vessels
**(D)** Transitional epithelium is well-adapted for absorptive functions

**103.** The basement membrane consists of which of the following components?

**(A)** A basal lamina
**(B)** Large collagen bundles and reticular fibers
**(C)** Plasma membrane of basal epithelial cells and large collagen bundles
**(D)** Reticular fibers and a basal lamina

**104.** Which of the following components is found within the intercellular spaces between epidermal cells?

**(A)** Cilia
**(B)** Melanocytes
**(C)** Meissner's tactile corpuscles
**(D)** Tonofilaments

**105.** Which of the following is found in osteons (haversian systems)?

**(A)** Canaliculi, through which blood is transported to the osteocytes
**(B)** Collagenous fibers, which, in adjacent lamellae, run parallel to each other
**(C)** Interstitial lamellae
**(D)** Osteoprogenitor cells, which occupy the osteon (haversian) canal

**106.** Which of the following is characteristic of both hyaline cartilage and bone?

**(A)** Avascular tissue
**(B)** Cells occupy lacunae
**(C)** Grow by both appositional and interstitial growth
**(D)** The intercellular matrix contains chondroitin sulfate

**107.** Cardiac muscle differs from adult skeletal muscle in that cardiac muscle possesses

**(A)** intercalated disks
**(B)** parallel fibers
**(C)** peripherally located nuclei
**(D)** Z lines

**108.** Which of the following actions occurs during excitation-contraction coupling in skeletal muscle?

**(A)** Calcium is released from mitochondria
**(B)** The A band shortens as a result of the interaction between thick and thin filaments
**(C)** The interaction of actin and myosin is regulated, at least in part, by troponin and tropomyosin
**(D)** A wave of membrane depolarization is carried into the depths of the muscle fiber by sarcoplasmic reticulum

**109.** Glial (neuroglial) cells are components of which of the following structures?

**(A)** Adrenal (suprarenal) medulla
**(B)** Pars distalis of the hypophysis
**(C)** Pars intermedia of the hypophysis
**(D)** Pineal gland

**110.** A drug that interferes with mitosis would be likely to directly affect the division of

**(A)** metamyelocytes
**(B)** myelocytes
**(C)** normoblasts
**(D)** reticulocytes

**111.** Which of the following statements about the respiratory system is true?

**(A)** Bronchioles possess hyaline cartilage
**(B)** Great alveolar (giant septal, pneumocyte II) cells produce surfactant
**(C)** Respiratory bronchioles possess smooth muscle and are lined with simple squamous epithelium
**(D)** The true vocal cords (folds) and the rest of the larynx are lined with pseudostratified columnar epithelium

**112.** Which of the following organs possesses submucosal glands?

**(A)** Colon
**(B)** Esophagus
**(C)** Fundus of the stomach
**(D)** Ileum

**113.** Which of the following statements about the liver is correct?

**(A)** Bile canaliculi are lined by simple squamous epithelium
**(B)** Discontinuities in the sinusoidal epithelium permit passage of some substances between the lumen and the perisinusoidal space
**(C)** Hepatocytes possess only smooth endoplasmic reticulum
**(D)** In the classic liver lobule, a mixture of arterial and venous blood flows toward the periphery of the lobule

**114.** Listed below are endocrine tissues matched with mechanisms involved in the regulation of the production and/or release of the hormone produced by that tissue. Of the tissues listed, which is matched with the correct mechanism?

**(A)** Adrenal medulla—stimulation of postganglionic sympathetic neurons
**(B)** Corpus luteum—stimulation of the posterior pituitary
**(C)** Pars distalis of the hypophysis—stimulated by releasing hormones from hypothalamic neurons via the hypophyseal portal system
**(D)** Thyroid—low levels of calcium in the blood

**115.** Which of the following cell–secretory product combinations is correct?

**(A)** Enteroendocrine cell—gastric lipase
**(B)** Fibroblasts—collagen precursor
**(C)** Mast cell—circulating antibodies
**(D)** Plasma cell—heparin

**116.** Which of the following cell–secretory product combinations is correct?

**(A)** Acidophil of the pars distalis—thyrotropin
**(B)** Chief cell of the stomach—serotonin
**(C)** Parietal cell of the stomach—hydrochloric acid
**(D)** Zona glomerulosa cell of the adrenal gland—glucocorticoids

**117.** A simple cuboidal or columnar epithelium with extensive basal infoldings of the plasma membrane is characteristically found in the

**(A)** distal convoluted tubules of the kidney
**(B)** intercalated ducts of the parotid gland
**(C)** lining epithelium of the oral cavity
**(D)** lining epithelium of the small intestine

**118.** Which of the following statements about the kidney is correct?

**(A)** The adjacent pedicels in the glomerulus are separated by a slit
**(B)** The collecting tubules are located only in the medulla
**(C)** The distal convoluted tubules have an extensive microvillous (brush) border
**(D)** The juxtaglomerular apparatus is a region of the distal tubule

**119.** Which of the following cells is correctly paired with its secretory product?

**(A)** Acidophils of hypophysis—follicle-stimulating hormone
**(B)** Basophils of hypophysis—prolactin (luteotropic hormone)
**(C)** Chief cells—parathyroid hormone
**(D)** Chromaffin cells of adrenal—aldosterone

**120.** Which of the following pairs is a target cell or target organ for the hormone indicated?

**(A)** Leydig cells—testosterone
**(B)** Prostate gland—testosterone
**(C)** Seminal vesicle—luteinizing hormone
**(D)** Spermatids—follicle-stimulating hormone

**121.** Which of the following statements about the testis is correct?

**(A)** Secondary spermatocytes are usually plentiful because they are relatively slow to divide into spermatids
**(B)** Sertoli cells help control passage of macromolecules to diploid cells
**(C)** Some spermatogonia proliferate and remain as stem cells, whereas others differentiate into primary spermatocytes
**(D)** The rough endoplasmic reticulum of Leydig cells is essential for testosterone production

**122.** The luteal (secretory) phase of the menstrual cycle is characterized by

**(A)** the absence of a functional corpus luteum
**(B)** coiling or sacculation of endometrial glands
**(C)** proliferation of ovarian follicles
**(D)** relatively low amounts of progesterone in the blood plasma

**123.** Which of the following statements about the membranous labyrinth of the ear is true?

**(A)** Ampullae are receptors for position sense
**(B)** Efferent nerves end on some of the hair cells
**(C)** Otoconia are components of the cristae
**(D)** Perilymph flow during head rotation produces forces on sensory hairs

**124.** Which of the following statements about the eye is correct?

**(A)** Aqueous humour passes, in sequence, through the anterior chamber, posterior chamber, trabecular meshwork (spaces of Fontana), canal of Schlemm, and veins
**(B)** Light striking the retina (excluding the fovea centralis and the optic papilla) encounters in order: rods and cones, bipolar cells, and ganglion cells
**(C)** The inner nuclear layer of the retina contains the nuclei of rod and cone cells
**(D)** The pathway of visual impulses in the retina is from rods and cones, to bipolar cells, to ganglion cells

**125.** In the processes leading to protein synthesis, which of the following occurs?

**(A)** Messenger RNA (mRNA) molecules are transcribed from exposed nitrogenous bases of DNA
**(B)** Only ribosomal RNA (rRNA) is involved in translating the mRNA message
**(C)** The genetic code is found in the nitrogenous base sequence of RNA
**(D)** The transfer RNA (tRNA) places an amino acid directly into the rough endoplasmic reticulum

**126.** Two months after complete ipsilateral destruction of the ventral roots of spinal nerves C5–C7, you would expect to see

**(A)** exaggerated myotatic (stretch) reflexes in the ipsilateral upper limb
**(B)** flaccid paralysis of muscles in the ipsilateral upper limb
**(C)** loss of superficial (cutaneous) abdominal reflexes
**(D)** minimal change in the muscle mass of the ipsilateral upper limb
**(E)** muscular symptoms concentrated primarily in the distal aspect of the ipsilateral upper limb

**127.** The pain and temperature pathway originating from receptors of the trunk and extremities

**(A)** ascends in the posterior funiculus of the spinal cord
**(B)** crosses to the opposite side at the level of the closed medulla
**(C)** includes central processes of primary (first-order) neurons whose cell bodies are located in the substantia gelatinosa
**(D)** includes third-order neurons whose axons project to the precentral gyrus
**(E)** is farther from the medial lemniscus in the medulla than at higher levels of the brainstem

**128.** Examination of a patient reveals loss of two-point discrimination and vibratory sense at T8 and below on the right; no response to pinprick at T9 and below on the left; deep tendon reflexes present in both lower limbs; and paralysis of the right lower limb. Which of the following lesions is most consistent with these findings?

**(A)** Bilateral destruction of the posterior gray at T9
**(B)** Blockage of the anterior spinal artery at T8
**(C)** Complete transection of the spinal cord at T9
**(D)** Destruction of the anterior white commissure at T8
**(E)** Hemisection of the spinal cord at T8

**129.** Which of the following structures projects directly to lower motor neurons?

**(A)** Cerebellum
**(B)** Globus pallidus
**(C)** Precentral gyrus
**(D)** Putamen
**(E)** Ventral lateral nucleus of the thalamus

**130.** Descending projections from the hypothalamus would most likely terminate on the

**(A)** general somatic efferent cell column in the spinal cord
**(B)** general visceral efferent cell column in the spinal cord
**(C)** special somatic afferent cell column in the brainstem
**(D)** special visceral afferent cell column in the brainstem
**(E)** special visceral efferent cell column in the brainstem

**131.** Which of the following statements regarding the thalamus is correct?

**(A)** Axons from many of its nuclei project to the cerebral cortex via the internal capsule
**(B)** The anterior nucleus projects fibers to the prefrontal cortex
**(C)** The pulvinar receives sensory information from the vestibular nuclei
**(D)** The thalamus is positioned lateral to the anterior limb of the internal capsule
**(E)** The ventral anterior nucleus receives fibers from the geniculate nuclei

**132.** Most of the fibers in the superior cerebellar peduncle originate from cell bodies in the

**(A)** dentate nucleus
**(B)** fastigial nucleus
**(C)** inferior olivary nucleus
**(D)** nucleus cuneatus
**(E)** ventral lateral nucleus of the thalamus

**133.** A unilateral lesion of the neocerebellum would most likely result in

**(A)** an intention tremor
**(B)** astereognosis
**(C)** athetoid motion at rest
**(D)** increased resistance to passive motion
**(E)** unilateral flaccid paralysis of the upper and lower limbs

**134.** A unilateral lesion of the neocerebellum would most likely result in

**(A)** bilateral involvement of both the upper and lower limbs
**(B)** contralateral involvement of primarily the lower limb
**(C)** contralateral involvement of primarily the upper limb
**(D)** contralateral involvement of both the upper and lower limbs
**(E)** ipsilateral involvement of both the upper and lower limbs

**135.** Disruption of corticobulbar fibers would occur with a lesion affecting the

**(A)** anterior limb of the internal capsule
**(B)** genu of the internal capsule
**(C)** posterior limb of the internal capsule
**(D)** retrolenticular part of the internal capsule
**(E)** sublenticular part of the internal capsule

**136.** The major source of efferent fibers from the basal ganglia is the

**(A)** caudate nucleus
**(B)** globus pallidus
**(C)** putamen
**(D)** substantia nigra
**(E)** subthalamic nucleus

**137.** Which of the following is characteristic of a spinal cord injury 1 month after the occurrence of the injury?

**(A)** Equal weakness (proximal and distal) of musculature throughout the musculature of the limbs
**(B)** Hyperreflexia of both superficial and deep reflexes
**(C)** Hypotonia
**(D)** Negative Babinski response
**(E)** Spastic paralysis of the involved muscles

**138.** Which of the following statements about the posterior spinocerebellar tract is true?

**(A)** It contains axons that arise from cell bodies in the nucleus proprius
**(B)** It enters the cerebellum via the inferior cerebellar peduncle
**(C)** It is found in the posterior funiculus of the spinal cord
**(D)** It is present at all levels of the spinal cord
**(E)** It transmits information detected by pain and temperature receptors in the skin

**139.** Your patient exhibits a spastic paralysis of the right upper and lower limbs and is unable to detect a pinprick in either the left upper or lower limb. The most likely site of a single lesion is the

**(A)** base of the midbrain on the left
**(B)** base of the pons on the left
**(C)** lateral funiculus of the spinal cord at the C2 level on the right
**(D)** left medullary pyramid
**(E)** posterior limb of the left posterior capsule

**140.** You have a patient whose uvula deviates to the left when he talks. The most likely location of a lesion is the

**(A)** medulla
**(B)** midbrain
**(C)** pons
**(D)** thalamus
**(E)** upper cervical spinal cord

**141.** The integrity of the nucleus ambiguus can be evaluated by testing the

**(A)** corneal reflex
**(B)** gag reflex
**(C)** jaw jerk reflex
**(D)** pupillary light reflex
**(E)** purse the lips

**142.** A patient has difficulty recognizing objects placed in the right hand, but she can feel both the texture and shape of the objects. The most likely location of a single lesion is the

**(A)** left medial lemniscus in the pons
**(B)** left parietal lobe unisensory association areas
**(C)** left postcentral gyrus
**(D)** right fasciculus cuneatus in the lower thoracic spinal cord
**(E)** right fasciculus cuneatus in the upper cervical spinal cord

**143.** Damage of the left motor cortex that involves corticobulbar neurons to the motor nucleus of the facial nerve would cause

**(A)** an inability to smile symmetrically
**(B)** deviation of the tongue to the right when it is protruded
**(C)** difficulty closing the right eye
**(D)** inability to clench the teeth
**(E)** ptosis of the right eye

**DIRECTIONS: Each of the numbered items or incomplete statements in this section is negatively phrased, as indicated by a capitalized word such as EXCEPT, LEAST, or NOT. Select the ONE lettered answer or completion that is best in each case.**

**144.** The pancreas contains each of the following EXCEPT

**(A)** ducts that demonstrate the presence of centroacinar cells
**(B)** exocrine glandular secretory end-pieces that are basophilic
**(C)** endocrine pancreatic islets
**(D)** myoepithelial cells

**145.** Infections in the pharynx can travel though open communications to each of the following areas EXCEPT the

**(A)** auditory tube
**(B)** inner ear
**(C)** mastoid air cells
**(D)** piriform recess
**(E)** tympanic cavity

**146.** Each of the following statements about the plasma membrane is true EXCEPT

**(A)** it contains glycolipids and glycoproteins
**(B)** it is coated by a glycocalyx on both its outer and inner surfaces
**(C)** it possesses a double layer of phospholipid molecules whose fatty acid components make up an intermediate hydrophobic zone
**(D)** it possesses transmembrane proteins that transport specific molecules into the cell

**147.** Each of the following statements about cytoplasmic organelles is true EXCEPT

**(A)** centrioles are similar in structure to basal bodies
**(B)** free ribosomes are mostly involved in the production of secretory proteins
**(C)** intermediate filaments are plentiful in epidermal cells
**(D)** vesicles coated with clathrin are involved in receptor-mediated endocytosis

**148.** Which of the following is NOT used in classifying the various types of epithelia?

**(A)** The number of layers of cells
**(B)** The amount of intercellular material relative to the cellular content
**(C)** The shape of the cells at only the free surface
**(D)** The terminal specialization or modification at the free surface

**149.** Which of the following hormones is NOT produced by a neuron?

**(A)** Antidiuretic hormone (ADH)
**(B)** Calcitonin
**(C)** Epinephrine
**(D)** Luteinizing hormone–releasing hormone (LHRH)

**150.** Which one of the following does NOT transport blood through its sinusoids?

**(A)** Bone marrow
**(B)** Liver
**(C)** Lymph node
**(D)** Spleen

**151.** Which of the following organs is LEAST likely to demonstrate mucus-secreting cells or glands?

**(A)** Cervix
**(B)** Colon
**(C)** Esophagus
**(D)** Vagina

**152.** Each of the following is a component of the mucosa EXCEPT

**(A)** enteroendocrine cells of the stomach
**(B)** lamina propria
**(C)** muscularis externa
**(D)** muscularis mucosa

**153.** Which of the following statements about the small intestine is NOT correct?

**(A)** Chylomicrons are often found in the intercellular spaces between absorptive cells
**(B)** Monoglycerides and fatty acids diffuse through the absorptive cell plasma membrane
**(C)** The nodules of Peyer's patches may extend into the submucosal layer
**(D)** The outer longitudinal layer of the muscularis externa is thickened into taenia coli

**154.** After the administration of radioactive glucose, which of the following structures or substances would NOT be labeled?

**(A)** Colloid in the thyroid follicle
**(B)** Glycocalyx covering the free surface of the intestinal lining cell
**(C)** Hyaline cartilage matrix
**(D)** Secretory product in plasma cells

**DIRECTIONS: Each set of matching questions in this section consists of a list of 4 to 15 options (some of which may be figures) followed by several numbered items. For each numbered item, select the ONE lettered option that is most closely associated. Each lettered option may be selected once, more than once, or not at all.**

*Questions 155–157*

**(A)** Anesthesia in the skin of the point of the shoulder
**(B)** Burning sensation in the dorsal aspect of the first web space (between thumb and index fingers)
**(C)** Difficulty identifying objects placed between the thumb, index, and middle fingers
**(D)** Inability to feel a pin prick in the hypothenar skin (medial part of the palm)
**(E)** Loss of sensation in the skin of the medial forearm
**(F)** Loss of sensation on the lateral aspect of the forearm
**(G)** Loss of sensibility in the skin over the insertion of the deltoid muscle

For each nerve, select the most likely sensory loss or disturbance that occurs if the structure is injured.

**155.** Superficial branch of the radial nerve

**156.** Median nerve in the carpal tunnel

**157.** Ventral ramus of spinal nerve C4

*Questions 158–161*

**(A)** Greatly reduced range of motion of opposition of the thumb
**(B)** Greatly weakened extension of the forearm
**(C)** Inability to adduct and abduct the four medial digits
**(D)** Inability to extend the thumb
**(E)** Loss of pronation of the forearm
**(F)** Loss of supination of the forearm
**(G)** Reduced biceps brachii reflex

For each nerve or nerve trunk indicated below, select the most likely muscular loss or disturbance that occurs if the structure is injured.

**158.** C5 spinal nerve

**159.** Ulnar nerve in the forearm

**160.** Radial nerve at the elbow

**161.** Median nerve in the carpal tunnel

*Questions 162 and 163*

**(A)** A lag in forward momentum at the push-off point in gait
**(B)** A noticeable foot slap at the heel-strike point in gait
**(C)** A noticeable lateral shift of the trunk, to the weight-bearing side, at heal strike and throughout stance
**(D)** A noticeable posterior shift of the trunk at the point of heel-strike
**(E)** At heel-strike, knee extension is maintained by extension of the thigh at the point of impact

For each of the muscular deficits indicated below, match the resulting change in gait.

**162.** Anterior muscles in anterior compartment of the leg

**163.** Weakness of the gluteus medius and minimus muscles

*Questions 164–166*

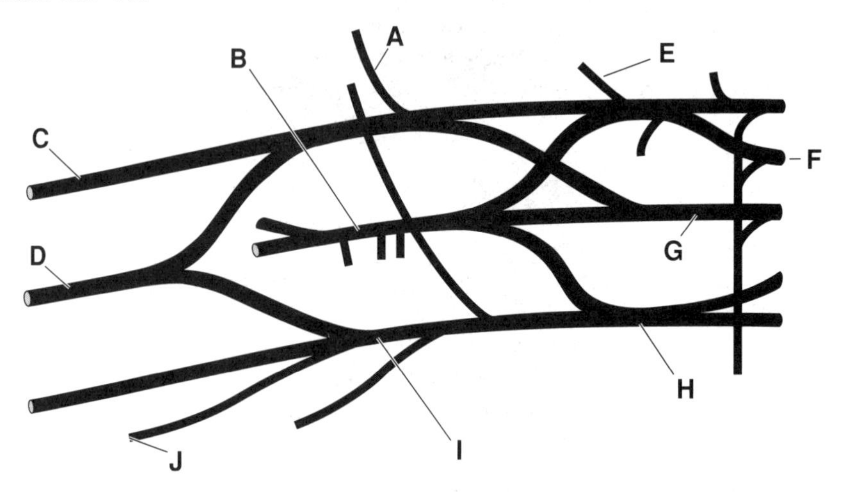


The illustration shown above is of the brachial plexus. Match each component below with its appropriate site (letter) on the figure.

**164.** Posterior cord

**165.** Suprascapular nerve

**166.** Ulnar nerve

*Questions 167–169*

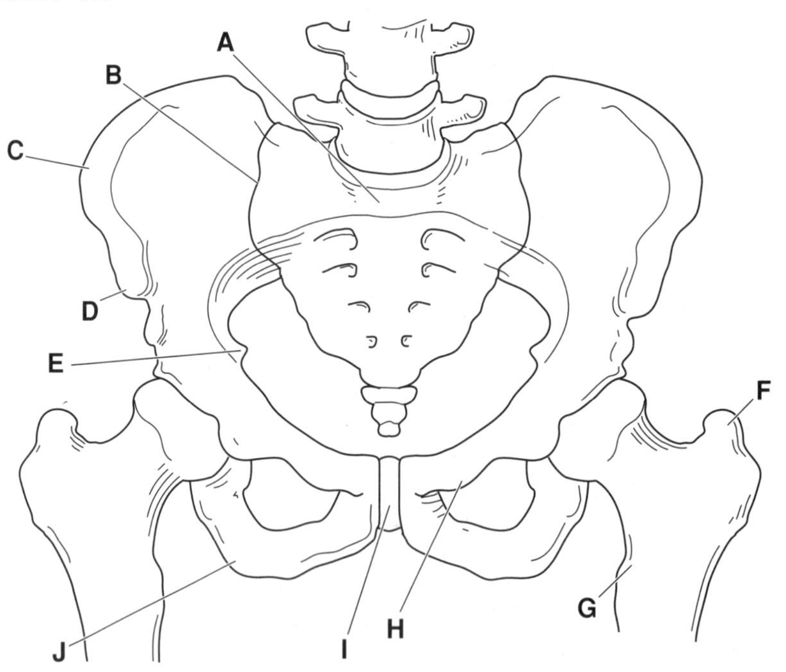


The illustration above shows the anterolateral view of the pelvis, lumbosacral spine, and proximal aspects of the femurs. Match each structure listed below with the appropriate site (letter) on the figure.

**167.** Sacroiliac joint

**168.** Ischial spine

**169.** Anterior superior iliac spine

*Questions 170–172*

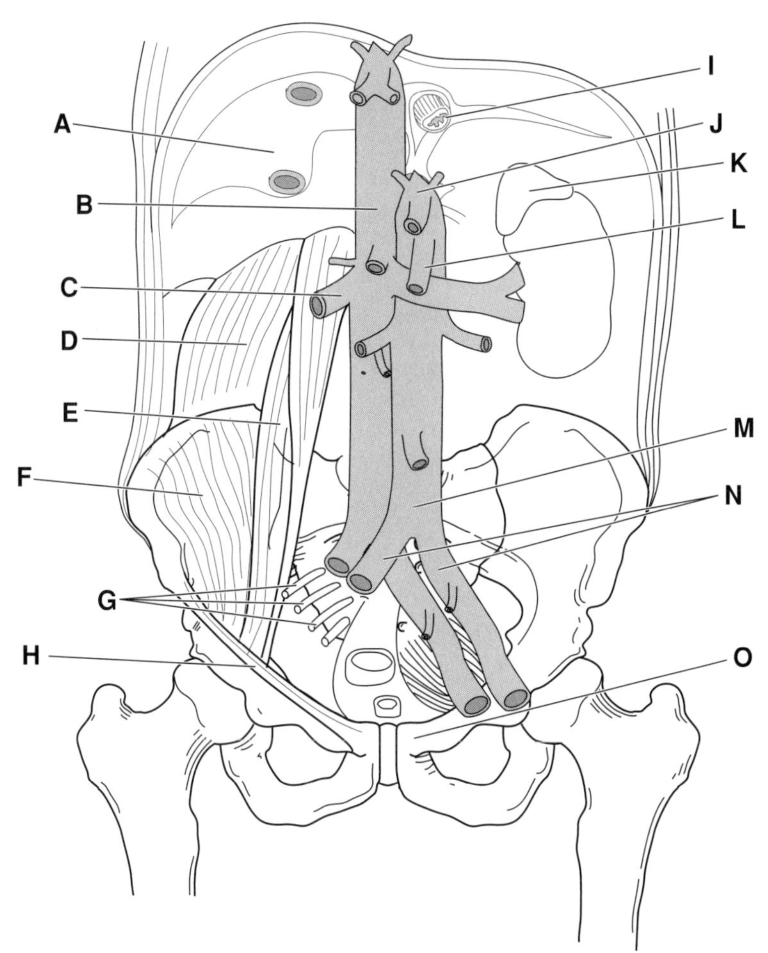


The above figure shows the posterior body wall of the abdomen and certain of the viscera and neurovascular structures. Match each structure listed below with the correct letter on the figure.

**170.** Superior mesenteric artery

**171.** Psoas major muscle

**172.** Central tendon of the respiratory diaphragm

*Questions 173–175*

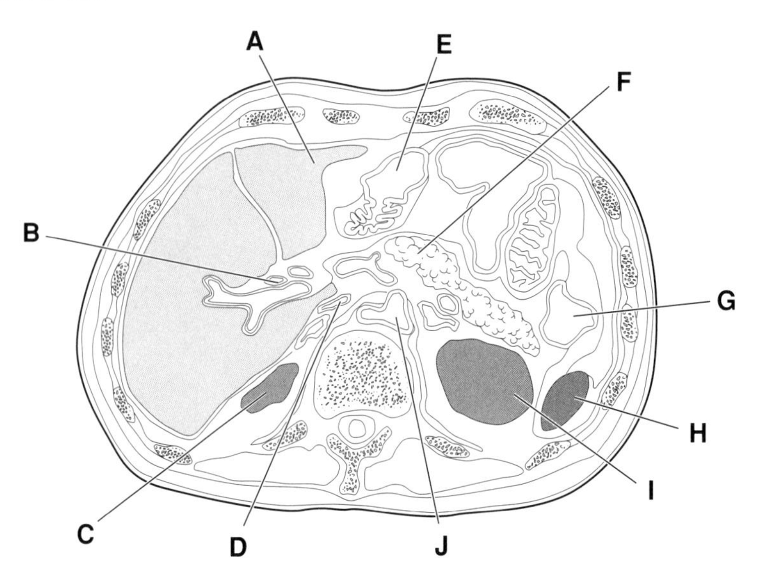


The figure above shows a cross section through the abdomen at vertebral level L1. It is oriented in the same way as a magnetic resonance image or computed tomogram. Match each structure listed below with the correct letter on the figure.

**173.** Abdominal aorta

**174.** Pancreas

**175.** Spleen

*Questions 176–178*

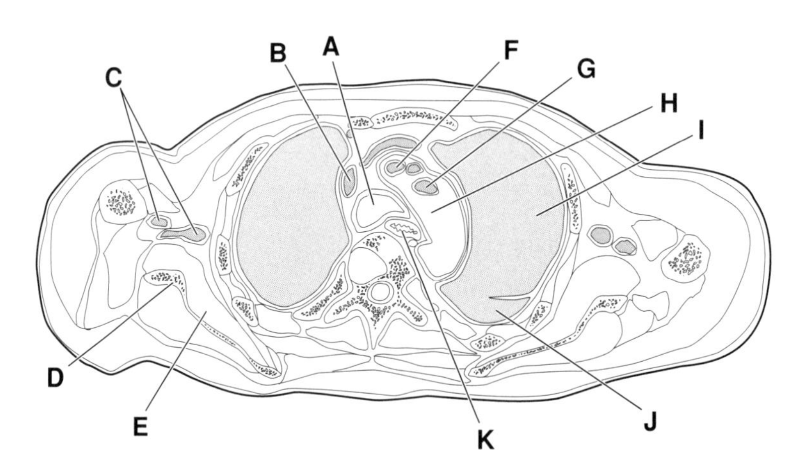


The figure shown above is a cross section through the thorax at the junction of the superior and inferior parts of the mediastinum. Identify each of the specific structures listed below with the correct letter on the figure.

**176.** Superior lobe, left lung

**177.** Brachiocephalic artery

**178.** Subscapularis muscle

*Questions 179–181*

**(A)** Carotid canal
**(B)** Foramen magnum
**(C)** Foramen ovale
**(D)** Foramen rotundum
**(E)** Foramen spinosum
**(F)** Hypoglossal canal
**(G)** Internal auditory meatus
**(H)** Jugular foramen
**(I)** Optic canal
**(J)** Superior orbital fissure

For each cranial nerve listed below, identify the opening in the base of the skull through which it passes.

**179.** Mandibular division of the trigeminal nerve

**180.** Abducens nerve

**181.** Facial nerve

*Questions 182–185*

**(A)** Neural crest
**(B)** Primitive node (Hensen)
**(C)** Primitive streak
**(D)** Third pharyngeal pouch
**(E)** Yolk sac entoderm

For each structure below, select the embryologic structure that most likely gives rise to it.

**182.** Sarcomere

**183.** Primordial sex cell

**184.** Notochord

**185.** Cardiac muscle

*Questions 186–189*

**(A)** Mesonephric duct
**(B)** Mesonephros
**(C)** Paramesonephric duct
**(D)** Urogenital sinus

For each structure below, select the embryologic structure that most likely gives rise to it.

**186.** Ovary

**187.** Vagina

**188.** Uterus

**189.** Ductus deferens

*Questions 190–193*

(**A**) Coated vesicle
(**B**) Lysosome
(**C**) Mitochondria
(**D**) Rough endoplasmic reticulum
(**E**) Smooth endoplasmic reticulum

For each cell function listed below, select the most likely organelle associated with that function.

**190.** Endocytosis

**191.** Phagocytosis

**192.** Steroid production

**193.** Generation of ATP

*Questions 194–197*

(**A**) Cortisol
(**B**) Follicle-stimulating hormone
(**C**) Glucagon
(**D**) Growth hormone
(**E**) Sodium- and bicarbonate-rich fluid

Match each cell below with the product it produces.

**194.** Adrenal zona fasciculata cell

**195.** Acidophils of the adenohypophysis

**196.** Alpha cell of pancreas

**197.** Striated duct

*Questions 198–202*

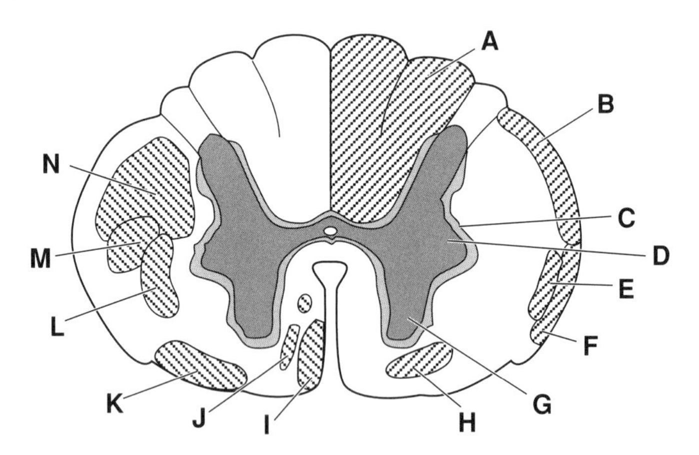


The locations of several spinal cord structures are indicated in the above illustration. Indicate the letter that depicts the location of each structure described below.

**198.** Concentration of interneurons

**199.** Location of cell bodies of preganglionic sympathetic neurons

**200.** Cell bodies of alpha and gamma motor neurons to axial musculature

**201.** A lesion in this area would most likely produce astereognosis

**202.** Lateral corticospinal tract

*Questions 203–206*

**(A)** Inferior cerebellar peduncle
**(B)** Ipsilateral posterior spinocerebellar tract
**(C)** Posterior limb of the internal capsule
**(D)** Superior cerebellar peduncle
**(E)** That part of the contralateral medial lemniscus containing fibers from cranial levels
**(F)** That part of the contralateral medial lemniscus containing fibers from sacral levels

Match each nucleus below with the correct location (tract, pathway, or fiber bundle) for the fibers whose cell bodies are located in that nucleus.

**203.** Nucleus dorsalis (Clarke)

**204.** Nucleus cuneatus

**205.** Ventral posteromedial and ventral posterolateral nuclei of the thalamus

**206.** Accessory cuneate nucleus

*Questions 207–211*

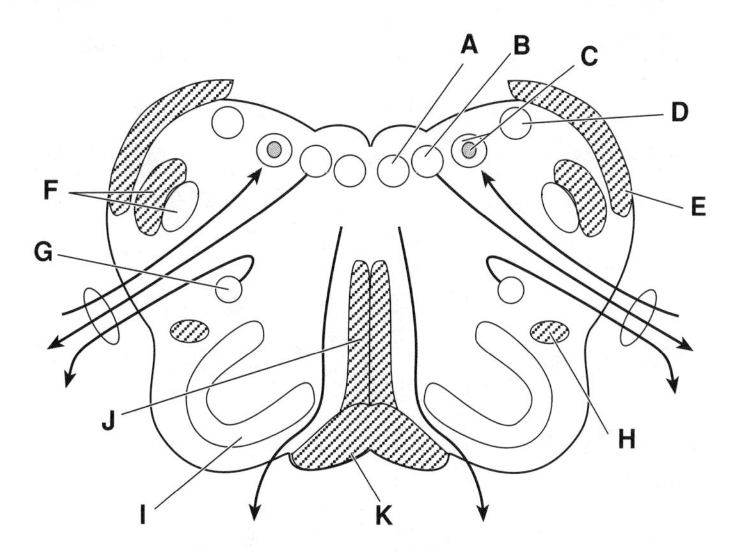


The locations of several medullary structures are indicated in the above figure of the high or open medulla. For each question below, indicate the letter identifying the location of each structure or the location of the lesion that causes the symptom described.

**207.** Medial lemniscus

**208.** Deviation of the tongue on protrusion

**209.** Inferior cerebellar peduncle

**210.** Unilateral paralysis of the larynx

**211.** Pyramid

*Questions 212–215*

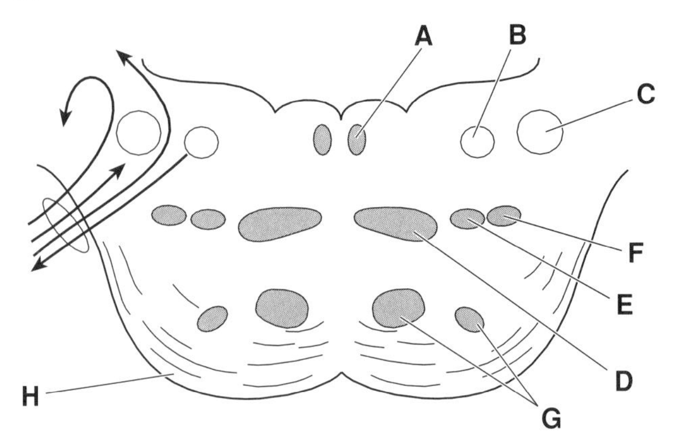


The locations of several pontine structures are indicated in the above illustration of the midpons. For each question below, indicate the letter identifying the location of each structure or the location of the lesion that causes the symptom described.

**212.** Corticospinal and corticobulbar fibers

**213.** Fibers from the cerebral cortex to the cerebellar cortex

**214.** Ipsilateral loss of general sensation on the face

**215.** Contralateral loss of two-point discrimination and vibratory sense

*Questions 216–219*

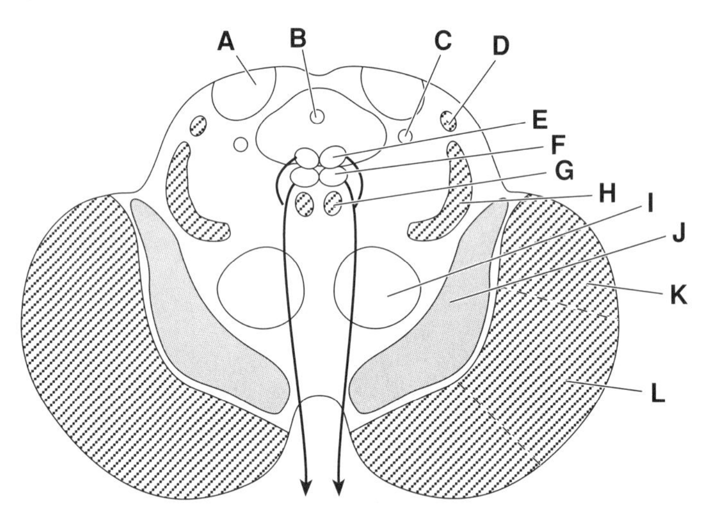


The locations of several midbrain structures are indicated in the above illustration of the upper midbrain. For each question below, indicate the letter identifying the location of the structure or the location of the lesion that causes the symptom described.

**216.** Functionally related to the basal ganglia

**217.** Cerebral aqueduct

**218.** Efferent fibers from cerebellar nuclei terminate here; other neurons then project to the cerebral cortex

**219.** Ipsilateral lateral strabismus

# ANATOMICAL SCIENCES ANSWERS AND DISCUSSION

**1—C (Chapter 2)** The cubital fossa is located anterior to the elbow and is bound by the pronator teres medially, the brachioradialis laterally, and a line interconnecting the medial and lateral humeral epicondyles proximally. The structures in this fossa, from lateral to medial, are the tendon of the biceps brachii muscle, brachial artery, and median nerve.

**2—C (Chapter 2)** The radial artery descends through the forearm in the anterior compartment. As it approaches the wrist, it is palpable just lateral to the tendon of the flexor carpi radialis muscle. The artery then crosses the wrist by passing laterally and dorsally, deep to the tendons of both the abductor pollicis longus and extensor pollicis brevis and into the anatomical snuff box.

**3—E (Chapter 2)** The axillary nerve and posterior circumflex humeral vessels curve horizontally around the posterior aspect of the humerus at the level of its surgical neck. The brachial artery is the continuation of the axillary artery, which is located medial to the humerus. The cephalic vein is a superficial vein, so it is separated from the humerus by considerable soft tissue, and the circumflex scapular artery curves posteriorly around the scapula.

**4—A (Chapter 2)** The annular ligament attaches to the edges of the radial notch of the ulna and surrounds the head of the radius; thus, it is ideally positioned to stabilize the proximal radioulnar joint. Of the other choices listed (i.e., biceps brachii muscle, pronator teres muscle, radial collateral ligament, or ulnar collateral ligament), only the pronator teres muscle interconnects the radius and ulna, but it is obliquely oriented and therefore unable to provide as much support as the annular ligament.

**5—D (Chapter 2)** Even though the muscles of the rotator cuff are located only on three sides (anterior, superior, and posterior) of the shoulder joint, they are the major supports of the joint. These muscles are supplied by spinal cord segments C5 and C6; they play no role in stabilizing the scapula, and they extend from the humerus to the scapula.

**6—B (Chapter 2)** The superficial palmar arch is located in the central compartment of the palm and formed primarily by the superficial branch of the ulnar artery; the superficial palmar branch of the radial artery usually completes the arch.

**7—E (Chapter 2)** The carpal tunnel is the connection between the anterior forearm and the palm. It is formed by the carpal bones medially, laterally, and dorsally, and the transverse carpal ligament (flexor retinaculum) anteriorly. It contains only the tendons of the long digital and thumb flexor muscles as well as the median nerve.

**8—D (Chapter 2)** The area just lateral to the acromion is occupied by the deltoid and supraspinatus muscles. The deltoid is superficial and its belly is located in that area. The supraspinatus is the deeper muscle; because it attaches to the greater tubercle of the humerus, it is tendinous just lateral to the acromion. In addition, it commonly degenerates in exactly that spot so it is the most likely structure to cause pain.

**9—B (Chapter 2)** Opposition is produced by the muscles in the thenar compartment, which are supplied by the recurrent branch of the median nerve. This branch is in a superficial position midway between the first metacarpophalangeal joint and the pisiform.

**10—B (Chapter 3)** The gluteus medius and minimus muscles interconnect the lateral aspect of the wing of the ilium and the greater trochanter of the femur, with the minimus positioned deep to the medius. They are rather large and form the majority of the soft tissue bulk in that area.

**11—D (Chapter 3)** The capsule of the hip joint is reinforced by three very strong ligaments (iliofemoral, ischiofemoral, and pubofemoral) that blend with the fibrous part of the capsule and provide the most support when the femur is extended. The capsule extends distally along the entire neck anteriorly, but only about two-thirds posteriorly. It attaches to the edge of the articular surface of the acetabulum, but to the trochanteric region (anteriorly) and neck (posteriorly) of the femur distally.

**12—B (Chapter 3)** The muscles in the anterior compartment of the thigh, which includes the quadriceps femoris, are supplied by the femoral nerve. Although the nerve contains fibers from L2 and L4, the primary segment is typically L3.

**13—B (Chapter 3)** From its attachment to the anterior intercondylar area of the tibia, the anterior cruciate ligament passes superiorly, posteriorly, and laterally to attach to the medial aspect of the lateral femoral condyle. This ligament prevents posterior displacement of the femur on the tibia (anterior displacement of the tibia on the femur).

**14—C (Chapter 3)** The posterior tibial artery enters the foot by passing approximately 2 cm posterior to the medial malleolus along with the tibial nerve and tendons of the tibialis posterior, flexor hallucis longus, and flexor digitorum longus.

**15—A (Chapter 3)** Plantar flexion at the ankle and inversion of the foot place stress on the lateral collateral ligament of the ankle, particularly the anterior talofibular ligament. The other ligaments are on the plantar or medial foot so that maneuver reduces their stress; the tibialis posterior tendon passes medial and posterior to the ankle, so it also is placed on slack.

**16—C (Chapter 3)** The bones of the foot are arranged into two longitudinal arches and one transverse arch. As a result, during quiet standing, weight is transferred to the floor posteriorly through the tuberosity of the calcaneus and anteriorly through the heads of the metatarsals.

**17—C (Chapter 1)** The ligamentum is a segmental ligament that interconnects adjacent lamina. With the lamina, it forms the posterior aspect of the vertebral canal. It is unusual in that it is composed of elastic cartilage. It is positioned medial to the zygapophyseal joints but does extend laterally, so it blends with the capsules of these joints and forms the medial aspect of the posterior wall of the intervertebral foramen.

**18—D (Chapter 1)** The intervertebral foramen is located posterior to the intervertebral disk. The vertical dimension of the foramen is greater than that of the disk at all levels, so the anterior boundary is completed by portions of the vertebral bodies both above and below the disk.

**19—E (Chapter 1)** The dorsal (primary) rami are branches of the spinal nerve, and they contain both motor and sensory fibers. These rami do not form a plexus; thus, they supply structures in a segmental manner. Specifically, they supply the deep muscles and medial skin of the back as well as structures of the vertebral column.

**20—C (Chapter 6)** The floor and anterior wall of the inguinal canal are formed by the aponeurosis of the external abdominal oblique muscle. The conjoined tendon is formed by the inferomedial aspects of the internal oblique and transversus abdominis muscles; it and the transversalis fascia form the posterior wall of the canal. The internal spermatic fascia is also derived from the transversalis fascia, and the deep inguinal ring is the beginning of that tubular sleeve of fascia.

**21—A (Chapter 6)** The descending colon is partially peritonealized in that it is covered on three sides with peritoneum, yet it has no mesentery. The ileum and transverse colon are (completely) peritonealized; the pancreas and kidney are retroperitoneal.

**22—D (Chapter 6)** The posterior aspect of the second part of the duodenum is related to the hilus of the right kidney, the renal vessels, and the ureter. The aorta is posterior to the third part of the duodenum; the superior mesenteric artery is anterior. The gallbladder is anterior to the first part of the duodenum. The right quadratus lumborum muscle is not related to the duodenum.

**23—B (Chapter 6)** Even though the taeniae coli are responsible for the formation of the haustra or sacculations of the colon, only the haustra can be identified when the colon is filled with a contrast medium. The epiploic appendices are characteristic of the colon but cannot be seen on a radiograph. The circular folds and vascular arcades are characteristics of the small intestine.

**24—C (Chapter 6)** The stomach is oriented obliquely. Its most posterior aspect, the fundus, is positioned posterolaterally to the left; from that point, the body and pylorus extend anteriorly and to the right. The lesser curvature is the superior border of the stomach, and the cardiac orifice is its communication with the esophagus.

**25—A (Chapter 6)** The fundus of the gallbladder is its most inferior part and the only part that projects beyond the inferior border of the liver. It is palpable at the tip of the ninth costal cartilage along the subcostal margin, which is where the midclavicular line intersects that margin.

**26—C (Chapter 6)** The head of the pancreas occupies the concavity formed by the first three parts of the duodenum, so the right border of the head is in contact with the second part of the duodenum. The spleen is related to its tail, and the splenic artery passes along its posterior surface. Neither the ascending colon nor left lobe of the liver is adjacent to the pancreas.

**27—A (Chapter 6)** The lymphatic drainage of the abdomen corresponds to the arterial supply. As a result, most of the lymph from the stomach drains through the celiac lymph nodes.

**28—E (Chapter 6)** The innervation to the viscera of the abdominopelvic cavity is provided by the aortic plexus; nerve fibers from its various subdivisions (superior mesenteric and so on) are distributed in periarterial plexus that follow the branches of the aorta. As a result, the junction of the large and small intestines is supplied by the superior mesenteric plexus.

**29—D (Chapter 6)** The parasympathetic innervation of the descending and sigmoid portions of the colon does not follow the pattern of innervation of most abdominal viscera. Rather than passing through periarterial plexus, these fibers pass directly to the viscera. They are derived from spinal cord segments S2, S3, and S4. They branch from those spinal nerves as pelvic splanchnic nerves and pass retroperitoneally directly to the descending and sigmoid portions of the colon.

**30—D (Chapter 5)** The surface landmarks that define the boundaries of the middle lobe of the right lung are oblique and horizontal fissures. The oblique fissure crosses the sixth rib in the midclavicular line, the fifth rib in the midaxillary line, and ends at the fourth rib in the scapular line. The horizontal fissure begins at the level of the fourth costal cartilage in the midclavicular line and ends where it meets the oblique fissure, which is the level of the fifth rib in the midaxillary line.

**31—A (Chapter 5)** The location and shape of the superior aspect of the lung or apex and the superior aspect of the pleura correspond closely. They both extend superiorly approximately 2.5 cm above the medial third of the clavicle.

**32—B (Chapter 5)** The costodiaphragmatic recess of the pleural cavity is the area of the pleural cavity inferior to the inferior border of the lung during quiet breathing. Projected to the surface, this recess is between the sixth and eighth ribs (intercostal spaces S6 and S7) in the midclavicular line; eighth and tenth ribs (intercostal spaces S8 and S9) in the midaxillary line; and the tenth and twelfth ribs (intercostal spaces S10 and S11) in the scapular line.

**33—A (Chapter 5)** The posterior surface or base of the heart consists primarily of the left atrium. It also is formed by the proximal aspects of the great vessels and part of the right atrium. This surface is related posteriorly to the esophagus, aorta, and thoracic duct.

**34—E (Chapter 5)** The majority of the diaphragmatic surface of the right and left ventricles is typically supplied by the posterior interventricular artery. The right coronary artery curves from right to left around the heart in the posterior part of the atrioventricular (coronary) sulcus before dividing into a small branch, which continues in that sulcus, and the posterior interventricular, which turns into the posterior interventricular sulcus and passes toward the apex of the heart.

**35—D (Chapter 5)** The fibrous pericardium encloses the heart and blends with the adventitia of the great vessels posteriorly and superiorly and the central tendon of the diaphragm inferiorly. It is lined by the parietal layer of serous pericardium, which reflects onto the heart as the visceral layer of serous pericardium. The reflection of visceral over the veins is in the form of an inverted U and is called the oblique sinus.

**36—E (Chapter 5)** The azygos vein is formed in the abdomen and ascends through the thorax to the right of the vertebral bodies (and thoracic duct) before passing posterior to the root of the right lung and arching anteriorly to join the superior vena cava. It drains most of the thoracic wall but no viscera.

**37—C (Chapter 5)** The left recurrent laryngeal nerve recurs around the arch of the aorta, and the right recurrent laryngeal nerve recurs around the right subclavian artery. The arch passes posteriorly and to the left and is connected to the left pulmonary artery. The thoracic duct is between the esophagus and vertebral bodies; the left brachiocephalic vein passes anterior to the brachiocephalic artery.

**38—B (Chapter 7)** The plane of least dimensions of the pelvis is the plane of the midpelvis. This plane extends from the inferior aspect of the symphysis pubis to the sacrum at the level of the ischial spines. The lateral diameter of this plane, between the ischial spines, is the smallest diameter of the pelvis.

**39—D (Chapter 7)** In the pelvis, the peritoneum covers the posterosuperior aspect of the uterus and continues posteriorly onto the anterior aspect of the rectum. Where it reflects from the uterus to the rectum, it forms a depression, the rectouterine pouch, which is separated from the posterior fornix of the vagina by only the thin muscular wall of the vagina and the peritoneum.

**40—A (Chapter 7)** The pudendal nerve branches from the sacral plexus in the pelvis. It passes from the pelvis to the perineum by curving around the posterior aspect of the pelvic diaphragm, which attaches posteriorly to the ischial spine.

**41—E (Chapter 7)** The membranous layer of subcutaneous tissue (Scarpa's fascia) in the abdominal wall is continuous with Colles' fascia of the perineum. Colles' fascia attaches to the posterior aspect of the urogenital diaphragm and thus is the inferior boundary of the superficial peroneal space. This fascial continuity makes the superficial perineal space and the facial plane deep to Scarpa's fascia continuous.

**42—E (Chapter 7, Fig. 7-1)** The central tendinous point is located in the midline at the junction of the urogenital and anal triangles. This muscular and tendinous mass is formed by the junction of the superficial transverse peroneus, external anal sphincter, and bulbocavernosus muscles. In the female, this point is positioned between the anus and vaginal opening.

**43—D (Chapter 7)** The major nerve of the perineum is the pudendal nerve. In addition, the anterior skin is supplied by the anterior labial (scrotal) nerves, which are branches of the ilioinguinal nerve.

**44—B (Chapter 4)** The carotid sinus is at the level of the bifurcation of the common carotid artery, which is between vertebral levels C3 and C4. The hyoid bone is at the C3 level and the thyroid prominence is at C4. The cricoid cartilage is at the C6 level.

**45—B (Chapter 4)** The majority of the pharyngeal mucosa is supplied by CN IX (glossopharyngeal); CN X (vagus) supplies its constrictor muscles. As a result, these two nerves are the sensory and motor limbs, respectively, of the gag reflex.

**46—D (Chapter 4)** The mandibular division of the trigeminal nerve supplies the muscles of mastication, which are responsible for the movement of the mandible. Because the x-ray indicates a fracture across the base of the skull and the symptoms implicate the mandibular nerve, the most likely location of the fracture is the middle cranial fossa (largely formed by the greater wing of the sphenoid), where the mandibular nerve exits the skull by passing through the foramen ovale.

**47—A (Chapter 4)** The buccinator muscle is supplied by the facial nerve and has an important masticatory function in that it ensures that food is kept between the teeth so it can be masticated. The medial and lateral pterygoid, masseter, and temporalis muscles are supplied by the mandibular division of the trigeminal nerve and are considered muscles of mastication because they control the mandible.

**48—D (Chapter 4)** The optic nerve (or chiasm) is immediately adjacent to the internal carotid, where it emerges from the cavernous sinus. The internal carotid artery emerges from the cavernous sinus at the anterior clinoid process, in the central part of the middle cranial fossa. The abducens and trochlear nerves pierce the dura in the posterior cranial fossa, and the mandibular and maxillary nerves are more lateral in the middle cranial fossa.

**49—E (Chapter 4)** Depression of the adducted eye is the function of the superior oblique muscle. This muscle is supplied by the trochlear nerve.

**50—A (Chapters 4 and 36)** The taste fibers to the anterior two-thirds of the tongue are conveyed to the tongue in the lingual nerve, which is a branch of the mandibular nerve. These fibers originate from the brainstem in the facial nerve and pass from the facial nerve to the lingual nerve via the chorda tympani nerve. Their cell bodies are located in the geniculate ganglion, which is the major sensory ganglion of the facial nerve.

**51—B (Chapter 4)** A lesion of the facial nerve as it exits from the brainstem would cause all of the symptoms listed in this question; however, when it exits from the facial canal via the stylomastoid foramen, it contains only those fibers that supply the muscles of facial expression, which includes the buccinator muscle. The fibers involved in the other functions listed branch from the nerve along its course in the temporal bone.

**52—D (Chapter 4)** The opening of the maxillary sinus is located on its medial wall well above the floor of the sinus. As a result of this location, it is necessary to lie on the opposite side (left) to ensure that the opening is the lowest part of the sinus.

**53—A (Chapter 9)** The ovum is released from the ovary and passes into the uterine tube. Fertilization generally occurs in the ampulla of the uterine tube. When the blastocyst reaches the uterine cavity, the optimal place of implantation is in the posterior wall of the body of the uterus. If the ovum is fertilized before it is carried into the uterine tube, it may implant in the rectouterine cavity, where it is known as an ectopic pregnancy.

**54—D (Chapter 9)** The layer of mesoderm that is associated with the endoderm is the splanchnopleuric mesoderm. The extraembryonic mesoderm splits to form the extraembryonic coelom. The layer of mesoderm that becomes associated with the ectoderm is called the somatopleuric mesoderm. The cytotrophoblast and the syncytiotrophoblast are the two layers of the trophoblast that contribute to the formation of the chorion and the extraembryonic membranes.

**55—A (Chapter 9)** Amniotic ectoderm, although connected to the surface ectoderm of the embryo, contributes only to the amniotic cavity. The intraembryonic paraxial mesoderm will give rise to the somites that form musculoskeletal structures. Mesenchyme is a general term for the loose tissue that condenses to form many of the deep body tissues. During development, the notochord forms as a midline rod that induces the overlying surface ectoderm to begin differentiation into the neural tube.

**56—C (Chapter 9)** Chorionic villi go through three stages of development in the placenta. The primary villi consist of a core of cytotrophoblast cells surrounded by a covering of syncytiotrophoblast. As the central core of the villus becomes invaded with mesenchyme, it becomes a secondary villus. Finally, when the mesenchyme gives rise to fetal capillaries in the villus, it is a tertiary or functional villus.

**57—C (Chapter 10)** The paraxial mesoderm of the somite differentiates into the sclerotome, which will form skeletal components; the dermatome, which will contribute to the formation of the dermis of the skin; and the myotome, which will give rise to primitive muscle cells, the myoblasts. The dorsal myoblasts become the epimere.

**58—D (Chapter 12)** Although the atria are initially separated from each other by the growth of septum primum, before the separation is complete, a foramen develops in the septum. This opening is called the foramen secundum. As the foramen secundum forms, a second septum grows from the atrial wall to cover the foramen secundum. This is called the septum secundum. It forms a one-way shutter valve over the foramen secundum. At birth, this shutter valve closes and the septum secundum fuses with the septum primum to partition the right and left atria from each other. The aorticopulmonary septum separates the aorta and the pulmonary artery from each other. The endocardial cushions form part of the cardiac skeleton.

**59—A (Chapter 12)** The fourth left aortic arch forms the arch of the aorta; the right aortic arch contributes to the development of the proximal part of the right subclavian artery. The pulmonary artery is largely derived from the sixth aortic arch. The internal carotid artery is *not* derived from any aortic arch.

**60—C (Chapter 15)** The primary palate, which is the anterior part of the hard palate, forms from the fusion of the medial nasal processes. The secondary palate develops from the fusion of the maxillary processes.

**61—D (Chapter 13)** Development of the lungs is divided into four phases. The first phase is the pseudoglandular period; during this time, the lung tissue resembles a gland. The second phase, from 16 to 25 weeks, is the canalicular period, during which primitive alveoli develop. The third phase is the terminal sac period, 24 weeks to birth, when type II pneumocytes develop and begin to secrete surfactant. The final phase, the alveolar period, is from birth to about 8 years, and is the time of development of mature alveoli.

**62—A (Chapter 14)** During development of the gastrointestinal tract, the primitive gut tube is divided into foregut, midgut, and hindgut. Although the ventral mesentery is present initially along the entire gut, it degenerates from the midgut and hind gut, persisting in the foregut region only. As a consequence, foregut derivatives are attached to the anterior body wall by way of remnants of the ventral mesentery. The spleen is not an outgrowth of the foregut but rather develops in the mesenchyme of the dorsal mesogastrium. The yolk sac is attached to the midgut, which receives its blood supply from the superior mesenteric artery. The celiac artery supplies the foregut.

**63—D (Chapter 14)** The pancreas develops by fusion of dorsal and ventral pancreatic buds. The ventral pancreatic bud contributes the proximal part of the pancreatic duct, and the dorsal pancreatic bud contributes the distal part. Occasionally the proximal part of the dorsal pancreatic duct remains attached to the duodenum as an accessory pancreatic duct.

**64—B (Chapter 14)** Initially, all parts of the gastrointestinal tract and its associated glands are intraperitoneal, suspended from the posterior body wall by way of the dorsal mesentery. As the abdominal contents rotate and become more crowded in the abdominal cavity, several parts of the gastrointestinal tract and the pancreas lose their mesentery and become retroperitoneal. The sigmoid and transverse colons remain intraperitoneal.

**65—D (Chapter 16)** Initially, the urogenital sinus and the hindgut open into a common area called the cloaca. The growth of the urorectal septum partitions the cloaca into a urogenital sinus, which contributes to the development of parts of the urinary system and the hindgut, which will develop into the rectum. The urachus is an embryonic structure that contributes to the wall of the urinary bladder.

**66—B (Chapter 19)** Cells that have high energy requirements must produce abundant amounts of adenosine triphosphate (ATP). Because ATP production is the primary activity of the mitochondrion, mitochondria are found in great numbers. Centrioles are needed for cell division and action of cilia. Rough endoplasmic reticulum is necessary for the production of secretory proteins. Peroxisomes are small membrane-bound vesicles containing oxidative enzymes.

**67—D (Chapter 20)** Secretory proteins are synthesized in the rough endoplasmic reticulum. They are further processed in the Golgi complex, where they are packaged for secretion. Proteins for intracellular use are produced on polyribosomes and free ribosomes. Cells producing steroid substances have abundant smooth endoplasmic reticulum.

**68—C (Chapter 19)** The nucleus is enclosed in a nuclear membrane similar in structure to the plasma lemma of the cell. The nuclear membrane possesses pores for the passage of material from the nucleus to the cytoplasm of the cell. Euchromatin is seen in areas of the nucleus where the DNA is actively transcribed and is electron lucent in electron micrographs. The areas of the nucleus that are more darkly stained are the areas of heterochromatin, which is inactive.

**69—A (Chapter 21)** Because radiation has its most potent effect on mitotic cells, the cardiac cells, which are postmitotic, would be *least* affected. The question describes a person who has received sufficient x-ray exposure to the entire body, causing cells to be destroyed as they attempt to divide.

**70—C (Chapter 22)** Mitochondria are the power generators for the cell and as such contain the enzyme systems for generating ATP in the Krebs cycle. Most of the ATP is generated through a system of electron transport enzymes that are localized in the mitochondrial matrix.

**71—B (Chapter 20)** Secretory proteins that are synthesized in the rough endoplasmic reticulum are transported to the Golgi complex, where they are concentrated and packaged in secretory vesicles. Carbohydrate moieties may be added in the Golgi complex. Polyribosomes produce proteins for intracellular use.

**72—B (Chapter 19)** Cilia are composed of nine sets of microtubules surrounding a central pair of microtubules. These insert into centrioles, which constitute the basal bodies that contribute to the movement of the cilia. Microvilli are absorptive modifications of the epithelial cell surface and contain a core of microfilaments. Stereocilia are very long microvilli and have a similar structure and are nonmotile.

**73—B (Chapter 19)** The gap junction represents a fusion of adjacent cell membranes with the presence of ionic channels that allow the cells to communicate with each other. The zonular adherens, also known as an occluding junction, represents a fusion of contiguous epithelial cells, thus providing a mechanism for preventing the passage of extracellular material between cells. Desmosomes and macula adherens provide similar, although less tight, attachments.

**74—D (Chapter 22)** Reticular fibers are connective tissue fibers, small in diameter, and their chemical composition is similar to collagen. Elastic fibers are composed of the protein elastin, and muscle fibers are made up of contractile proteins, actin, and myosin. Neurofibrils are protein structures found in neurons and represent aggregation of neurofilaments and neurotubules, cytoskeletal elements of the neuron.

**75—B (Chapter 22)** Tendons are classified as dense, regular connective tissue and consist largely of parallel arranged bundles of collagen fibers with intervening fibroblasts. Tendons tend to be inelastic and therefore contain no significant amount of elastic fibers.

**76—C (Chapter 22)** Some of the upper respiratory passages contain hyaline cartilage as a skeletal element. Developing bone is also composed of hyaline cartilage, whereas the inner ear is embedded in bony tissue. The larynx, which needs strength plus resilience, contains elastic cartilage.

**77—A (Chapter 22)** Chondroitin sulfate is a major matrix component of cartilage. It contributes stiffness to the tissue without making it brittle.

**78—A (Chapter 22)** Bone remodeling is a lifelong process that involves the removal of bone by osteoclasts. New bone is generally formed on the inner surfaces of osteons. Calcified cartilage is present only during the development of endochondral bone formation. Increased levels of blood calcium or stress are factors that contribute to bone remodeling.

**79—C (Chapter 22)** Gap junctions are a component of the intercalated disks of the heart and are also found between contiguous smooth muscle cells. Both skeletal and cardiac muscle have high energy requirements and are capable of quick contractions. Smooth muscle is capable of sustained or tonic contractions. Cardiac muscle cells and smooth muscle cells in the wall of the gastrointestinal tract must contract in concert with each other. As a result, these cells communicate with each other via gap junctions.

**80—A (Chapter 22)** The neuropil of the central nervous system (CNS) consists of the many processes of glial cells, among which are the oligodendrocytes that produce myelin on the axons of the CNS. Neurons of the autonomic ganglia are multipolar. Pseudounipolar neurons are characteristic of the sensory ganglia. Nissl substance is formed by rough endoplasmic reticulum and is found in the cell body and dendrites of neurons, but not in the axons.

**81—B (Chapter 22)** Neuromuscular spindles are specialized sensory organs that regulate the tonus of skeletal muscles and contribute to controlling contraction. They consist of afferent endings attached to small intrafusal muscle fibers. The majority of the muscle fibers that make up a skeletal muscle are the larger extrafusal muscle fibers.

**82—C (Chapter 24)** Myeloblasts differentiate into promyelocytes. The platelet is derived as a fragment of the mature megakaryocyte. The reticulocyte is the last developmental stage before the mature erythrocyte. The neutrophilic metamyelocyte is a late stage of the granulocytic cell series.

**83—D (Chapter 24)** In a normal human blood smear, the most numerous white blood cells are the neutrophils. The second most numerous are lymphocytes and monocytes. Eosinophils are fewer in number, and basophils are the least numerous.

**84—D (Chapter 24)** Myelocytes are cells that have "committed" to becoming specific granulocytes. Initially they possess nonspecific granules, but as they develop, they become more granulation-specific. They give rise to the metamyelocytes, which have an indented nucleus and are amitotic.

**85—C (Chapter 22)** The innermost layer of the heart is the endocardium, which is thicker in the atria than in the ventricles. The middle or muscular layer is the myocardium, which is thicker in the ventricles than in the atria. The outer layer is the epicardium, which includes the visceral pericardium. Purkinje cells are located in the endocardium and consist of modified cardiac muscle cells, which are part of the cardiac conduction system. The sinoatrial node is also a part of the cardiac conduction system and similarly consists of modified cardiac muscle. Papillary muscles attach to the valve cusps via the chordae tendinea. These are made of cardiac muscle as well.

**86—A (Chapter 26)** Although the thymus, Peyer's patches, and the spleen are all considered lymphoid organs, only the lymph node has a direct position along the lymphatic vessels and as such serves as a "way station" for the flow of lymph fluid. Afferent lymph vessels enter each lymph node on its surface. Efferent lymph vessels leave each lymph node at its hilum.

**87—C (Chapter 27)** The liver has sinusoidal capillaries that are lined by fenestrated epithelium. The space of Disse is outside the endothelial cells and contains macrophages called Kupffer cells and cytoplasmic processes of the hepatocytes. Blood enters the liver through the hepatic artery and portal vein and mixes in the sinusoids. From the sinusoids, the blood is collected into central veins for drainage from the liver.

**88—C (Chapter 29)** Intralobular arteries arise from the arcuate arteries. Eighty percent of glomerular filtrate is resorbed in the proximal tubule. Renin must activate angiotensin to influence smooth muscle.

**89—A (Chapter 32)** During spermiogenesis, or differentiation of a spermatozoan, the acrosome contains enzymes that allow the spermatozoa to penetrate the zona pellucid and ovum cell membrane. They are packaged in the Golgi complex of the spermatid.

**90—A (Chapter 32)** Spermatozoa that are produced in the seminiferous tubules of the testis are stored and mature in the epididymis. They reach the epididymis by passing through the rete testis and then the efferent ductules. During ejaculation, the sperm are propelled up the ductus deferens by the smooth muscle in its wall and mixed with secretions from the seminal vesical and prostate gland. The seminal vesical, which also has smooth muscle in its wall, contributes fructose to the seminal fluid.

**91—D (Chapter 31)** In the female reproductive system, uterine glands secrete a carbohydrate-rich substance and also are necessary for regeneration of the surface epithelium of the uterus during the menstrual cycle.

**92—C (Chapter 31)** The syncytiotrophoblast secretes estrogen and progesterone and helps to maintain the pregnancy. Glucagon is produced in the alpha cells of the pancreatic islets. Follicle-stimulating hormone is produced by basophils. The zona glomerulosa secretes aldosterone.

**93—A (Chapter 31)** On the twenty-seventh day of the menstrual cycle, the coiled arteries constrict, thereby cutting off blood supply to the functional layer, resulting in necrosis and sloughing off of the endometrium.

**94—D (Chapter 31)** Although both the ovary and the testis are endocrine and exocrine in their functions, the cells of the ovary complete their second meiotic division after they are shed from the ovary. The pancreas is also both an endocrine and exocrine organ, but none of its cells undergoes meiosis. The liver is an exocrine gland.

**95—B (Chapter 27)** Only the parotid gland fits all the criteria listed in the question (i.e., possesses serous acini, intercalated ducts, striated ducts, and has no mucous alveoli). The sebaceous gland, which functions by holocrine secretion, has a simple duct system. The pancreas does not possess striated ducts. The sublingual gland has both serous and mucous acini.

**96—C (Chapter 27)** The tongue possesses all of the characteristics listed in the question (i.e., stratified squamous nonkeratinized epithelium, serous and mucous glands, skeletal muscle, and special visceral afferent nerve fibers). The esophagus possesses all of these characteristics except the special visceral nerve fibers, which convey the sensation of taste, and the presence of serous glands. The anal canal has skeletal muscle and mucous glands as well as nonkeratinized stratified squamous epithelium. The vagina has neither glands nor special visceral afferent nerve fibers.

**97—D (Chapter 24)** Macrophages are commonly a feature of regions where materials can pass from the outside easily into the body.

**98—A (Chapter 33)** Bipolar neurons in the spiral cochlear ganglion each have a peripheral process that ends on hair cells and a central process that is a component of the cochlear nerve. Perilymph fills the scala tympani and scala vestibuli. The stapes is located in the oval window.

**99—B (Chapter 33)** The blind spot produced by the optic disc is medial to the visual axis. Accommodation results in contraction of the ciliary muscle. The pupil increases in diameter because of sympathetic stimulation. Visual pigments are located in the rods and cones.

**100—B (Chapter 19)** Cells of the zona fasciculata of the suprarenal gland that produce steroid hormones typically contain abundant amounts of smooth endoplasmic reticulum and mitochondria with tubular cristae. Cells producing protein secretory products tend to have large accumulations of rough endoplasmic reticulum.

**101—B (Chapter 27)** Kupffer cells are phagocytic cells located in the space of Disse in the liver. They serve a phagocytic function and ingest particulate matter that circulates through the liver. Lymphocytes and plasma cells are components of the lymphatic system and generally are not phagocytic. Fibroblasts produce collagen in connective tissue.

**102—A (Chapter 22)** Simple squamous epithelium lines blood vessels and is referred to as endothelium. Transitional epithelium lines the bladder and parts of the urinary system and is of variable thickness, depending on the state of distention of the organ. Columnar epithelium is generally absorptive. A limitation to the thickness of epithelial layers is the fact that epithelium is avascular and the cells must derive nutrition by diffusion.

**103—D (Chapter 19)** The basement membrane is a structure that has been observed in light microscopy and is associated with the basal surface of epithelia. It can be colored with the periodic acid–Schiff reaction. At the electron micrograph level, the basement membrane can be resolved into two substructures: the basal lamina, believed to be secreted by the epithelial cells, and some reticular fibers, believed to be produced by the underlying connective tissue cells.

**104—B (Chapter 22)** Melanocytes, which provide pigmentation to the skin, have cytoplasmic processes interposed between epithelial cells. There is very little extracellular space between contiguous epithelia cells. Laterally, epithelial cells may be united by a variety of junctions, including desmosomes, an important component of which is the intracellular presence of tonofilaments. Cilia are important specializations of the surface of epithelial cells. Meissner's corpuscles are sensory endings found in the skin and located in the papillary layer of the dermis, deep to the epithelium.

**105—D (Chapter 22)** Osteons consist of concentric lamellae of collagenous tissue arranged perpendicular to each other. Interstitial lamellae are found between adjacent osteons. Within the osteon, blood vessels are limited to the canal of the osteon (haversian canal), which also contains osteoprogenitor cells. Osteocytes in the more peripheral layers of the osteon communicate with and gain nutrients from cells in the inner lamellae by way of cytoplasmic processes contained within minute canaliculi.

**106—B (Chapter 22)** The cells of both bone and cartilage occupy spaces called lacunae. Cartilage is an avascular tissue; bone is not. Cartilage may grow by both interstitial and appositional growth, whereas bone grows only by appositional growth. The matrix of cartilage contains chondroitin sulfate.

**107—A (Chapter 22)** Although both cardiac muscle and adult skeletal muscle are striated, they differ in several respects. Skeletal muscle fibers are elongated and contain many peripherally placed nuclei. Individual cardiac muscle cells are branched and contain a single, centrally placed nucleus. In addition, cardiac muscle cells are electrically connected by way of intercalated disks, which contain gap junctions. Z lines in both types of muscle anchor contractile proteins.

**108—C (Chapter 22)** During excitation contraction coupling in skeletal muscle, the T tubules carry the excitation deep into the muscle fiber, producing a rapid release of calcium ions from the sarcoplasmic reticulum. Calcium binds to troponin molecules on the actin filaments, changing the configuration of the molecule and exposing myosin binding sites on the actin; thus, actin and myosin bind and contraction occurs, shortening the I band.

**109—D (Chapter 30)** The pineal gland develops as a dorsal outgrowth of the diencephalon and can be considered to be an extension of the central nervous system. It does contain neural elements and glial cells in the form of astrocytes. Both the pars intermedia and the pars distalis of the hypophysis consist of glandular components derived from Rathke's pouch and therefore are devoid of neural elements. Although the adrenal medulla is neural in its function, neuroglial cells are not present.

**110—B (Chapter 24)** As cells in the developing blood cell line become more differentiated, they tend to loose their ability to divide. The earliest cells, such as the myelocytes, would be most affected. Reticulocytes, which represent very late stages in the development of red blood cells, have no nuclei.

**111—B (Chapter 28)** The true vocal cords are covered by stratified squamous epithelium, whereas the false vocal cords possess epithelium typical of the upper respiratory tract (pseudostratified ciliated columnar). Bronchioles contain no cartilage in their walls, but rather have a layer of spirally arranged smooth muscle cells. Respiratory bronchioles have a lining of simple cuboidal epithelium. Within the alveoli, the type I pneumocytes contribute to the wall of the alveolus; the type II pneumocytes produce surfactant.

**112—B (Chapter 27)** Submucosal glands are a feature of the esophagus and duodenum. In other parts of the gastrointestinal tract, the glands are limited to the mucosa.

**113—D (Chapter 27)** In the hepatic lobule, a mixture of venous and arterial blood flows toward the center of the lobule. Bile canaliculi are lined by hepatocytes, which synthesize both proteins and steroids and thereby possess both smooth and rough endoplasmic reticulum. In the liver, the discontinuities in the endothelial cells of the sinusoids permit large molecular weight substances to come into contact with hepatocytes in the space of Disse. Plasma cells produce antibodies that are proteins.

**114—C (Chapter 30)** Releasing hormones reach the pars distalis from hypothalamic neurons by way of the hypophyseal portal system and therefore regulate its secretory activity. Parafollicular cells in the thyroid respond to high levels of calcium in the blood. Cells in the adrenal medulla are the equivalent of postganglionic sympathetic neurons and therefore respond to preganglionic sympathetic neurons. The corpus luteum of pregnancy is maintained by hormones released initially by the anterior pituitary and then by the developing placenta.

**115—B (Chapter 26)** Fibroblasts secrete a collagen precursor that assembles into collagen in the extracellular space. Enteroendocrine cells elaborate hormones that are secreted into the bloodstream and affect digestive activity locally. Gastric lipase is a digestive enzyme secreted by chief cells in the gastric glands. Mast cells secrete heparin. Plasma cells secrete circulating antibodies.

**116—C (Chapter 27)** Parietal cells of the stomach produce hydrochloric acid. Acidophils of the pars distalis produce growth hormone and prolactin. Chief cells of the stomach secrete digestive enzymes such as pepsinogen; serotonin is secreted by enteroendocrine cells. Thyrotropin is secreted by basophils. Zona glomerulosa cells of the adrenal gland secrete mineralocorticoids. Glucocorticoids are secreted by the zona fasciculata.

**117—A (Chapter 27)** The epithelium of the distal convoluted tubules of the kidney is modified for active ion transport and therefore demonstrates basal infoldings of the plasma lemma. Intercalated ducts of the parotid gland possess low cuboidal epithelium, although intercalated ducts possess basal infoldings because they are involved in ion transport. The lining epithelium of the oral cavity is stratified squamous epithelium. The lining epithelium of the small intestine is simple columnar tissue.

**118—A (Chapter 29)** Adjacent pedicels in the glomerulus are separated by a slit bridged by a very thin slit membrane. The macula densa is a region of the distal tubule that is closely associated with the juxtaglomerular apparatus, a region of the afferent glomerular arteriole. Collecting tubules are located in both the medulla and the cortex. Proximal convoluted tubules have an extensive microvillous (brush) border.

**119—C (Chapter 30)** Chief cells of the parathyroid secrete parathyroid hormone. Acidophils secrete prolactin (luteotropic hormone). Basophils of hypophysis secrete follicle-stimulating hormone. Aldosterone is secreted by the cells of the zona glomerulosa of the adrenal.

**120—B (Chapter 32)** The prostate gland responds to testosterone as does the seminal vesicle. Leydig cells produce testosterone and are influenced by luteinizing hormone from the anterior pituitary. Follicle-stimulating hormone promotes the growth of the follicles in the ovary and the production of spermatocytes in the seminiferous tubules.

**121—C (Chapter 32)** Some spermatogonia proliferate and remain as stem cells, whereas others differentiate into primary spermatocytes. Because the Leydig cells are responsible for testosterone production, a steroid hormone, they have abundant smooth endoplasmic reticulum. Sertoli cells help control passage of macromolecules to haploid cells. Secondary spermatocytes are usually difficult to observe, because they are relatively quick to divide into spermatids.

**122—B (Chapter 31)** The luteal (secretory) phase of the menstrual cycle is characterized by a coiling or sacculation of endometrial glands, relatively high amounts of progesterone in the blood plasma, the presence of a functional corpus luteum, and atresia of some ovarian follicles.

**123—B (Chapter 33)** Efferent nerves end on some of the hair cells, thereby sharpening the afferent messages. In the membranous labyrinth of the ear, maculae are receptors for position sense and contain minute concretions called otoconia. Perilymph flow during head rotation produces forces on crista ampullaris that cause them to fire nerve impulses.

**124—D (Chapter 33)** The pathway of visual impulses in the retina is from rods and cones, to bipolar cells, to ganglion cells. Aqueous humor passes, in sequence, through the posterior chamber, anterior chamber, trabecular meshwork (spaces of Fontana), canal of Schlemm, and veins. Light striking the retina (excluding the fovea centralis and the optic papilla) encounters, in sequence, ganglion cells, bipolar cells, and rods and cones. The inner nuclear layer of the retina contains the nuclei of bipolar cells, horizontal cells, and amacrine cells.

**125—A (Chapter 20)** Messenger RNA (mRNA) molecules are transcribed from exposed nitrogenous bases of DNA. The genetic code is found in the nitrogenous base sequence of DNA. Both ribosomal RNA (rRNA) and transfer RNA (tRNA) are involved in translating the mRNA message. The tRNA places an amino acid directly into the appropriate place within the protein molecule.

**126—B (Chapter 41)** The symptomology would be concentrated primarily in the proximal aspect of the ipsilateral upper limb. Generally, the upper plexus supplies proximal aspects of the limb, and the lower plexus supplies distal aspects of the limb. Lower motor neurons are damaged so the stretch reflex would be diminished or lost, muscle atrophy would occur, and the paralysis would be flaccid. The abdominal wall is supplied by branches of lower thoracic and upper lumbar spinal nerves.

**127—E (Chapter 35)** The lateral-spinothalamic tract is next to the medial lemniscus at pons and mesencephalon levels, but in the postolivary sulcus in the medulla. First-order neuron cell bodies are in the dorsal root ganglia, second-order neurons cross in the spinal cord, and third-order neurons project to the postcentral gyrus and parietal lobe immediately above the lateral fissure. This system passes through the lateral funiculus of the spinal cord.

**128—E (Chapter 34)** Paralysis of only one lower limb rules out transection, and involvement of the posterior and lateral funiculi indicates hemisection. Destruction of the anterior spinal artery and anterior white commissure would result in an inability to detect a pinprick, and bilateral loss of the posterior gray would produce local bilateral symptoms.

**129—C (Chapter 37)** The cerebellum and globus pallidus project to brainstem nuclei other than motor nuclei of cranial nerves. The putamen projects to other structures of the basal ganglia, and the ventral lateral nucleus projects to the precentral gyrus. Fibers from cell bodies in the precentral gyrus may synapse directly on anterior horn cells or motor nuclei of cranial nerves.

**130—B (Chapter 35)** Because the hypothalamus is the highest subcortical center for autonomic regulation, it likely projects to the general visceral efferent cell columns in both the spinal cord and brainstem. The special visceral efferent cell column contains motor neurons that supply skeletal muscles of branchial arch origin.

**131—A (Chapter 35)** The anterior nucleus projects primarily to the cingulate gyrus, and the pulvinar receives sensory information from the geniculate nuclei. The globus pallidus projects to the ventral anterior nucleus, and the thalamus is located medial to the posterior limb of the internal capsule.

**132—A (Chapter 38)** The superior cerebellar peduncle consists primarily of fibers originating in the dentate nucleus and terminating in the thalamus. This is part of the large feedback system between the cerebral cortex and the cerebellum (neocerebellum).

**133—A (Chapter 41)** An intention tremor, a tremor that occurs when an active motion is attempted, is a classic sign of lateral or neocerebellar disease. Flaccid paralysis is the result of a lower motor neuron lesion; involuntary motion and rigidity are associated with basal ganglia disease; and astereognosis is a sensory disturbance secondary to involvement of the posterior white columns.

**134—E (Chapter 41)** The symptoms resulting from a unilateral lesion of the lateral or neocerebellum are unilateral and occur on the same side as the lesion. This is because of the feedback system between one side of the neocerebellum and the opposite side of the cerebral cortex.

**135—B (Chapter 39)** Corticobulbar fibers are restricted to the genu of the internal capsule. Auditory fibers pass through the sublenticular portion of the capsule; optic fibers pass through both the sublenticular and retrolenticular portions. Corticospinal fibers and projections to the parietal lobe occupy the posterior limb; the anterior limb contains only projections to the prefrontal cortex.

**136—B (Chapter 39)** The subthalamic nucleus and substantia nigra are not strictly part of the basal ganglia, but are functionally related and involved in feedback circuits with the globus pallidus (subthalamic nucleus) and striatum (substantia nigra). The putamen and caudate nucleus receive most of the afferent fibers that project to the basal ganglia.

**137—E (Chapter 41)** Even though there is usually an initial period of flaccidity, a lesion of an upper motor neuron is characterized by a spastic paralysis. With a lesion of an upper motor neuron, the deep tendon reflexes are exaggerated, but the superficial reflexes are lost. There is a positive Babinski reflex and hypertonia, and the distal musculature is typically more affected than the proximal musculature.

**138—B (Chapter 37)** The posterior spinocerebellar tract is found at all levels of the spinal cord above L3. Its fibers originate from cell bodies in the nucleus dorsalis, ascend in the ipsilateral lateral white funiculus, and then enter the cerebellum via the inferior cerebellar peduncle. It conveys proprioceptive information that originates in neuromuscular spindles and Golgi tendon organs.

**139—C (Chapter 34)** The spastic paralysis in this patient indicates involvement of both the corticospinal and either corticoreticular or reticulospinal tracts. The loss of pain perception indicates a lesion of brainstem tegmentum or lateral white funiculus of the spinal cord. A lesion of the medullary pyramid or base of the midbrain or pons would produce only motor deficits. A lesion of the internal capsule would cause symptoms on the contralateral side only.

**140—A (Chapter 35)** The levator veli palatini is the primary mover of the uvula and is supplied by the vagus nerve. Because the uvula moves to the left with phonation, the left vagus nerve is intact but the right is not. This means that the lesion involves the nerve itself or its origin, which is the nucleus ambiguus of the medulla.

**141—B (Chapter 36)** The nucleus ambiguus is composed of special visceral efferent neurons and provides fibers to CN IX, CN X, and CN XI (glossopharyngeal, vagus, and accessory nerves, respectively). The gag reflex tests CN IX and CN X. The corneal reflex tests CN V and CN VII (trigeminal and facial, respectively); the pupillary light reflex tests CN II and CN III (optic and oculomotor, respectively); and the jaw jerk reflex tests CN V. The lips are pursed by the muscles of facial expression, which are supplied by CN VII.

**142—B (Chapter 39)** Stereognosis is dependent on intact unisensory association areas (5, 7, and 40) in the parietal lobe. There would be perception of neither texture nor shape with loss of the postcentral gyrus, fasciculus cuneatus in the upper cervical cord, or the medial lemniscus. The fasciculus cuneatus is not present inferior to the middle thoracic spinal cord; in addition, fibers from the hand pass through the brachial plexus.

**143—A (Chapter 37)** Corticobulbar fibers to the motor nucleus supplying the lower facial muscles are primarily crossed, and those to the nucleus to the upper facial muscles are bilateral. As a result, the right side of the mouth droops when a smile is attempted, and there is no trouble closing either eye. The muscles of the tongue are supplied by the hypoglossal nerve, and the teeth are clenched by the muscles of mastication (mandibular, V). Ptosis results from a lesion of the oculomotor nerve or the sympathetic supply of the orbit.

**144—D (Chapter 27)** The pancreas is both an endocrine gland and an exocrine gland. The exocrine portion of the gland consists of glandular end pieces that contain a considerable amount of rough endoplasmic reticulum, making them very basophilic. The acini of the exocrine part of the pancreas are characterized by the presence of centroacinar cells, which represent the initial portion of the duct system. The endocrine part of the pancreas resides in the pancreatic islets and is responsible for the production of digestive hormones. Myoepithelial cells are found in the mammary gland.

**145—B (Chapter 4)** The auditory or eustachian tube interconnects the middle ear or tympanic cavity. The mastoid air cells occupy the mastoid process, also part of the temporal bone, and are in direct communication with the tympanic cavity. The piriform recess is part of the laryngopharynx.

**146—B (Chapter 22)** The plasma membrane separates the inside from the outside of the cell and is composed of a phospholipid bilayer with the hydrophobic ends directed centrally. The plasma lemma also has numerous integral membrane proteins that contribute to its structure. The outer surface of some cells possesses a glycocalyx.

**147—B (Chapter 20)** Cells that produce proteins for secretion have an abundance of rough endoplasmic reticulum. Polyribosomes are found in cells that produce protein for intracellular use.

**148—B (Chapter 22)** Epithelial cells are classified according to the number of layers of cells, the shape of the surface cells, and certain surface specializations of the cells. Connective tissues are generally classified on the basis of the composition and amount of their intercellular matrix.

**149—B (Chapter 22)** Neurons are capable of elaborating a number of protein substances that they secrete. Antidiuretic hormone is produced by neurons in the supraoptic nucleus of the hypothalamus and transported via axons into the posterior pituitary gland, where it is secreted into the bloodstream. Epinephrine is modified from the amino acid tyrosine and serves as a neurotransmitter in many neural pathways. Cells in the hypothalamus also produce luteinizing hormone–releasing factor, which is secreted into the hypothalamohypophyseal portal system, where it influences the activity of cells in the anterior pituitary. Calcitonin is produced by parafollicular cells in the thyroid gland.

**150—C (Chapter 25)** The sinusoids of the liver, spleen, and bone marrow are modified capillaries of the blood vascular system and, as such, transport blood. The sinusoids of lymph nodes, on the other hand, are part of the system of lymphatic vessels and therefore transport only lymph.

**151—D (Chapter 22)** The vagina is lined with nonkeratinized stratified squamous epithelium and contains no glands. Both the esophagus and cervix have mucous glands associated with their walls. The colon has numerous goblet cells in its wall.

**152—C (Chapter 27)** The walls of the alimentary canal consist of an inner mucosa, which includes the mucous membrane and attendant intraepithelial glands and enteroendocrine cells, a connective tissue layer, the lamina propria, and a layer of smooth muscle, the muscularis mucosa. Beneath the mucosa is the submucosa, which is also connective tissue and contains the submucosal or Meissner's plexus of nerves. Deep to the submucosa is the muscularis externa, consisting of alternating layers of circular and longitudinal smooth muscle. The outermost layer of the wall is the serosa or adventitia, which depends on the peritoneal relationships of the organ.

**153—D (Chapter 27)** The major digestive role of the small intestine is the absorption of the digestive products and transport of these substances to the bloodstream or, in the case of chylomicrons, to the lacteals and then to the lymphatic system. Columnar absorptive cells have the means to absorb these substances both actively and passively. A feature of the distal small intestine, the ileum, is the presence of aggregates of lymphatic tissue known as Peyer's patches. These are localized in the lamina propria but may extend into the submucosa as well. In the colon, the outer longitudinal muscle layer is condensed to form three parallel bands of smooth muscle called the tenia coli.

**154—D (Chapter 20)** The fuzz (glycocalyx) covering the free surface of the intestinal lining cell, the colloid in the thyroid follicle, and the matrix of hyaline cartilage all contain complex carbohydrates and, therefore, will incorporate radiolabeled glucose. Secretory product in plasma cells will not incorporate radiolabeled glucose.

**155—B (Chapter 2)** The superficial branch of the radial nerve supplies the dorsal aspects of the three-and-a-half lateral digits and corresponding dorsum of the hand; thus, if the nerve is injured, a burning sensation in the dorsal aspect of the first web space (between the thumb and index fingers) occurs. Because of potential variation, involvement of the central part of that area ensures involvement of that nerve.

**156—C (Chapter 2)** The digital branches of the median nerve supply the skin of the volar aspects of the three-and-a-half lateral digits and corresponding part of the palm. Difficulty identifying objects held between the thumb, index, and middle fingers would accompany a median lesion, such as carpal tunnel syndrome.

**157—A (Chapter 2)** The skin of the point of the shoulder is supplied by spinal cord segment C4 via the supraclavicular nerves.

**158—G (Chapter 2)** Because the biceps brachii is one of the muscles in the anterior compartment of the arm, it is supplied by the musculocutaneous nerve. The musculocutaneous nerve, one of the terminal branches of the lateral cord, contains fibers from spinal cord segments C5 and C6.

**159—C (Chapter 2)** The ulnar nerve supplies all of the interossei muscles, which produce adduction and abduction of the four medial digits. These muscles are supplied specifically by the deep branch of the ulnar nerve, which begins at the wrist. As a result, injury of the ulnar anywhere proximal to the wrist will result in loss of the interossei muscles.

**160—D (Chapter 2)** The radial nerve supplies all of the muscles in the posterior compartment of the forearm. Branches to the majority of the superficial muscles arise proximal to the elbow. Branches to the deep muscles, which include the extensor pollicis longus and brevis and which extend the thumb, arise distal to the forearm.

**161—A (Chapter 2)** The median nerve branches into its digital and recurrent (motor, thenar) branches at the distal end of the carpal tunnel. The muscles in the thenar compartment, which includes the opponens pollicis, are supplied by the recurrent branch of the median nerve. Advanced carpal tunnel syndrome commonly causes thenar atrophy and reduced thenar function, which is opposition.

**162—B (Chapter 3)** The muscles in the anterior compartment of the leg, particularly the tibialis anterior, are active at heel-strike. At that point, they counteract the tendency for the ball of the foot to slap the ground.

**163—C (Chapter 3)** During the stance phase, there is a tendency for the pelvis to drop to the non-weight-bearing side. This tendency is counteracted by the hip abductors, the gluteus medius and minimus, which act to limit adduction of the thigh on the weight-bearing side.

**164—B (Chapter 2)** The posterior cord is found posterior to the second part of the axillary artery. It is formed by the posterior divisions of the superior, middle, and inferior trunks.

**165—E (Chapter 2)** The suprascapular nerve is the only branch from the trunks of the brachial plexus. It branches from the superior trunk and contains fibers from spinal cord segments C5 and C6.

**166—J (Chapter 2)** The ulnar nerve is one of the terminal branches of the medial cord. The other branch joins a similar branch from the lateral cord to form the median nerve.

**167—B (Chapter 3)** The bones of the pelvis are joined at the symphysis pubis and two sacroiliac joints. The sacroiliac joint is a synovial joint that is formed between the lateral aspect of the sacrum and the body of the ilium.

**168—E (Chapter 3)** The ischial spine projects posteromedially from the body of the ischium and separates the greater and lesser sciatic foramina. The distance between these spines is the transverse diameter of the midpelvis and the smallest diameter of the entire pelvis.

**169—D (Chapter 3)** The anterior superior spine of the ilium is the prominent anterior limit of the iliac crest. This spine is readily palpable and serves as the origin of the sartorius muscle.

**170—L (Chapter 6)** The superior mesenteric artery is the second unpaired artery that branches from the abdominal aorta. It branches at about vertebral level L1, just inferior to the celiac trunk, and passes anteriorly between the pancreas and third part of the duodenum.

**171—E (Chapter 3)** The psoas major muscle forms part of the posterior body wall just lateral to the lumbar vertebral bodies. This muscle is the major flexor of the femur; it extends from the anterolateral aspects of the lumbar vertebral bodies and transverse processes to the lesser trochanter of the femur.

**172—A (Chapter 5)** The flat central tendon of the respiratory diaphragm forms the central portion of the dome-shaped structure and serves as the insertion of the muscular portion of the diaphragm.

**173—J (Chapter 6)** The abdominal portion of the aorta descends through the abdominal cavity anterior and slightly to the left of the lumbar vertebral bodies.

**174—F (Chapter 6)** The pancreas is located between the stomach and posterior abdominal wall and is retroperitoneal. Its head occupies the "C" formed by the duodenum; from that point, it extends to the left, superiorly and posteriorly, where its tail is in contact with the spleen.

**175—H (Chapter 6)** The spleen is found in the left upper quadrant of the abdomen, where it is against the posterior body wall and deep to ribs 9, 10, and 11.

**176—I (Chapter 5)** The left lung has only two lobes, the superior and inferior, which are separated by the oblique fissure. This fissure is oriented so it descends from posterior to anterior. As a result, the superior lobe is anterior and superior, and the inferior lobe is posterior and inferior. In a cross section, the superior lobe is anterior to the inferior lobe.

**177—F (Chapter 5)** The arch of the aorta curves posteriorly and to the left and has three branches: the brachiocephalic, left common carotid, and left subclavian arteries.

**178—E (Chapter 2)** The subscapularis muscle is one of the rotator cuff muscles. It arises from the entire deep or anterior surface of the scapula and inserts on the lesser tubercle of the humerus.

**179—C (Chapter 4)** The mandibular division of the trigeminal nerve branches from the semilunar ganglion in the middle cranial fossa. It then passes inferiorly through the foramen ovale into the infratemporal fossa.

**180—J (Chapter 4)** The abducens nerve branches from the brainstem at the pons-medulla junction in the posterior cranial fossa. It passes anteriorly and pierces the dura on the basilar portion of the occipital bone, passes through the cavernous sinus, and then into the orbit through the superior orbital fissure.

**181—G (Chapter 4)** The facial nerve exits from the brainstem at the pons-medulla junction. It passes laterally through the posterior cranial fossa and enters the internal auditory meatus, which is continuous with the facial canal. It leaves the facial canal via the stylomastoid foramen.

**182—C (Chapter 10)** Sarcomeres are the functional subunits of skeletal and cardiac muscle. These tissues develop from mesoderm, which initially forms by the migration of cells in the primitive streak during gastrulation.

**183—E (Chapter 16)** Primordial sex cells first appear in the wall of the yolk sac, from which they migrate to seed the gonad, which develops from the mesonephros.

**184—B (Chapter 9)** Cells of the primitive node (Hensen) migrate forward in the embryo and condense to form the notochord.

**185—A (Chapter 11)** Ganglia of the autonomic nervous system as well as the dorsal root and sensory ganglia develop from the neural crest.

**186—B (Chapter 16)** In both the male and female, the gonad develops from the mesonephros.

**187—D (Chapter 16)** The vagina develops from the urogenital sinus.

**188—C (Chapter 16)** The uterus develops from the paramesonephric duct.

**189—A (Chapter 16)** The ductus deferens develops from the mesonephric duct.

**190—A (Chapter 19)** After endocytosis, the ingested material is generally contained within a coated vesicle.

**191—B (Chapter 19)** Lysosomes fuse with phagosomes or endocytotic vesicles to destroy or digest ingested material in the cell.

**192—E (Chapter 19)** Cells that engage in the production of steroid hormones generally possess abundant smooth endoplasmic reticulum and mitochondria demonstrating tubular cristae.

**193—C (Chapter 19)** Mitochondria are often called the "powerhouses of the cell" because of their role in ATP production.

**194—A (Chapter 30)** The three zones of the adrenal cortex are generally considered to produce different hormones. The zona glomerulosa produces mineralocorticoids; the zona fasciculata produces glucocorticoids, including hydrocortisone or cortisol; and the zona reticularis produces sex hormones.

**195—D (Chapter 30)** Acidophils of the adenohypophysis are responsible for the production of growth hormone.

**196—C (Chapter 30)** Alpha cells of the pancreatic islets produce glucagon, and beta cells produce insulin.

**197—E (Chapter 30)** Striated ducts are modified for the absorption of bicarbonates from glandular secretions. They possess abundant infoldings of the basal membranes of the cells.

**198—C (Chapter 34)** The fasciculus proprius or spinal-spinal system consists of a concentration of interneurons or internuncials. It forms a thin strip that surrounds the gray matter of the spinal cord.

**199—D (Chapter 34)** The cell bodies of preganglionic autonomic neurons are found in the lateral horn (Rexed's lamina, VII) of the gray matter. The sympathetic neurons are in thoracic and upper lumbar levels, and the parasympathetic are in segments S2, S3, and S4.

**200—G (Chapter 34)** The cell bodies of the alpha and gamma motoneurons are found in the anterior gray horn. The more medially positioned neurons supply the more axial musculature; the laterally positioned supply the more distal musculature.

**201—A (Chapter 37)** The descriptive information used to identify an object that is placed in the hand is carried through the spinal cord by the posterior white columns.

**202—N (Chapter 37)** The lateral corticospinal tract descends through the posterolateral aspect of the lateral funiculus of the spinal cord.

**203—B (Chapter 37)** Afferent fibers conveying position sense from segments below C8 terminate in the nucleus dorsalis. Cell bodies in this nucleus give rise to the fibers that form the ipsilateral posterior spinocerebellar tract.

**204—E (Chapter 37)** The nuclei gracilis and cuneatus give rise to the fibers that form the medial lemniscus. The first-order neurons above spinal cord level T6 terminate in the nucleus cuneatus.

**205—C (Chapter 35)** The ventral posterolateral and posteromedial nuclei of the thalamus are sensory relay nuclei. They give rise to fibers that project to the postcentral gyrus of the cerebral cortex via the posterior limb of the internal capsule.

**206—A (Chapter 37)** The accessory cuneate nucleus receives first-order neurons conveying position sense that enter the spinal cord above level C8. Fibers from this nucleus then project to the cerebellum through the inferior cerebellar peduncle.

**207—J (Chapter 35)** In the high medulla, the medial lemniscus is located medially and just dorsal to the pyramids.

**208—A (Chapter 36)** The muscles of the tongue are supplied by the hypoglossal nerve; its nucleus, consisting of general somatic efferent neurons, is the most medial structure in the floor of the fourth ventricle in the upper medulla.

**209—E (Chapter 35)** The inferior cerebellar peduncle is located dorsolaterally and lateral to the fourth ventricle and the spinal tract of CN V in the open medulla.

**210—G (Chapter 36)** The muscles of the larynx are innervated by the special visceral efferent fibers of the vagus nerve. These fibers begin in the nucleus ambiguus, which is in the reticular formation between the inferior olivary nucleus and the spinal nucleus of CN V.

**211—K (Chapter 35)** The pyramid is the most medial structure in the anterior aspect of the open medulla. It is prominent on the surface and delineated by the midline and the preolivary sulcus.

**212—G (Chapter 35)** The corticospinal and corticobulbar fibers descend through the basilar part of the pons in varying numbers of bundles.

**213—H (Chapter 35)** Fibers from the cerebral cortex to the cerebellar cortex first descend to pontine nuclei, where they synapse. The axons of the second-order neurons in this chain cross the midline in the middle cerebellar peduncle en route to the cerebellar cortex.

**214—B (Chapter 36)** The skin of the face is supplied by the trigeminal nerve (CN V). The fibers conveying particularly touch, and some pain and temperature, terminate in the principal sensory nucleus of the CN V. This nucleus is found in the central part of the tegmentum at the midpons level.

**215—D (Chapter 37)** The modalities of vibratory sense and two-point discrimination are transported via the posterior white columns. This tract crosses the midline in the medulla, and then ascends as the medial lemniscus. At the midpons level, the medial lemniscus is located medially at the junction of the tegmentum and basis pontis.

**216—J (Chapter 39)** Anatomically, the substantia nigra is not grouped with the basal ganglia; however, it is interconnected with them and considered a functional part.

**217—B (Chapter 35)** The midbrain part of the ventricular system is the iter or cerebral aqueduct of Sylvius.

**218—I (Chapter 38)** The red nucleus is part of the feedback system between the cerebral cortex and the neocerebellum. Fibers from the dentate nucleus project to the red nucleus; fibers originating in the red nucleus then project to the cerebral cortex.

**219—F (Chapter 38)** The general somatic efferent fibers of the oculomotor nerve (CN III) supply several extraocular eye muscles including the medial rectus. Loss of this muscle causes the eye to deviate laterally, a lateral or external strabismus. The motor nucleus of CN III is located medially, just anterior to the central gray at the level of the superior colliculus.

# Anatomical Sciences

## Must-Know Topics

The following are must-know topics discussed in this review. It would be useful for you to formulate outlines on these subjects because knowledge of the related material will be key to your understanding of the subject and material and for performing well on the examination.

## *Gross Anatomy*

### Back

- Structures traversed during lumbar puncture
- Length of spinal cord versus vertebral column
- Formation and major branches of a spinal nerve
- Distribution of the dorsal and ventral rami
- Formation of the intervertebral foramen; cervical versus lumbar levels
- Composition and function of an intervertebral disk

### Upper Limb

- Formation and location of the brachial plexus
- Formation of the scalene groove (triangle); structures that traverse this groove
- Injury of the upper plexus (superior trunk) versus injury of the lower plexus (inferior trunk)
- Rotator cuff muscles and their function
- Course and distribution of median nerve; muscular and cutaneous losses after injury
- Course and distribution of ulnar nerve; muscular and cutaneous losses after injury
- Course and distribution of radial nerve; muscular and cutaneous losses after injury
- Course and distribution of axillary nerve; muscular and cutaneous losses after injury
- Course and distribution of musculocutaneous nerve; muscular and cutaneous losses after injury
- Dermatomes of the upper limb
- Segmental motor evaluation of the upper limb

*(continued)*

- Intrinsic muscles (thenar, hypothenar, lumbrical, and interossei) of the hand and their functions
- Carpal tunnel syndrome and symptomology
- Clavicular fractures and potential complications involving closely related structures
- Fractures of the wrist (Colles) and scaphoid and potential involvement of closely related structures

## Lower Limb

- Femoral neck fractures and potential complications
- Femoral triangle and contents
- Knee joint and function of the collateral and cruciate ligaments and the menisci
- Formation of lumbar and sacral components of lumbosacral plexus
- Course and distribution of obturator nerve; muscular and cutaneous losses after injury
- Course and distribution of femoral nerve; muscular and cutaneous losses after injury
- Courses and distribution of the superior and inferior gluteal nerves; muscular and cutaneous losses after injury
- Courses and distribution of the components (tibial and common peroneal) of the sciatic nerve; muscular and cutaneous losses after injury
- Courses of vessels and nerves through gluteal region; hazards for injection
- Medial versus lateral stability of the ankle
- Femoral, popliteal, posterior tibial, and dorsalis pedis pulse points
- Dermatomes of the lower limb
- Segmental motor evaluation of the lower limb

## Head and Neck

- Location and contents of carotid sheath
- Other structures located at the level of the hyoid bone, laryngeal prominence, and cricoid cartilage
- Courses and functions of the right and left recurrent (inferior) laryngeal nerves
- Primary muscles of facial expression; their functions and innervation
- Distribution and functions of the trigeminal nerve and its branches
- Foramina in the anterior, middle, and posterior cranial fossae; structures that traverse each opening
- Meningeal layers and related spaces
- Dural venous sinuses
- Muscles of mastication; their functions and innervation
- Extraocular eye muscles; their functions, innervation, and how tested
- Muscles of the tongue and their innervation
- Blood supply of the nasal cavity
- Paranasal sinuses and best position for drainage of each
- Muscles of the palate and nerve supply

## Thorax

- Projection of the lungs and pleural cavities to surface of the thorax
- Lobes of each lung and how separated on the surface of the thorax

*(continued)*

- Lobes versus bronchopulmonary segments
- Projection of the outline of the heart to the surface of the thorax
- Projection of the cardiac valves to the surface of the thorax
- Coronary arteries and the regions of the heart each supplies
- Parts of the mediastinum and contents of each
- Attachments and shape of the respiratory diaphragm; structures that pass through the diaphragm
- Pattern of lymphatic drainage of the structures of the thorax

## Abdomen

- Layers of abdominal wall
- Formation of inguinal canal; abdominal wall structures contributing to each part
- Anastomoses between portal and systemic venous systems
- Surface locations of abdominal viscera
- Layers of spermatic cord and structures contributing to each layer
- Courses of direct and indirect inguinal herniae
- Structures supplied by vagal versus sacral parasympathetics
- Relationships of abdominal viscera
- Pattern of lymphatic drainage of each of the abdominal viscera
- Greater versus lesser omental bursae
- Peritoneal relationships of each of the abdominal viscera
- Autonomic plexuses; their inputs and distribution
- Blood supply of each of the abdominal viscera

## Pelvis and Perineum

- Relationships of the male and female pelvic organs
- Walls and diameters of the pelvis
- Location and boundaries of the perineum
- Broad ligament and its components
- Posterior fornix of the vagina and its relationships
- Blood supply to pelvis versus perineum
- Course of pudendal nerve
- Autonomic supply of male and female perineal structures
- Formation of pelvic floor
- Urogenital versus anal triangles of perineum
- Superficial versus deep perineal spaces (pouches)
- Continuity between perineum and abdominal wall

# *Embryology*

- Development of the heart and congenital malformations
- Fetal circulation and changes in circulation at birth
- Derivatives of the foregut, midgut, hindgut, and their mesenteries
- Structure and derivatives of the branchial arches
- Development of the tongue, palate, and face
- Structure and derivatives of a typical somite
- Development of the male and female reproductive tracts and the homologues of each

# Histology

- Functions and structure of cell organelles
- Process of mitosis and meiosis
- Protein synthesis and secretion
- Functional modifications of epithelial cells
- Cell types of connective tissue
- Methods of gland secretion
- Structure of the neuron and peripheral nerves
- Microanatomy of skin
- Structure of lymphoid organs
- Histology of the respiratory tract
- Histology of the nephron
- Structure of the wall of the gastrointestinal tract and variations from region to region

# Neuroanatomy

## Spinal Cord

- Location of cell columns and Rexed's lamina in the gray matter of the spinal cord
- Location of the major ascending and descending tracts in a cross section of the spinal cord
- Areas of the spinal cord supplied by the anterior and posterior spinal arteries

## Brainstem

- Major differences between the upper (closed) and lower (open) levels of the medulla
- Cranial nerve nuclei located in the medulla, pons, and midbrain
- Component differences between the tegmentum and basilar portions of the pons
- Functions of the superior and inferior colliculi
- Functions of the autonomic, neuroendocrine, and olfactory nuclei of the hypothalamus
- Functional differences of the medial, lateral, and anterior thalamic nuclei

## Cranial Nerves

- Fiber types (functional components) found in each cranial nerve
- Lesions of the visual pathways
- Visual reflexes
- Cranial nerve conveying preganglionic parasympathetic fibers from the brain stem
- Symptoms resulting from lesions of cranial nerves III, IV, and VI
- Symptoms resulting from an injury of cranial nerve VII
- Cranial nerves in the gag reflex

## Ascending and Descending Pathways

- Courses of the axons forming the lateral spinothalamic tract; symptoms of injury above and below decussation
- Symptoms of injury of fasciculus gracilis versus fasciculus cuneatus
- Course and termination of posterior spinocerebellar tract
- Difference in functions of corticospinal and corticobulbar tracts
- Functional roles of vestibulospinal, reticulospinal, and rubrospinal tracts

## Cerebellum

- Functional differences of archicerebellum (vestibulo-), paleocerebellum (spino-), and neocerebellum (ponto-)
- Circuitry of the cerebellar cortex

## Telencephalon

- Areas of the cerebral cortex supplied by the anterior, middle, and posterior cerebral arteries
- Primary receptive areas of the cerebral cortex and their functional roles
- Motor, premotor, frontal eye fields, prefrontal, and Broca's area of the cerebral cortex and their functions
- Components of the anterior and posterior limbs, and the genu of the internal capsule
- Location and components of the basal ganglia
- Circuitry of the basal ganglia

## Motor Control Mechanisms

- Regulation and centers involved in gross stereotyped movement
- Feedback system between the cerebral cortex and cerebellum
- Symptoms and location of lesions of the basal ganglia
- Symptoms and location of lesions of the neocerebellum versus the vestibulocerebellum
- Difference between upper and lower motor neuron lesions
- Deficits resulting from destruction of various areas of the cerebral cortex

# Index

Note: Page numbers followed by *f* refer to figures; page numbers followed by *t* refer to tables.